To my father

István Berta, M.D.

who turned my mind onto medical sciences, and who so successfully
planted his ambitions in his three children.

AUTHOR

András Berta, M.D., C.Sc., is Clinical Associate Professor of Ophthalmology, Senior Lecturer in Ophthalmology, and Head of the General Diagnostic and Research Laboratory in the Department of Ophthalmology at the University Medical School of Debrecen, Hungary.

Dr. Berta graduated summa cum laude and received his M.D. degree in 1979 from the University Medical School of Debrecen, Hungary. He obtained his C.Sc. degree (a Ph.D. equivalent scientific degree of the Hungarian Academy of Sciences) in 1987 based on a thesis entitled, *Diagnostic Tear Protein Determinations.*

As a student he worked as a part time university lecturer, taught general biology for medical students, and did research work in the Department of Biophysics. He was one of the organizers and a leader of the Students' Scientific Society at the University Medical School of Debrecen. Since 1979 he has been working at and has held different positions at the Department of Ophthalmology, University Medical School of Debrecen. After 4 years of residency and a qualifying examination in 1984 in General Ophthalmology he specialized in anterior segment surgery of the eye. Besides his medical practice and teaching activity he has been the head of the General Diagnostic and Research Laboratory since 1984. In 1989 he did a research fellowship in the Wilmer Eye Institute at The Johns Hopkins University, Baltimore, MD (USA). In 1990 he worked for a year as a visiting professor at the Dry Eye Institute in Lubbock, TX (USA).

Dr. Berta is a member of the Hungarian Ophthalmolgical Society, European Society of Lacrimology, International Society for Eye Research, and the European Association for Eye Research. He is a founding member of the International Society of Dakryology. He was the secretary of the Organizing Committee of the "First Meeting of the International Society of Dakryology" held in 1987 in Budapest, Hungary. He is an elected member of the Advisory Board of the International Society of Dakryology. He was a member of the Scientific Committee of the "Sixth International Symposium on the Lacrimal System" held in 1990 in Singapore.

He received the Weszprémi Award and the Pro Universitate Award of the Medical University of Debrecen, Hungary, the Lacrima Award of the Dry Eye Institute, Lubbock, Texas (USA); the Papolczy Award of the Papolczy Founadation, Budapest, Hungary; twice the First Prize of the Hungarian Ophthalmological Society, Budapest, Hungary; and the Chibret Travel Fellowship Award of Chibret International, Zurich, Switzerland.

He presented more than 70 papers at scientific congresses, and is the author of 45 papers published in international journals and of 2 book chapters published in the United States. His major scientific interest, besides tear research and the diagnosis and therapy of dry eyes, includes corneal transplantation, ophthalmic laser therapy, and the radiotherapy of intraocular tumors.

PREFACE

The total protein concentration of tears is about 10% of that of the plasma. Up to 80 polypeptide components have been detected in tears by two-dimensional electrophoresis. The number of identified proteins is above 30 and nearly half of them are enzymes. One of these, lysozyme accounts for approximately 30% of the total protein concentration; the others are present in varying amounts. Not only are the number and the quantity of enzymes in the tear fluid high but they are the most active, therefore, probably functionally the most important tear proteins. Some of the tear enzymes are secreted by the lacrimal glands; others are produced by or released from the epithelial cells of the cornea and the conjunctiva; still others originate from the plasma and appear in tears only in cases with increased permeability of the conjunctival vessels.

Lysozyme is the most studied tear enzyme, and is also the most studied enzyme in general. We know much about lysozyme, including its conformation, the exact mechanism of its enzymatic action, and many other characteristics, but its specific role (or roles) in tears is still poorly understood. It is a bacteriolytic substance, but it actually acts on a very limited number of bacteria that are mostly apathogenic saprophytes. The antibacterial effect of lysozyme might not be the only cause why the ability to keep such high concentration in tears has not been lost in the process of the evolution of species.

The activity of lactate dehydrogenase (LDH) in the tears is higher than that in the plasma. In tears, unlike in plasma, LDH isoenzymes built up mainly of M (muscle)-type subunits are predominant, suggesting the local production of this enzyme. Different diseases that affect corneal epithelium increase LDH activity and change LDH isoenzyme pattern in tears. The amount or rather the ratio of enzymes involved in the aerobic and anaerobic pathways of energy-producing mechanisms are thought to reflect the general metabolic status of corneal epithelial cells in health, disease, and contact lens wear.

Various tear enzymes are used in the diagnostics of genetically determined enzymopathies. The determination of β-hexosaminidase and α-galactosidase in tears is important in heterozygote screening in Tay-Sachs' and Fabry's diseases. In both diseases heterozygotes have about one half of the normal enzyme activity in tears, whereas it is virtually absent in homozygotes.

Proteolytic enzymes are some of the most important tear enzymes playing a decisive role in the pathogenesis of a number of corneal diseases, including bacterial ulcers, caustic injuries, herpetic keratitis, persistent or recurrent erosions, contact lens-associated lesions of the cornea, etc. The detection of collagenase in the tears of patients with corneal ulcers raised great enthusiasm and hope concerning the use of collagenase inhibitors in the treatment of these devastating diseases. The use of collagenase inhibitor eyedrops on humans provided only limited results, suggesting that other enzymes and enzyme inhibitors may be involved in the degradation and healing of the cornea.

Recently the plasminogen activator-plasmin system of tears has been shown to initiate a proteolytic cascade resulting in the activation of various proteases capable of degrading corneal tissue. A complex mechanism regulating the secretion, activation, and inhibition of different enzymes of this cascade is thought to be responsible for maintaining a homeostasis in the healthy eye. Any change in the balance of inhibitors and activators, some of which are also active enzymes, may result in the formation of tissue defect, the propagation of microorganisms in the otherwise compact corneal tissue, or the defective healing of some of the above-mentioned corneal diseases. The studies of plasminogen activators, plasmin, plasminogen activator inhibitors, and plasmin inhibitors are some of the most promising fields in the diagnosis and therapy of corneal diseases.

A large number of studies concerning various enzymes of the tear fluid have been performed and published; various authors studied tear enzymes from different aspects. I have attempted to review the literature and put together the small pieces of information available in ophthalmological and biochemical literature. While preparing this manuscript it was my intention to cover various fields of clinical and experimental tear research, to evalute different views, and to integrate the results produced by ophthalmologists and biochemists. I also did have the opportunity to summarize the results of our own clinical and experimental studies, express my personal opinions, and make suggestions for the solution of unsolved problems. The aim of this work is to provide clinical and biochemical information about tear enzymes both for ophthalmologists and for research scientists who are interested in the clinical and the experimental aspects of tear enzymology.

András Berta

ACKNOWLEDGMENTS

I am most grateful to my professor, Béla Alberth, M.D., D.Sc., who started my carrier at the Department of Ophthalmology University Medical School of Debrecen, Hungary, as an ophthalmologist, university lecturer, and research scientist in 1979, and ever since helped my work with his valuable advice and with personal sympathy.

I greatly acknowledge the help of Micheal Berman, Ph.D., and Frank J. Holly, Ph.D., in the experimental work during my stays at the Wilmer Eye Institute in Baltimore and at the Dry Eye Institute in Lubbock.

I would like to thank the Johns Hopkins University, the University Medical School of Debrecen, the Hungarian Academy of Sciences, the Dry Eye Institute, and the International Society of Dakryology, whose grants and financial support made possible to perform my experimental and clinical studies and to prepare this manuscript.

TABLE OF CONTENTS

Chapter 1

ANATOMY AND PHYSIOLOGY OF THE LACRIMAL SYSTEM

I. INTRODUCTION

The exposed surface of the eyeball is covered by a thin layer of tears. This layer is formed and maintained by the lacrimal apparatus. A continuous tear film provides the cornea with a surface of high optical quality, serves as a lubricant ensuring the smooth movement of the lids during blinking, owing to its antibacterial properties protects the eye from infections, and helps to maintain the well being of the corneal and conjunctival epithelium by keeping the surface wet and providing oxygen and nutrients for the superficial epithelial cells.[1]

The lacrimal system is also capable of protecting the eye by a flushing and cleansing action resulting from an increased secretion of tears. The secretion rate of the lacrimal glands can increase by a hundredfold or more in a very short time due to a reflex started by mechanical or chemical irritation of the eye or the nasal mucosa. A sudden increase in tear secretion is highly effective in removing minor foreign bodies and flushing and diluting contaminants or noxious chemicals from the surface of the eye.[1]

The lacrimal apparatus consists of three parts: the secretory part, the distributory part, and the excretory part.

II. THE SECRETORY SYSTEM

Aqueous tears forming the overwhelming majority of the tear secretion are produced by the main and accessory lacrimal glands. Palpebral glands, producing lipids, and conjunctival goblet cells secreting mucus also contribute to the composition of tears.

The main lacrimal gland is an almond-shaped secretory gland located in the upper outer orbital region above the eyeball. The lacrimal gland consists of two parts: a large orbital or superior portion and a small palpebral or inferior portion, divided by the upper part of the orbital septum. The orbital part is lodged in the fossa glandulae lacrimalis of the frontal bone on the anterior and lateral part of the roof of the orbit. This part of the gland cannot be examined directly. In order to reach this portion during surgery one has to penetrate the skin of the upper lid, the orbicularis muscle, and the orbital septum. The palpebral portion of the lacrimal gland lies anterior to the lateral portion of the upper part of the orbital septum. When the patient's eye turns nasally and down, and the upper lid is everted, this part of the gland can be seen through the upper fornix of the conjunctiva.

Lipids forming a continuous layer on the surface of the aqueous tears are secreted by palpebral glands. The main source of lipid secretion are the meibomian glands. They are located in the tarsus of the upper and lower lids. Their openings are located along the lid margins just behind the gray line. Their secretions supply the outer portion of the tear film, which prevents rapid tear evaporation and tear overflow, and provide tight eyelid closure. Some lipids are also secreted by the glands of Zeis, located at the palpebral margin of each eyelid, and by the glands of Moll, found at the roots of the eyelashes. Meibomian glands are the most important secretors of the tear lipids. The other palpebral glands help with their secretions, as similar glands elsewhere in the body to prevent the hair (eyelashes) from becoming dry and brittle.

Most mucous material covering the superficial epithelium under the layer of the aqueous tears originates from the conjunctival goblet cells. These are large one-cell mucous glands in the superficial layers of the conjunctiva. Their secretion is distributed over the ocular surface by lid motion. The epithelial cells of the conjunctiva and the cornea are also capable of secreting glyco-proteins that can serve as the foundation of the mucous layer covering the epithelial surface. In adddition, small amounts of glycoproteins are secreted by the lacrimal glands, and under pathological conditions plasma glycoproteins passing the blood-tear barrier also contribute to the mucous content of tears.[2,3]

III. THE DISTRIBUTIONAL SYSTEM

The eyelids have many functions such as protecting the eye, regulating the light, and covering the eye during sleep. Its paramount functions are, however, to distribute the lacrimal fluid on the anterior surface of the eye, to regulate the evaporation, to expel superfluous quantities of tears, and to form a stable pre-corneal tear film. The eyelids are closed by the simultaneous contraction of the upper and lower parts of the orbicularis muscle. There is also a horizontal component of the lid movement in closing, and the lid is stretched as it moves medially when the eye is closed. The upper lid is raised by the levator palpebrae superioris muscle, and the lower lid is retracted by the inferior rectus muscle (capsulopalpebral part). The lid retractors have a reciprocal innervation with the protractors, the upper and lower palpebral part of the orbicularis muscle. When the eye is closed pressure against the globe increases. When the eye is open the pressure decreases.[4]

Every time the lids pass over the exposed surface of the eyeball, the lipid layer is compressed and the tearfilm-air interface is eliminated. The shear forces acting across the thin aqueous tear layer between the moving lid and the eyeball rejuvenate the mucous layer by removing lipid-laden mucus and redis-tributing mucus freshly expressed from the goblet cells.[1]

The tear volume in the normal open eye is approximately 8 μl. This volume consists of three parts: 1 μl the actual volume of the precorneal tear film, 3 μl in the tear menisci, and 4 μl in the fornices. The tear film is so thin that gravitation has no effect on it. Practically no hydraulic flow occurs in it over the ocular surface. The only flow that takes place is in the upper and lower tear menisci.[5]

IV. THE EXCRETORY SYSTEM

The lacrimal excretory system consists of the upper and lower lacrimal ducts (canaliculi), the tear sac, and the nasolacrimal (or lacrimonasal) duct. The excretory system is the sink of the lacrimal apparatus. Tears enter the drainage system through the lacrimal puncta, which are small round openings of the lacrimal ducts located on slight elevations (lacrimal papillae) of the lid margins near the medial canthus. Both puncta point backward and immerse in the lacrimal lake (lacus lacrimalis), a local thickening of the tear film around the lacrimal caruncle and near the semilunar plica. Normally the puncta are not visible unless the lid is pulled away from the globe or the lid margin is everted. If the puncta is visible without such a maneuver it is usually a sign of the eversion of the lacrimal punctum, or of the malposition of the lid or of the lid margin. Either of these situations prevents tears from entering the drainage system and usually causes epiphora. The puncta lead into the canaliculi. Each canaliculus consists of a vertical and a horizontal part. The vertical portion is about 2 mm, the horizontal about 8 mm long. The upper and lower canaliculi lead either directly or meet in a short common canaliculus into the lacrimal sac. The lacrimal ducts are located in the palpebral parts of the orbicularis muscle. The contraction of this muscle provides a peristaltic action propelling the lacrimal fluid through the excretory passages. A bunch of fibers belonging to the same muscle is attached to the upper part of the lacrimal sac pulling it on each contraction, creating a negative pressure inside the sac. The suction created by this mechanism also contributes to the propagation of tears through the lacrimal ducts into the lacrimal sac. From the sac, the tears drain into the inferior nasal meatus. This latter movement is brought about by the gravitation and may be assisted by increased pressure within the sac.

A. Basal Tear Secretion — Reflex Tear Secretion

According to Jones[6] the tear secretory system consists of basic secretors and reflex secretors. The former are the accessory lacrimal glands of Krause and Wolfring (lacrimal secretors); the goblet cells, and the glands of Manz (mucin secretors); the meibomian glands, the glands of Zeis and Moll (oil secretors); while the latter are the main lacrimal glands (orbital and palpebral portion). He proposed that the two systems function separately and are responsible for tear secretion under basically different conditions. In this concept in sleep, when the eyelids are closed, the basic secretors alone produce tears. During the waking hours reflex secretors secrete various amounts of tears to provide enough hydration.[7]

The idea of basic and reflex tear production was described in one of Jones' early papers.[6] In this paper he refers to the results of another investigator de Roeth (de Rötth), Sr.[8] de Roeth measured tear secretion with the Schirmer test without and with stimulation in dry eye patients[8] and in normal subjects.[9] de Roeth did not provide data directly proving the existence of basic tear flow. No such data can be found in Jones' original paper either, except for a

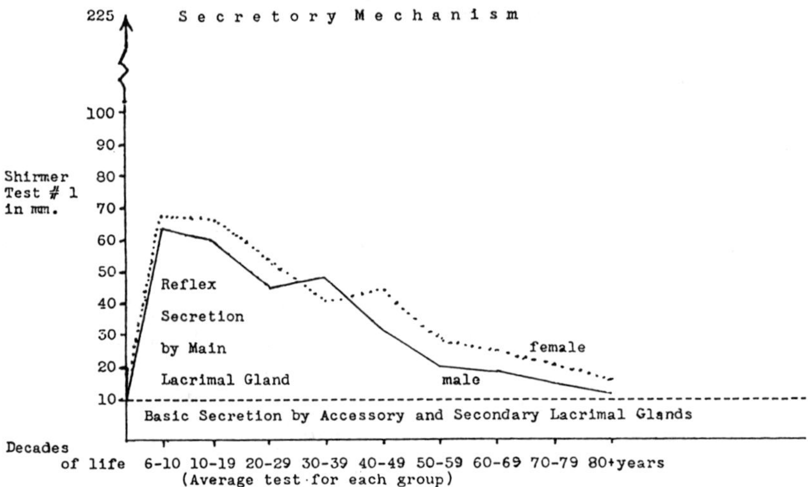

FIGURE 1. Reflex and basic secretion of tears in millimeters pf wetting in Shirmer No. 1 test (based on a study of 827 persons with normal eyes by A. deRoeth, Sr., 1953). (From Jones, L. T., *Arch. Ophthalmol.,* 66, 137, 1961. With permission.)

horizontal dotted line drawn under the curves plotted from the data of de Roeth (Figure 1).

In spite of the dearth of data supporting the idea of basically different basal and reflex tear secretory systems and secretory mechanisms, Jones' concept became widely accepted. Much effort was devoted to study basal tear flow and the composition of basal tears trying to minimize the stimuli caused by the measurements or by tear sample collection. No direct data proving the existence of separate basal and reflex tear secretion have been published until now; it has become clear, however, that the composition and the osmolarity of tears changes with increasing secretion rates.[10-13]

Jones' idea inspired a number of clinical and experimental studies and had a clearly beneficial effect on the development of dakryology. Other opinions and explanations, however, were also published[14-16] arguing that both the main lacrimal glands and the rest of the tear secretory glands (accessory lacrimal glands, meibomian glands, palpebral glands, conjunctival glands) function under the influence of continuously acting stimuli. The effects of the regulating neuroendocrine system, and those of external stimuli, though are different under various circumstances, never can be completely eliminated. Neither of these glands function in a "black box", and it can hardly be accepted that any of these glands are completely turned off when the eye is closed. Various physical and psychic stimuli most likely reach all tear-secreting glands in sleep, too. The proper function of all tear glands seems to be necessary for the maintainance of the three-layered tear film and for the sufficient hydration of the ocular surface both when the eye is closed and when the eye is open.

This latter view may also be supported by the fact that, in diseases and in pathological states where the so-called basic secretors are missing or not functioning, no obvious signs of dryness develop in sleep; on the contrary, all disabling symptoms are present during the day when the "reflex secretors" are functioning. In diseases with decreased lacrimal gland secretion like Sjögren syndrome, on the other hand, the lack of sufficient hydration during the night most likely play a role in the development of epithelial lesions due to the temporary adhesions of the surface of the cornea and the inner surface of the upper lid.[17]

Based on the concept of the two basically different secretory systems and separate secretory mechanisms, Jones[7] also developed a test to measure basic tear secretion. The basic tear secretion test is a modification of the original Schirmer test, and is performed after the topical anesthesia of the conjunctiva and "gentle drying" the lower cul de sac with a cotton applicator. To decide whether such a test is suitable to measure basic tear flow eliminating the effects of external stimuli is beyond the scope of this book. This test certainly gives lower values for tear secretion than the original Shirmer test.

B. Normal Tear Secretion and the Turnover of Tears

Schirmer estimated the normal tear flow to be 0.6 to 0.8 µl/min, by measuring the time required to produce epiphora in patients with obstruction of the lacrimal puncta. He also developed a test, bearing his name, to measure tear secretion by placing a filter paper strip in the eye with a 5-mm long bent portion placed between the globe and the lower eyelid, and evaluating the length of the wetted part of the paper in 5 min.[18]

Kirschner measured the rate of tear secretion by instilling fluorescein into the conjunctival sac, and evaluating the change in the fluorescence of the tear samples collected periodically from the eye. He found values for tear flow close to those reported by Schirmer.[19]

Mishima et al.,[5] using a quantitative fluorophotometer, measured the concentration of fluorescein in and the disappearance of this dye from the lower tear meniscus. The average tear flow with this method was 1.2 µl/min, with a range from 0.5 to 2.2 µl/min. Sorensen[20] measured tear flow rate to be 0.6 µl/min using a technetium 99 tracer and a gamma counter. Both Mishima and Sorenson emphasized that measurements of tear flow should be made without causing reflex lacrimation, trying to exclude one of the important drawbacks of the Schirmer test. Mishima and colleagues have shown, however, that the instillation of 1 µl fluorescein solution also causes an increase in tear secretion, lasting up to 5 min in some eyes.[5]

Lamberts et al.[21] reported the results of a study of normal tear secretion performed with calibrated Schirmer strips. They found a tear flow of 2.51 ± 1.18 µl/min in unanesthetized eyes and 1.52 ± 1.00 µl/min in anesthetized eyes.

The estimated turnover rate of tears determined by Mishima et al.[5] with *in situ* fluorophotometry was 16%/min. Puffer et al.[22] using the same method found a mean value of 15% for the tear elimination coefficient in normal subjects.

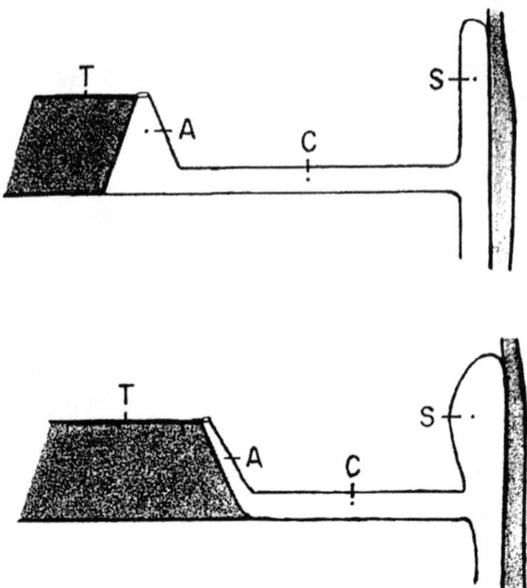

FIGURE 2. Schematic drawing of the Jones' lacrimal pump theory. Upper, lids open. Lower, lids closed. T, tarsus; A, ampulla; C, canaliculus; S, fundus of the tear sac. (From Jones, L. T., *Arch. Ophthalmol.*, 66, 137, 1961. With permission.)

C. Blinking and the Drainage of Tears

In addition to providing mechanical protection for the cornea and reforming the precorneal tear film, physiological movements of the lids during blinking play a crucial role in the drainage of the tear fluid through the excretory passages.

The mechanism of tear fluid drainage is most often explained by the "lacrimal pump" theory of Jones. According to this theory, during lid closure the canaliculi shorten and move medially at the same time as the lacrimal sac distends (Figure 2). The expansion of the sac would create negative pressure that would draw tear fluid into the sac via the canaliculi at the time of lid closure.[23,24]

Another theory published by Breinen and Snall[25] suggested that a pressure increase in the conjunctival sac at the moment of lid closure would provide "a positive, direct force" that would drive the tear fluid though the drainage channels.

Neither of these theories has taken into consideration the early works of Rosengren[26] and of Freiberg[27] concerning pressure changes in the lacrimal sac. Rosengren inserted a catheter to the lacrimal sac via the nasolacrimal duct and measured pressure changes during blinking. He detected a pressure rise in the lacrimal duct during the closing phase of the blink and a continuously increasing pressure following a series of spontaneous blinks. The pressure rose consistently with each lid closure until it was sufficient to cause the lacrimal sac to remain distended and full of fluid. He suggested that during each lid closure tear fluid is forced into the sac, and a valve-like mechanism prevents fluid from regurgitating back into the canaliculi during the opening phase of the blink.[26,28]

Doane[29] studied the interaction of the eyelids and tears in the dynamics of normal human eyeblink. He described a mechanism for tear drainage through the secretory pathways based on high-speed motion pictures of the eyelid movements and the precise analysis of changes in the tear film and tear menisci. During each blink cycle, the upper lid descends over the eye. The globe is pushed posteriorly as the lid descends, and returns to its anterior position as the lid rises. High-speed photography showed a 1.5- to 2.0-mm retropulsion of the globe during each blink.[29] It is notewothy that there is no upward rotation (no Bell's motion) of the globe during normal blinks. An average velocity of 190 mm/s was found for the upper lid during the blinks. The upper lid accelerates rapidly, reaching its maximum velocity by the time it crosses the visual axis, then decelerates to a stop, often reversing its motion before actually making contact with the lower lid. The opening (rising) phase of the upper lid's motion is much slower, with lid velocities about 100 mm/s; the last third of the opening movement is particularly slow. An average total blink time for natural, nonforced blinks is about one quarter of a second.[30]

The photography showed that the lacrimal papillae containing the punctal openings extend and rise from the marginal edge of the lids as the upper lid begins its downward sweep. The puncta are located near the medial junction of the upper and lower lids. These portions of the lid margins meet forcefully by the time the blink is less than half complete. Besides this abutting effect there is a muscular contraction around the proximal ends of the canaliculi closing the lacrimal puncta and preventing a back flow in the canaliculi.

Doane[30] suggested that the tear drainage process is driven by blinking, with the closing of the lids being associated with compression of the canaliculi and the lacrimal sac. As the lids close, the punctal openings are occluded, by either the abutting lid margins or a sphinchter-like closure of the puncta themselves. Further closure of the lids squeezes fluid from the canaliculi into the sac. As the lids open, pressure on the canaliculi is relaxed, allowing the elastic walls to expand to their relaxed shape. Since both ends of the canaliculi are functionally occluded at this point a negative pressure is created (there is a one-way valve at the distal end of the canaliculi near the sac). When the lids open sufficiently to expose the punctal openings, the negative pressure within the canaliculi draws in fluid from the tear menisci. This continues until the canaliculi are filled, or until the level of fluid in the menisci falls to a point where contact with the punctal openings is lost. The canaliculi full of fluid are ready to start the cycle again with the next blink.[30]

The tear drainage process is capable of transporting fluid at a much higher rate than that at which the tears are normally secreted. It is only during periods of crying, characterized by very high tear secretion rate, that the drainage system is overloaded and incapable of handling the fluid flow. Freiberg[27] showed that each blink could result in the transport of up to 2 µl of fluid through the canaliculi. Since a normal blink rate is about 16 blinks per minute, and the generally accepted normal rate of tear secretion is about 1 µl/min, it is easy to calculate that the drainage system is usually working at a small portion of its capacity. Maurice suggested that under normal circumstances the low amounts of tear fluid entering the lacrimal sac are probably absorbed by the mucosal walls of the sac and nasolacrimal duct, and only in the case of copious reflex tearing do tears actually enter the nasal cavity.[31]

REFERENCES

1. **Holly, F. J., Lamberts, D. W., and Buesseler, J. A.,** The human lacrimal apparatus: anatomy, physiology, pathology, and surgical aspects, *Plast. Reconstr. Surg.,* 74, 438, 1984.
2. **Berta, A. and Török, M.,** Soluble glycoproteins in aqueous tears, in *The Preocular Tear Film in Health, Disease, and Contact Lens Wear,* Holly, F. J., Ed., Dry Eye Institute, Lubbock, TX, 1986, 506.
3. **Berta, A. and Török, M.,** Tear glycoprotein determinations in the diagnosis and differential diagnosis of dry eyes, *Scand. J. Rheumatol.,* 61(suppl.), 228, 1986.
4. **Jones, L. T.,** Anatomy of the tear system, *Int. Ophthalmol. Clin.,* 13(1), 3, 1973.
5. **Mishima, S., Gasset, A., Klyce, S. D., Jr., and Baum, J. L.,** Determination of tear volume and tear flow, *Invest. Ophthalmol.,* 5, 264, 1966.
6. **Jones, L. T.,** An anatomical approach to problems of the eyelids and lacrimal apparatus, *Arch. Ophthalmol.,* 66, 137, 1961.
7. **Jones, L. T.,** The lacrimal secretory system and its treatment, *Am. J. Ophthalmol.,* 62, 47, 1966.
8. **de Rötth, A.,** Low flow of tears: the dry eye, *Am. J. Ophthalmol.,* 35, 782, 1952.
9. **de Roeth, A.,** Sr., Lacrimation in normal eyes, *Arch. Ophthalmol.,* 49, 185, 1953.
10. **Berta, A.,** Standardization of tear protein determinations. The effects of sampling, flow rate, and vascular permeability, in *The Preocular Tear Film in Health, Disease, and Contact Lens Wear,* Holly, F. J., Ed., Dry Eye Institute, Lubbock, TX, 1986, 418.
11. **Farris, R. L., Stuchell, R. N., and Mandel, I. D.,** Basal and reflex human tear analysis. I. Physical measurements: osmolarity, basal volumes, and reflex flow rate, *Ophthalmology,* 88, 852, 1981.
12. **Gilbard, J. P. and Dartt, D. A.,** Changes in lacrimal gland fluid osmolarity with flow rate, *Invest. Ophthalmol.,* 23, 804, 1982.
13. **Stuchell, R. N., Farris, R. L., and Mandel, I. D.,** Basal and reflex human tear analysis. II. Chemical analysis: lactoferrin and lysozyme, *Ophthalmology,* 88, 858, 1981.
14. **Baum, J. L.,** Clinical implications of basal tear flow, in *The Preocular Tear Film in Health, Disease, and Contact Lens Wear,* Holly, F. J., Ed., Dry Eye Institute, Lubbock, TX, 1986, 646.
15. **Berta, A.,** Collection of tear samples with or without stimulation, *Am. J. Ophthalmol.,* 96, 115, 1983.
16. **Jordan, A. and Baum, J.,** Basic tear flow. Does it exist?, *Ophthalmology,* 87, 920, 1980.
17. **Holly, F. J.,** Ocular surface disease update: diagnosis and treatment, manuscript.

18. **Schirmer, O.,** Studien zur Physiologie und Pathologie der Tranenabsonderung und Tranenabfuhr, *Graefes Arch. Klin. Ophthalmol.,* 56, 197, 1903.
19. **Kirschner, C.,** Untersuchungen über das Ausmass der Tranensekretion beim Menschen, *Klin. Monatsbl. Augenheilk.,* 144, 412, 1964.
20. **Sorensen, T.,** Determination of tear flow using a radioactive tracer, *Acta Ophthamol.,* 125 (Suppl.), 43, 1975.
21. **Lamberts, D. W., Foster, C. S., and Perry, H. D.,** Schirmer test after topical anesthesia and the tear meniscus height in normal eyes, *Arch. Ophthalmol.,* 97, 1082, 1979.
22. **Puffer, M. J., Neault, R. W., and Brubaker, R. F.,** Basal precorneal tear turnover in the human eye, *Am. J. Ophthalmol.,* 89, 369, 1980.
23. **Jones, L.,** Epiphora II. Its relation to the anatomic structures and surgery of the medial canthal region, *Am. J. Ophthalmol.,* 43, 203, 1957.
24. **Jones, L. T., Marquis, M. M., and Vincent, N. J.,** Lacrimal function, *Am. J. Ophthalmol.,* 73, 558, 1972.
25. **Brienen, J. A. and Snall, C. D.,** The mechanism of the lacrimal flow, *Ophthalmologica,* 159, 223, 1969.
26. **Rosengren, B.,** Zur Frage der Mechanik der Tranenableitung, *Acta Ophthalmol.,* 6, 367, 1928.
27. **Freiberg, T.,** Einige physiologische probleme der menschlichen Tranenabflusswege, *Z. Augenheilkd.,* 67, 1, 1929.
28. **Rosengren, B.,** On lacrimal drainage, *Ophthalmologica,* 164, 409, 1972.
29. **Doane, M. G.,** Interaction of eyelids and tears and the dynamics of the normal human eyeblink, *Am. J. Ophthalmol.,* 89, 507, 1980.
30. **Doane, M. G.,** Tear spreading, turnover and drainage, in *The Preocular Tear Film in Health, Disease, and Contact Lens Wear,* Holly, F. J., Ed., Dry Eye Institute, Lubbuck, TX, 1986, 652.
31. **Maurice, D.,** The dynamics and drainage of tears, *Int. Ophthalmol. Clin.,* 13(1), 103, 1973.

Chapter 2

THE PRECORNEAL TEAR FILM

The basic role of the lacrimal system is the formation and maintenance of the tear film covering the exposed, anterior surface of the eyeball. The functions of the tear film are (1) optical, (2) protective, (3) moisturizing, and (4) nutritive.

The average tear volume in the eye, as determined by dye dilution techniques, is 6 to 7 µl.[1] The distribution of this volume in the eye is as follows: 1 µl in the precorneal tear film, 2 to 3 µl in the tear menisci (tear strips by the lid margins), and 3 to 4 µl in the fornices. The secretion of the lacrimal glands gets first into the tear volume contained in the upper conjunctival fornix, then runs along the lid margins and is mixed with tears in the upper and lower menisci. The tear fluid forming the precorneal tear film is mixed with the secretion of the lacrimal glands during blinking; the lower fornix is thought to be the destination of the secreted tears. A considerable portion of the secreted tears never gets into the lower fornix because it is drained out from the marginal fornices by the suction of the upper and lower tear canaliculi before proper mixing occurs. The tear menisci are much smaller or seem to be missing in severe dry eyes. The total tear volume as determined by Doane and Dohlman is decreased only by 25% in patients with keratoconjunctivitis sicca suggesting that tear menisci may serve as a "reservoir" for aqueous tears.[2]

The turnover of the tear film depends on the rate of blinking. There is no exchange of fluid between the marginal tear menisci and the precorneal tear film, except during blinking. Blinking action is very efficient in mixing tear components in the precorneal tear film. A single blink can quite uniformly spread a 1.0-µl drop of fluorescein over the exposed surface of the cornea and the conjunctiva.

The preocular tear film is an approximately 10-µm thick fluid layer covering the anterior surface of the cornea and the bulbar conjunctiva. The tear film consists of three layers (Figure 1). The superficial lipid layer comprises about 1%, the middle aqueous layer over 98.5%, and the inner mucus layer less than 0.5% of its total thickness.[3]

The lipid layer consists mainly of nonpolar lipids: waxy and cholesterol esters. These lipids are fluid at body temperature despite their high cholesterol content and molecular weight. The melting point of most of the the tear lipids is around 35°C.[4] Polar lipids — triglycerides, free fatty acids, and phospholipids — are present only in negligible amounts.[1] The majority of tear lipids is produced by the meibomian glands of the tarsus of the lids and are secreted through orifices along the lid margins. The glands of Zeis at the palpebral margin and the glands of Moll at the roots of the eyelashes secrete lipids that contribute to the outermost lipid layer of the tear film. This oily layer is capable of retarding the evaporation of the aqueous layer, and also increases the

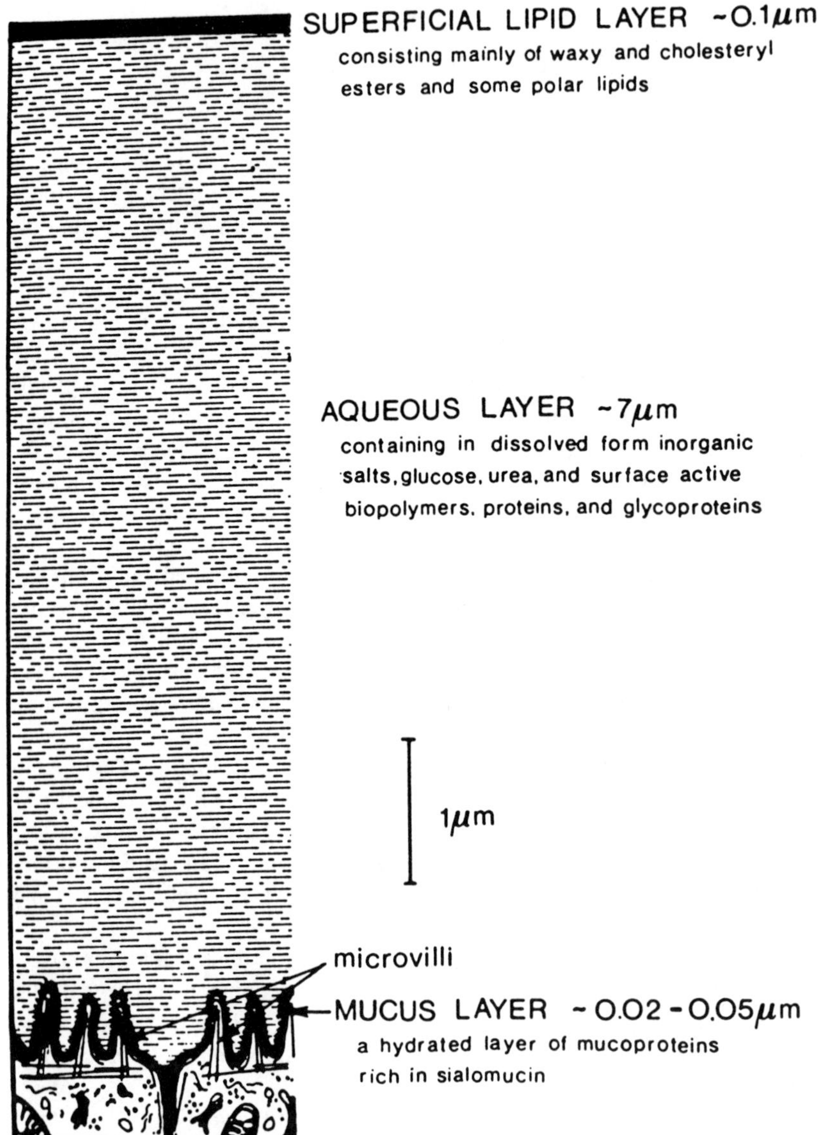

FIGURE 1. Structure and composition of the precorneal tear film. (From Holly, F.J. and Lemp, M.A., *Surv. Ophthalmol.*, 22, 69, 1977. With permission.)

stability of the tear film. Another important role of the superficial lipid layer may be the protection of the surface of the tear film from the invasion of highly polar lipids originating from the sebaceous glands of the skin surounding the palpebral fissure. Polar lipids, if they get into the tear film, almost instantaneously cause its rupture resulting in the formation of dry spots. The thickness of the superficial lipid layer is about 0.1 μm in a normal open eye.[6] Its thickness depends on the palpebral fissure width and changes during blinks.[7]

The aqueous layer is an aqueous solution of macromolecules, inorganic salts, and small molecular weight organic substances. It is known to be secreted by the main and accessory lacrimal glands. In humans the basic tear secretion was suggested to be due to the continuous secretion of the accessory lacrimal glands, while the main lacrimal glands are thought to produce aqueous secretion on different stimuli. Maurice[8] suggested that water flow across the epithelium that accompanies the active transport of inorganic ions may also contribute to the aqueous layer of the tear film. Some water flow through the walls of the conjunctival vessels is also possible. Transudation is more significant in cases with irritation and inflammation of the conjunctiva. There are considerable fluid movement and flow within the aqueous layer. The secretion of the lacrimal glands runs around the lid margins in the tear menisci. This movement is a hydraulic flow called the "lacrimal river" driven by pressure gradients between different parts of the tear film and the tear menisci. The precorneal tear film is so thin that surface forces are determining the fluid movements and the effect of gravitational forces is negligible. The aqueous layer also has a high hydraulic resistance. That is why there is no spontaneous bulk flow within the tear film, but substantial mixing occurs during blinks. Aqueous tears contain inorganic salts, small molecular weight organic components, and different proteins. The latter are functionally the most important components. The pH of the aqueous phase is normally slightly alkaline (about 7.35). A change of pH in the range 6.6 to 7.8 is tolerated without discomfort. Bicarbonate and proteins serve as the major buffer systems of the tear film. In spite of their action a shift toward slight acidity often occurs in different forms of inflammation and when there is an accumulation of lactate due to anaerobic energy production in the surrounding tissues. The osmolarity of aqueous tears is normally around 300 mOsmol, close to the osmolarity of physiologic NaCl solution. It has been suggested that hyperosmolarity of the tear film causes epithelial damage and may be an important causative factor in the development of the surface disturbances developing in dry eyes.

The inner, mucous layer of the tear film is composed of acidic glycoproteins, mainly sulfated glycoproteins (sulfomucins), but contain sialomucins, too. Mucus on the surface of the eye forms a gel-like layer, coating the cornea and the conjunctiva, being adherent to the microvilli of the superficial epithelial cells. The mucin is known to be secreted by the goblet cells of the conjunctiva. The density of goblet cells in the bulbar conjunctiva is $8.8/mm^2$. The mucous layer has been suggested to play a double role. One of its functions is to make the epithelial surface wettable for aqueous tears. The other function of this layer is to trap lipid contamination, thereby maintaining tear film stability. Besides the adsorbed mucous, solved glycoproteins and hydrated mucous particles are also dispersed in the aqueous phase of the tear film. On each blink these mobile mucous components are thought to help in recovering the mucin sheet at the dry, hydrophobic areas of the surface of the eyeball.[2,9]

The tear film covering the exposed surface of the eyeball is formed by the action of the moving eyelids during blinking. Aqueous tears are secreted by the main and accessory lacrimal glands through the orifices located in the upper fornix of the conjunctival sac. The tear fluid is spread on the mucous-covered epithelial surface as a result of lid movements. The aqueous layer is covered by a sheet of polar lipids. This layer is very thin, practically a monomolecular layer when the eye is wide open. During blinking the superficial lipid layer is compressed, becoming several times thicker, while the aqueous layer remains under the eyelid providing sufficient lubrication so that the lids do not injure the surface epithelium. The inner edge of the moving upper lid margin is thought to play a role in the distribution of the mucus secreted by the goblet cells of the conjunctiva, as well as in the removal of the debris, consisting of desquamated epithelial cells and lipid-contaminated mucus. When the eye is open again the compressed lipid layer will spontaneously spread over the aqueous layer at a very high speed. That is why an aqueous tear surface uncoated by lipids never exists.[3,10]

The dynamics of eyelid movements were determined by high-speed photography. The upper lid moves downward at a velocity of 250 mm/s in normal, and more than 400 mm/s in forced blinking. The stationary phase (where the upper lid changes direction from upward to downward) lasts 2 ms in normal blinking. The lower lid transverses in a nasally oriented vertical motion totally synchronized with the upper lid. The total globe is pushed 1 to 2 mm backward during each blink, and there is little or no tendency for the upward rotation of the globe (Bell's phenomenon) during normal blinks. The normal blinking rate varies considerably by age, sex, condition of the eye, and the effects of external stimuli. The average interval between blinks is 4 to 5 s.[11]

The precorneal tear film is thermodynamically unstabile. Processes leading to the rupture of the continuous fluid layer covering the cornea start right from the moment when the tear film is formed. Local thinnings develop in the formerly even fluid layer resulting in the formation of dry spots on the surface of the cornea. The time interval between the last blinking and the appearance of the first dry spot is called breakup time (BUT). The value of the BUT is normally 20 to 40 s.

Most of the problems of tear film stability and tear film breakup can be explained on the basis surface chemical considerations. The surface chemical equivalent of precorneal tear film is a thin fluid layer covering a solid hydrophilic surface. Such a fluid film will remain stable only if the sum of the surface tension of the fluid and the interfacial tension at the solid-fluid boundary is smaller than the surface tension of the bare solid. The surface tension of normal aqueous tears is 40 to 42 dyn/cm, as determined by Holly and Lemp.[12] The surface tension of aqueous tears can be attributed to some of its macromolecules of glycoprotein nature. The surface tension of the tear film *in situ* is to a large extent determined by the film pressure of the superficial lipid layer. The lipid film pressure is largely dependent on the actual palpebral slit width. The more the lipid layer is compressed during blinking the lower the surface tension becomes. The other factor determining tear film stability is the tear epithelium interfacial tension that is kept low by the mucus layer that makes the otherwise hydrophobic epithelial surface of the cornea to a large extent hydrophilic. Any disturbance lowering the surface tension of the tears either by changing its glycoprotein composition or lowering the film pressure of the superficial lipid layer, just like local effects diminishing the tear-epithelial interfacial tension, will cause the rupture of the precorneal tear film.[3,12]

The mechanism of tear film rupture, as suggested by Holly,[7] involves the migration of lipid molecules through the aqueous layer to the mucus sheet covering the cornea. As a result of this the otherwise hydrophilic epithelial surface locally and temporarily becomes hydrophobic and the rupture occurs. Factors leading to early tear film break up are (1) local thinning of the tear film due to circumscribed elevation of the epithelial surface, or irregularity of the corneal curvature; (2) thinning due to local Marangoni effect (surface tension gradient-driven liquid flow inside the tear film); (3) locally inadequate mucus layer generally occurring at sites where the microvilli of the surface epithelial cells or the epithelial cells themselves are missing due to erosions, local epitheliopathy, scars, squamous metaplasia, presence of vesiculi or bullae, etc.; (4) unevenly formed tear film due to lid deformities and/or lid movement disturbances, subepithelial calcified particles, or foreign bodies on the inner surface of the upper eyelid; (5) a pathological increase in the polar fraction of the superficial lipid layer (e.g., in meibomianitis, in chronic blepharitis, in rosacea dermatitis involving the lid margins); and (6) foreign bodies in the tear film or on the surface of the cornea (corneal foreign bodies, contact lenses, etc.).[3,13]

The original method for the determination of the BUT is based on measuring the time interval between the last blink and the appearance of the first dry spots on the cornea under slitlamp, using a cobalt blue filter, holding the lids of the patient to prevent blinking, following the instillation of a drop of fluorescein in the patient's eye. This method has been shown to be very variable and poorly reproducible. Several modifications have been advocated and attempts have been made for its standardization. The suggestions intending to reduce the variability of the BUT test include the establishment of uniform starting conditions by repeated (not enforced) blinkings, the use of small volumes of concentrated fluorescein drops without preservatives and surface-active contaminants, the avoidance the use of local anesthetics, and the elimination of external stimuli (intensive light, draft, etc.). These modifications did improve the reproducibility of the test but could not eliminate the fact that any intervention (adding something to the tear film, holding the lids of the patient) surely changes the situation in the eye and the measurements reflect tear film stability under these changed, nonphysiological conditions.

Different methods have been developed for the "noninvasive", "*in situ*", "physiological" measurement of tear film BUT. These methods eliminated most but not all the disturbing factors; to measure tear film stability under physiological conditions, sophisticated methods and expensive equipment are needed; this is why they have not yet been accepted for general use in ophthalmological practice.

According to Holly and Lemp[2] tear film abnormalities can be divided into five main groups. They were originally based on theoretical and experimental considerations but have been confirmed by clinical observations, too. These groups are as follows: (1) aqueous deficiencies, (2) mucus deficiencies, (3) tear lipid anomalies, (4) tear film abnormalities due to inadequate blinking, and (5) corneal epithelial surface diseases.

Aqueous tear deficiencies are by far the most common among the possible causes of the tear film abnormalities. There are a number of conditions known to lead to the decreased secretion of the lacrimal glands. They can be either congenital or acquired. Congenital aqueous tear deficiencies are rare diseases in which the dysfunction of the lacrimal gland is a part of genetically determined multiple disorders usually affecting the functions of several excretory glands. Riley-Day syndrome, anhydrotic ectodermal dysplasia, alacrima congenita, Cri du chat syndrome, Adie syndrome, and congenital familial anhydrosis are some of these congenital diseases. Among the acquired forms trauma, surgical removal, toxic drug effects, neuroparalytic conditions, tumors, and infiltrative systemic diseases (leukemias, sarcoidosis, amyloidosis) have to be taken into consideration. Besides these diverse etiologies one clinical entity, Sjögren's syndrome, is the most common and the most important among the possible causes of aqueous tear deficiency. Sjögren's syndrome is considered to be an autoimmune disease that is characterized by chronic inflammation in the lacrimal, salivary, and other excretory glands that are affected and progressively damaged by activated autoaggressive lymphocytes.

Mucus deficiencies form a more compact group of diseases. Practically all of the diseases causing mucus tear deficiency do that by affecting the secretion of or destroying the goblet cells of the conjunctiva. Goblet cell dysfunction is also known to develop in vitamin A deficiency. The most common causes of goblet cell destruction are alkali burns, cicatriceal conjunctival pemphigoid, Stevens Johnson syndrome, Lyel syndrome, and trachoma. There are drug-induced conditions like practolol and echothiopathe iodide intoxication that may also lead to damage of the conjunctiva. These conditions are usually characterized by a decrease of goblet cell density, which can be determined by filter paper imprint technique. The level of mucus glycoproteins was also shown to decrease in the tears of these patients. Changes in the composition of tear glycoproteins is probably even more informative in the diagnosis and differential diagnosis of these diseases.[9]

Lipid abnormalities are usually divided into quantitative and qualitative disorders of the secretion of the meibomian glands and other sebaceous excretory glands of the upper and lower lids. Severe anhydrotic ectodermal dysplasia is the only known condition leading to total tear lipid deficiency. This disease is characterized, among others, by the congenital absence of the excretory glads of the lids. Much more often occurring disorders are the so-called quantitative tear lipid abnormalities. The most common causes for changes in the composition of tear lipids are the different types of blepharitis and blepharoconjunctivitis. The bacteria that invade the meibomian glands are capable of secreting lipases that produce free fatty acids from the lipids of the normal meibomian gland secretion. Free fatty acids, like other polar lipids, are extremely surface active and can cause the rupture of the otherwise stable tear film. Besides decreasing tear film stability fatty acids have been suggested to have a direct toxic effect on corneal epithelium. Both of these mechanisms may play a role in the formation of punctate superficial corneal epithelial defects commonly seen in blepharitis.

To the fourth group of tear film anomalies belong disorders that are brought about by the impaired resurfacing function of the eyelids. Disturbances in the form, position, or movement of the eyelids, breaking the apposition of the lid margins to the ocular surface or hampering its physiologic movements during blinking, can potentially produce areas of dryness. The most common conditions leading to inadequate blinking are lagophthalmus and proptosis, which may lead to exposure keratitis, as well as ectropium, entropium, certain forms of symblepharon, large lid notches, tumors, scars, and other major lid deformaties, like Vth or VIIth nerve palsies, that cause the inhibition either of the afferent or of the efferent line of the blinking reflex.

Major deformities of the ocular surface, like deforming scars, bullous keratopathy, pterygia, ectatic corneal dystrophies, erosions, ulcers, dellen formation, etc., may also impair the resurfacing function of otherwise normal eyelids. Such irregular, elevated, or depressed areas will predispose the tear film to break up instantly at that spot. The role of surface epitheliopathy, as a primary disease, in the development of tear film abnormalities is still debated. Epithelial defects, shown by punctate or filiform staining of the corneal surface, sooner or later develop in the course of the disease irrespective of the specific cause. It has been suggested, however, that multiple minor defects of the epithelial surface occurring in epithelial or endoepithelial dystrophies, in superficial forms of keratitis, Thygeson keratitis, etc., may lead to the development of or may exaggerate the symptoms of the already existing dry eyes.

REFERENCES

1. **Mishima, S., Gasset, A., Klyce, S. D., Jr., and Baum, J. L.,** Determination of tear volume and tear flow, *Invest. Ophthalmol.,* 5, 264, 1966.
2. **Holly, F. J. and Lemp, M. A.,** Tear physiology and dry eyes, *Surv. Ophthalmol.,* 22, 69, 1977.
3. **Holly, F. J.,** Tear film physiology and contact lens wear. I. Pertinent aspects of tear film physiology, *Am. J. Optom. Physiol. Opt.,* 58, 324, 1981.
4. **Brown, S. I. and Dervichian, D. G.,** The oils of the meibomian glands: physical and surface characteristics, *Arch. Ophthalmol.,* 82, 537, 1969.
5. **Andrews, J. S.,** The meibomian secretion, *Int. Ophthamol. Clin.,* 13(1), 23, 1973.

6. **McDonald, J. E.,** Surface phenomena of tear films, *Trans. Am. Ophthalmol. Soc.,* 66, 905, 1968.
7. **Holly, F. J.,** Formation and rupture of the tear film, *Exp. Eye Res.,* 15, 515, 1973.
8. **Maurice, D. M.,** The dynamics and drainage of tears, in *The Preocular Tear Film and Dry Eye Syndromes,* Holly, F. J. and Lemp, M. A., Eds., *Int. Ophthamol. Clin.,* 13(1), 103, 1973.
9. **Berta, A. and Török M.,** Tear glycoprotein determinations in the diagnosis and differential diagnosis of dry eyes, *Scand. J. Rheumatol.,* 61(suppl.), 228, 1986.
10. **Berger, R. E. and Corrsin, S.,** A surface tension gradient mechanism for driving precorneal tear film after blink, *J. Biomech.,* 7, 225, 1974.
11. **Doane, M. G.,** Tear spreading, turnover and drainage, in *The Preocular Tear Film in Health, Disease, and Contact Lens Wear,* Holly, F. J., Ed., Dry Eye Institute, Lubbock, TX, 1986, 652.
12. **Holly, F. J. and Lemp, M. A.,** Surface chemistry of the tear film; implications for dry eye syndromes, contact lenses, and ophthalmic polymers, *J. Contact Lens Assoc. Am.,* 5, 1, 1971.
13. **Holly, F. J.,** Tear film physiology, *Am. J. Optom. Phys. Opt.,* 57, 252, 1980.

Chapter 3

TEAR PROTEIN DETERMINATIONS*

The collection of a small tear sample and its analysis with readily available biochemical methods seem to be a simple way of obtaining information on pathological events occurring in the anterior segment of the eye. The relative simplicity and the noninvasive nature of tear sampling and its subsequent analysis have motivated researchers to study the composition of normal and pathological tears during the past 150 years. Several types of compounds have been detected in human tears; however, tear proteins seemed to be the most valuable for the purposes of clinical diagnosis.

The development of useful clinical tear protein tests theoretically is not a complex problem. Tear protein concentrations must be determined in normal tears and changes in their levels correlated with clinical signs and symptoms of various eye diseases. However, anybody who has attempted this task knows that this is not as simple as it appears. The accuracy and reproducibility of tear protein determinations are influenced by a variety of factors.

The first group of these factors results from the complexity of the structure and function of the lacrimal system. These may be termed as physiological factors:

1. Tears in the conjunctival sac consist of the secretions of the main and accessory lacrimal glands, the conjunctival goblet cells, and the palpebral meibomian glands.
2. Tears arise from a heterogeneous (multiphase) system. The multilayered film consists of a lipid, an aqueous, and a mucus phase. Furthermore, even within one phase (layer) the mixing is incomplete as the composition of tears was shown to be different at different areas of the eye.
3. The ever-changing rate of tear production and protein synthesis, the incomplete mixing of tear constituents, and the effects of the varying rates of drainage, evaporation, and absorption by the corneal and conjunctival tissue further complicate matters.

Other important factors are pathological changes caused by inflammation, cell destruction, the presence of microorganisms, specific and nonspecific reactions to trauma, physical or chemical insults, or the invasion of microorganisms:

---

* This chapter is based on a paper presented by the author at the 1st International Tear Film Symposium (Nov. 7–10, 1984. Lubbock, TX) and was published in the proceedings of the symposium, *The Preocular Tear Film in Health, Disease and Contact Lens Wear*, Holly, F.J., Ed., Dry Eye Institute, Lubbock, TX. 1986, 418.

19

1.   The various tear film-producing glands change their function in response to both different pathological states (ocular and systemic diseases) and locally or systematically administered drugs.
2.   Any change in the permeability of conjunctival vessels facilitates an influx of water, salts, small molecular weight organic compounds, and macromolecules like proteins from the serum to the tears.
3.   Cell destruction and physiological desquamation of epithelial cells may cause an accumulation of intracellular compounds (e.g., lysosomal enzymes) in tears.
4.   There is growing evidence that various compounds such as glycoproteins and immunologic factors may be produced locally by the epithelial cells and by immune-competent cells in subconjunctival follicles.

Different methods and conditions of tear sample collection change the results of tear protein determination to a large extent. Tear sample collection can be performed

1.   With or without stimulation
2.   With filter papers, sponges, or capillary tubes
3.   At various tear flow rates
4.   With or without touching (damaging) the conjunctiva
5.   At different locations in the eye (lower fornix, tear meniscus, near the outer or the inner canthus, etc.)
6.   Under various external and internal conditions (light, temperature, air currents, under different physical or emotional stresses)

The use of various methods for protein determinations makes the comparison of the results even more difficult. The following factors may alter protein quantitation:

1.   Cells and mucus threads present in tear samples (if not removed) may change the result of protein determinations.
2.   Storage,[1] especially repeated freezing and melting, may cause the denaturation of tear proteins.
3.   When an absorbing porous medium (sponge, filter paper) is used for tear collection, the extracted tear composition is affected by the absorbed tear-absorbing medium weight ratio.
4.   Contamination of samples during handling, concentration by evaporation, loss of proteins during concentrating, and diluting procedures.
5.   Differences due to the use of various spectrophotometric, electrophoretic, and immunological methods.
6.   Differences in staining, calibration, and evaluation.
7.   The lack of generally accepted procedures, conditions, and standards makes the results of different authors incomparable.

The results of tear protein concentration determinations are influenced by a number of factors. Three of them appear to be the most important. These are

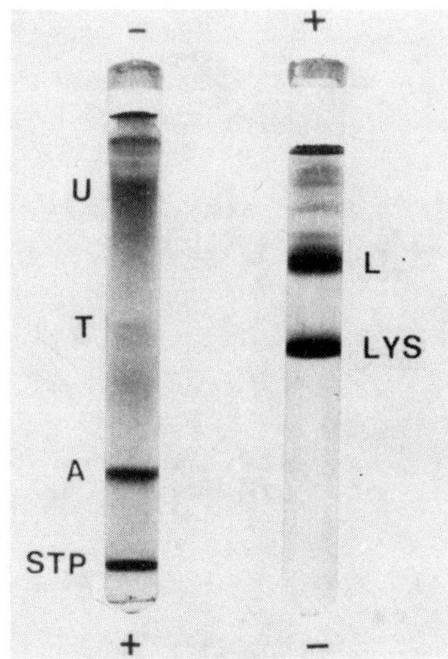

FIGURE 1. Polyacrylamide-gel electrophoretic pattern of normal tears: (a) in the nondenaturing (nondissociating) electrophoretic system of Ornstein and Davis; (b) in the system of Reisfeld. Abbreviations: STP, specific tear prealbumin; A, serum albumin; T, transferrin; U, unidentified major tear protein; L, lactoferrin; LYS, lysozyme. (From Berta, A., *The Preocular Tear Film in Health, Disease and Contact Lens Wear*, Holly, F. J., Ed., Dry Eye Institute, Lubbock, TX, 1986, 418. With permission.)

(1) tear secretion rate, (2) vascular permeability, and (3) type of the sampling method. To study the effects of these factors we investigated the protein content and composition of normal and pathological tear samples collected with different methods, at various flow rates, and under different conditions characterized by changes in the permeability of conjunctival vessel walls. The protein concentration in the tear samples was determined with the folin phenol reagent according to the method of Lowry.[2] Polyacrylamide-gel electrophoresis was performed under nondenaturing conditions in the system of Ornstein[3] and Davis[4] at pH 8.9, and in the system of Reisfeld[5] at pH 4.5, as well as under reducing conditions according to Laemmli[6] (Figures 1 and 2). Gels were stained with Coomassie brillant blue,[7] with Amido black,[8] or with Fast green.[9] The concentrations of major tear proteins were determined densitometrically using standards run, stained and evaluated in the same electrophoretic systems.

Total protein concentration values were plotted against tear secretion rates. The protein concentration in tear samples collected at low flow rates was significantly higher than that of tears collected at higher secretion rates. The protein content of tears remained relatively constant between flow rate values 10 and 60 μl/min. Tear protein concentration values appeared to decline with increasing secretion rates greater than 60 μl/min, but the change was not statistically significant (Figure 3).

The concentrations of major tear protein components were determined by the densitometric evaluation of the gels. The level of specific tear prealbumin

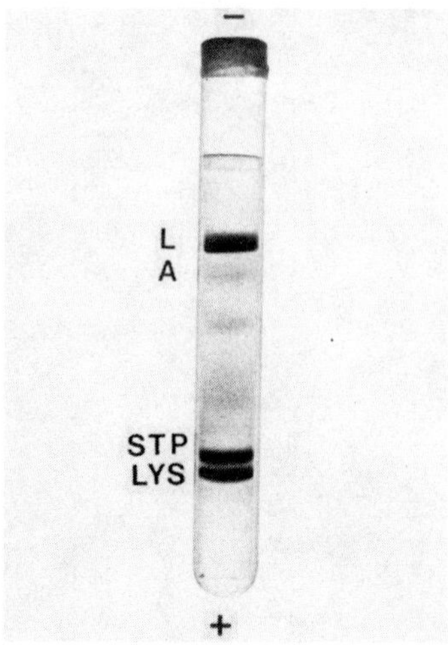

FIGURE 2. Polyacrylamide-gel electrophoretic pattern of normal tears in the denaturing (dissociating, SDS) electrophoretic system of Laemmli. Abbreviations: STP, specific tear prealbumin; A, serum albumin; L, lactoferrin; LYS, lysozyme. (From Berta, A., *The Preocular Tear Film in Health, Disease and Contact Lens Wear*, Holly, F. J., Ed., Dry Eye Institute, Lubbock, TX, 1986, 418. With permission.)

was 1.84 (±0.43) mg/ml, that of lysozyme 2.01 (±0.68) mg/ml, and that of lactoferrin 2.24 (±0.84) mg/ml. The concentration values did not change with flow rates up to 60 µl/min. There was a decrease in these values at very high flow rates but the change was marginal (Figure 3). Serum albumin was found in all of our samples in considerable amounts. Its concentration was higher at low flow rates, then it decreased to the level of 0.25 mg/ml at a rate of 20 µl/min, where it remained invariant even at very high flow rates (Figure 4). Two minor components, transferrin and immunoglobulin G (IgG) were determined in tears with radial immunodiffusion. Transferrin was found in normal tears only at low flow rates. We could demonstrate IgG in tears collected at high flow rates, but at a lower concentration than its level at low secretion rates (Figure 4).

The concentration of serum albumin, transferrin, and Ig was found higher in irritative states when the permeability of conjunctival vessels was increased. There was a decrease in the concentration of tear-specific proteins in these cases probably due to the dilution of tears through transudation. Any mild trauma that irritated the conjunctiva resulted in the rise of the levels of serum proteins in tears. The total protein concentration of these samples was also higher. A decrease was observed in the level of lysozyme, lactoferrin, and tear-specific prealbumin in samples collected at high (more than 70 µl/min) and at nearly basal flow rates (less than 5 µl/min). Very few minor components were detected in the former case, while minor components were present in significant amounts in tears from the latter case.[10,11]

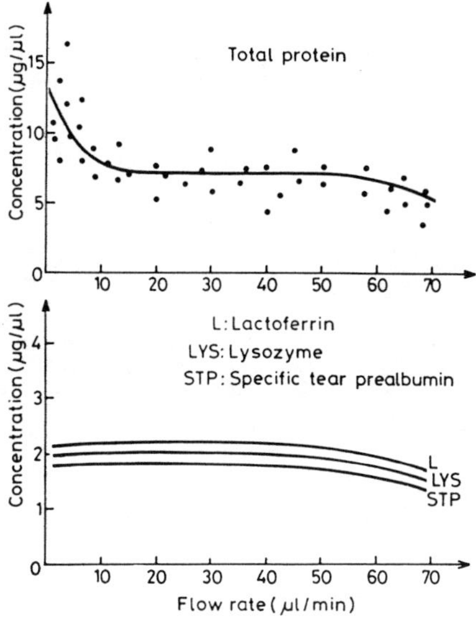

FIGURE 3. The total protein concentration of normal tears collected with capillary tubes at different flow rates (upper part). The levels of tear-specific major protein fractions in normal tears collected with capillary tubes at different flow rates (lower part). (From Berta, A., *The Preocular Tear Film in Health, Disease and Contact Lens Wear,* Holly, F. J., Ed., Dry Eye Institute, Lubbock, TX, 1986, 418. With permission.)

Tears are an ever-changing mixture of the secretions of different tear-secreting glands that vary their rate and probably the composition of their secretions under the effects of external and internal (neuroendocrine) stimuli. The main lacrimal gland has a unique property: it can increase its rate of production several hundredfold within seconds. Thus, the relative amount of lacrimal fluid from the main lacrimal gland in the mixture of secretions increases considerably after stimulation. As a result, the secretions of other tear-producing glands are diluted to different extent by lacrimal fluid at various flow rates. Therefore, the rate of tear secretion must be determined during sample collection so that concentration values measured in samples that were collected at different flow rates can be made comparable to each other.

An increase in the permeability of the conjunctival vessels changes the protein content and composition of tear samples to a great extent. The amount of serum proteins increases; the relative and absolute amount of tear proteins may decrease due to diluting effects. Therefore, vascular permeability has to be taken into consideration when evaluating and interpreting the results of tear protein determinations.

The tear collection method and its conditions must be defined precisely as different sampling methods influence the results of concentration determinations.

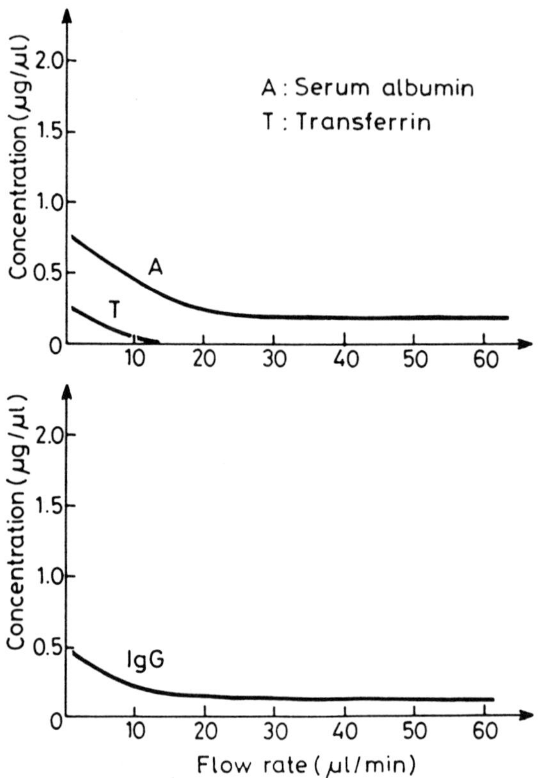

FIGURE 4. The levels of serum albumin, transferrin, and IgG in normal tears collected with capillary tubes at different flow rates. (From Berta, A., *The Preocular Tear Film in Health, Disease and Contact Lens Wear,* Holly, F. J., Ed., Dry Eye Institute, Lubbock, TX, 1986, 418. With permission.)

It is important to know the origin of protein fractions in tears because various influencing factors have different effects on the concentration of proteins with different origin. The levels of serum proteins are higher, and the levels of tear-specific proteins may be lower in case of increased permeability. The concentration of proteins originating from the conjunctiva decreases with increasing flow rates, while the concentration of proteins produced by the lacrimal gland appears to be unchanged at high flow rates. Lysozyme, lactoferrin, tear-specific prealbumin, and secretory IgA are known to be secreted by the main lacrimal glands. They were demonstrated in human lacrimal glands by immunohistological techniques.[12,13] They were shown to be secreted by the lacrimal gland in tissue culture.[14] A high degree of correlation was found between changes in the concentration of lysozyme and other tear-specific proteins in persons with normal lacrimal gland function and in patients with keratoconjunctivitis sicca.[15,16]

Avisar et al.,[17] studying the lysozyme content of tears in external eye infections, found lower levels at high flow rates. Stuchell et al.[18] found the concentration of lysozyme significantly greater in reflex tears than in basal tears. Others found no differences in the lysozyme content of tears collected at different flow rates and with different sampling methods.[19-23] The same has been found for lactoferrin by Kijlstra et al.[24] The levels of serum albumin, transferrin, and IgG were shown to be higher when tears were collected with filter papers, probably due to the irritating effect of papers.[23] Intracellular enzymes were demonstrated in tears collected with Schirmer strips but were absent from tears obtained with capillary tubes.[25]

The total protein concentration in the secretion of the main lacrimal gland was studied in rabbits at different flow rates by Dartt and Botelho.[1] Protein concentration was found to be higher at low flow rates (2 to 10 μl/min) than at flow rates above 10 μl/min. The osmolarity of rabbit lacrimal gland fluid was found to be high at extremely low flow rates and was shown to decrease at higher flow rates.[26]

The total protein concentration of tear samples collected at low flow rates (less than 10 μl/min) were shown to be higher than at higher flow rates in our experiments. A number of minor components were demonstrated in tears at low flow rates, which were not present in stimulated tears. The levels of the three tear-specific major components did not change with flow rate in the range of 2 to 50 μl/min. The high total protein concentration in basal tears may be due to the presence of minor components that are not the products of the main lacrimal gland but are probably produced by other lacrimal, conjunctival, and palpebral glands. Some components may be derived from serum or epithelial cells.[11]

These observations suggest that the composition of the main lacrimal gland fluid does not change at different rates of tear production. The secretions of other tear-producing glands (with high total protein concentration) are probably mixed with small volumes of the secretion of the main lacrimal gland (with relatively lower protein concentration) at low flow rates and with greater volumes at high flow rates. The contribution of other glands to the protein content of tear samples is so small in reflex tears that protein concentration does not change with flow rates. A decrease in total protein concentration and in the levels of lysozyme, lactoferrin, and tear-specific prealbumin was observed at extremely high flow rates that may be the sign of "exhaustion" of the lacrimal gland. Protein synthesis may not be able to keep pace with a secretion of that magnitude.

We observed a decrease in the levels of tear-specific proteins at extremely low flow rates, which may be the result of the dilution of small volumes of lacrimal gland secretion by other fluids. A similar effect was observed in cases of increased permeability when tears were diluted through transudation from the serum.

Serum albumin was reported by Janssen and van Bijsterveld[14] to be absent from stimulated tears collected with capillary tubes. Serum albumin is usually present in all tear samples, even those collected at extremely high flow rates. In a study we found that serum albumin level in tears was higher at low flow rates but it was always less than 1.0 mg/ml in normal cases. Its concentration was around 2.0 mg/ml when tears were collected with filter papers and was generally above 2.0 mg/ml in irritative cases (conjunctivitis, iritis, postoperative uveitis, etc.).[11] Transferrin was shown to be present in normal tears collected with capillary tubes at flow rates lower than 10 μl/min. IgG was found in tears at a concentration lower than 0.2 mg/ml except for flow rates lower than 10 μl/min when its concentration was between 0.2 and 0.5 mg/ml. The levels of transferrin and IgG were also elevated in cases with increased vascular permeability. The results of protein determinations may differ greatly because of the use of various methods. This can be best illustrated by the variability of tear lysozyme levels produced by different authors with different methods in different laboratories.[20,27,28] The exact circumstances of sampling and protein determinations must be well defined and kept constant so that the results of different authors can be compared.

The same is true for electrophoretic-densitometric protein determinations. The conditions of the complex procedure must be kept constant to have reliable and reproducible data. There is an important advantage of this method: the concentration of the four major tear protein components (lysozyme, lactoferrin, specific tear prealbumin, and serum albumin) and the level of a few minor components can be determined simultaneously from a small sample of 1 to 2 μl.[10] In our opinion, sample collection always causes undesired and uncontrolled stimulation to some degree. Any change in tear production rate or in the permeability of conjunctival vessels alters the results of tear protein determinations. It is better to use well-controlled stimulation methods than to try avoiding irritation.[29] Protein concentration values must be evaluated as a function of flow rates, corrected to the degree of dilution, or expressed in relative amounts compared either to the total protein content or to the concentration of serum protein present in the tears.

Our tear samples were collected with stimulation (one drop of nasally applied 70% ethanol). This proved to be a very convenient method for stimulation. Ethanol has no unpleasant smell; therefore, it could be used in outpatient departments, wards, and operating rooms. There was no problem with dosage. It was easily accepted by the patients (unlike tear gases). It did however cause an unpleasant sensation lasting for about 30 s. This was well tolerated by the patients when they were prepared in advance for the discomfort that would last only a few seconds and would cause no harm.

We have shown that changes in flow rate, vascular permeability, and the use of different methods for sampling and biochemical analysis alter the results of tear protein determinations to a great extent. Generally accepted methods, circumstances, and standards of sampling and protein determinations are needed so that the results obtained in the future can be better defined, more precise, and thus more comparable.

Normal tears collected at different flow rates and with different sampling methods and tears of patients with external eye diseases were studied by SDS polyacrylamide-gel electrophoresis. Different conditions of sampling and biochemical analysis altered the results of protein determinations to a great extent.

The total protein concentration (determined with the method of Lowry[2] or the method of Bradford[30]) is usually higher at low flow rates than in reflex tears. The levels of lysozyme, lactoferrin, and specific tear prealbumin (determined by the densitometric evaluation of polyacrylamide gels) are constant in the flow rate range between 2 and 50 µl/min. A decrease in their levels was observed at extreme low and high flow rates. Serum albumin and IgG was found in considerable amounts in normal tears even at high flow rates. Their concentration (together with transferrin, which was present in normal tears only at low flow rates) was higher in tears when irritation was present (acute conjunctivitis, iritis, postoperative uveitis, etc.). The levels of specific tear proteins were shown to be lower in cases with increased vascular permeability.

We suggest that flow rate determinations are necessary during sample collection, the extent of vascular permeability must be taken into consideration, and the exact circumstances of sample collection and protein determinations must be defined and kept constant so that the published results of various authors can be compared.

REFERENCES

1. **Dartt, L. A. and Botelho, S. Y.,** Protein in rabbit lacrimal gland fluid, *Invest. Ophthalmol. Vis. Sci.,* 18, 1207, 1979.
2. **Lowry, O. H., Rosenbrough, N. J., Farr, A. L., and Eandall, R. J.,** Protein measurement with the folin phenol reagent, *J. Biol. Chem.,* 193, 265, 1951.
3. **Ornstein, L.,** Disc electrophoresis. I. Background and theory, *Ann. N.Y. Acad. Sci.,* 121, 321, 1964.
4. **Davis, B. J.,** Disc electrophoresis. II. Method and application to human serum proteins, *Ann. N.Y. Acad. Sci.,* 121, 404, 1964.
5. **Reisefeld, R. A., Lewes, U. J., and Williams, D. E.,** Disc electrophoresis of basic proteins and peptides on polyacrylamide gels, *Nature,* 195, 281, 1962.
6. **Laemmli, U. K.,** Cleavage of structural proteins during assembly of the head of bacteriophage T4, *Nature,* 227, 680, 1970.
7. **Chrambach, A., Reisfeld, R. A., Wyckoff, M., and Zeccari, J.,** A procedure for rapid and sensitive staining of protein fractionated by polyacrylamide-gel electrophoresis, *Anal. Biochem.,* 20, 150, 1967.
8. **Johns, E. W.,** The electrophoresis of histones in polyacrylamide-gel and their quantitative determination, *Biochem. J.,* 104, 78, 1967.
9. **Gorovsky, M. A., Carlson, K., and Rosenbaum, J. L.,** Simple method for quantitative densitometry of polyacrylamide gels using fast green, *Anal. Biochem.,* 35, 359, 1970.
10. **Berta, A.,** A polyacrylamide-gel electrophoretic study of human tear proteins, *Graefe's Arch. Clin. Exp. Ophthalmol.,* 219, 95, 1982.
11. **Berta, A.,** Standardization of tear protein determinations. The effects of sampling, flow rate, and vascular permeability, in *The Preocular Tear Film in Health, Disease and Contact Lens Wear,* Holly F.J., Ed., Dry Eye Institute, Lubbock, TX, 1986, 418.
12. **Franklin, R. M., Kenyon, K. R., and Tomasi, T. B., Jr.,** Immunohistologic studies of human lacrimal gland: localization of immunoglobulins, secretory component and lactoferrin, *J. Immunol.,* 110, 984, 1973.

13. **Gillette, T. W., Allansmith, M. R., Greiner, J. V., and Janusz, M.,** Histologic and immunohistologic comparison of main and accessory lacrimal tissue, *Am. J. Ophthalmol.,* 89, 724, 1980.
14. **Janssen, P. T. and van Bijsterveld, O. P.,** Origin and biosynthesis of human tear fluid proteins, *Invest. Ophthalmol. Vis. Sci.,* 24, 623, 1983.
15. **Janssen, P. T. and van Bijsterveld, O. P.,** A simple test for lacrimal gland function: a tear lactoferrin assay by radial immunodiffusion, *Graefe's Arch. Clin. Exp. Ophthalmol.,* 220, 171, 1983.
16. **Janssen, P. T. and van Bijsterveld, O. P.,** The relations between tear fluid concentrations of lysozyme, tear specific prealbumin and lactoferrin, *Exp. Eye Res.,* 36, 773, 1983.
17. **Avisar, R., Menache, R., Shaked, P., and Savir, H.,** Lysozyme content of tears in some external eye infections, *Am. J. Ophthalmol.,* 92, 555, 1981.
18. **Stuchell, R. N., Farris, R. L., and Mandell, I. D.,** Basal and reflex tear analysis. II. Chemical analysis: lactoferrin and lysozyme, *Ophthalmology,* 88, 858, 1981.
19. **Copeland, J. R., Lamberts, D. W., and Holly, F. J.,** Investigation of the accuracy of tear lysozyme determination by the Quantiplate TM method, *Invest. Ophthalmol. Vis. Sci.,* 22, 103, 1982.
20. **Grabner, G., Formanek, I., Dorda, W., and Luger, T.,** Human tear lysozyme. A comparison of electro-immunodiffusion, radial immunodiffusion and spectrophotometric assay, *Graefe's Arch. Clin. Exp. Ophthalmol.,* 218, 265, 1982.
21. **de Konig, E. W. J. and van Bijsterveld, O. P.,** Schirmer test values and lysozyme content of tears in acute dendritic keratitis, *Invest. Ophthalmol. Vis. Sci.,* 25, 55, 1984.
22. **Mackie, I.A. and Seal, D.V.,** Quantitative tear lysozyme assay in units of activity per microliters, *Br. J. Ophthalmol.,* 60, 70, 1976.
23. **Stuchell, R. N., Feldman, J. J., Farris, R. L., and Mandel, I. D.,** The effect of collection technique on tear composition, *Invest. Ophthalmol. Vis. Sci.,* 25, 374, 1984.
24. **Kijlstra, A., Jenrissen, S. H. M., and Koning, K. M.,** Lactoferrin levels in normal human tears, *Br. J. Ophthalmol.,* 67, 199, 1983.
25. **van Haeringen, N. J. and Glasius, E.,** The origin of some enzymes in tear fluid, determined by comparative investigation with two collection methods, *Exp. Eye Res.,* 22, 267, 1976.
26. **Gilbard, J. P. and Dartt, D. A.,** Changes in rabbit lacrimal gland fluid osmolarity with flow rate, *Invest. Ophthalmol. Vis. Sci.,* 23, 804, 1982.
27. **Ensink, F. T. E. and van Haeringen, N. J.,** Pitfalls in the assay of lysozyme in human tear fluid, *Ophthal. Res.,* 9, 366, 1977.
28. **Huemmer, P. and Rieger, G.,** Zur Problematik der Standardisierung des Lysozymgehaltes in der Tranenflussigkeit, *Klin. Mbl. Augenheilk.,* 184, 43, 1984.
29. **Berta, A.,** Collection of tear samples with or without stimulation, *Am. J. Ophthalmol.,* 96, 115, 1983.
30. **Bradford, M. M.,** A rapid and sensitive method for the quantitation of microgram quantities of protein utilizing the principle of protein-dye binding, *Anal. Biochem.,* 72, 248, 1976.

Chapter 4

ENZYMES:
STRUCTURE, FUNCTION, AND CLASSIFICATION

The enzymes make up the largest and most highly specialized class of proteins. They catalyze thousands of chemical reactions that collectively constitute the intermediary metabolism of cells and are the primary instruments for the expression of gene action. Over a thousand different enzymes have been identified so far, and there are genetic grounds for suspecting that many more remain to be discovered. Many of these have been isolated in pure homogeneous form, and over 200 have been crystallized.[1]

Enzymes are macromolecules of protein nature and of globular achitechture. Enzymes are built up of one or more polypeptide chains in which amino acids are bound together by peptide bonds (primary structure). Depending on the amino acid sequence the enzyme molecule contains varying amounts of α-helix and β-conformation (secondary structure). X-ray analysis of the three-dimensional conformation of the enzyme molecules revealed that their chains are folded forming a globular (tertiary) structure, leaving little space in the interior for water molecules. The globular form of the enzyme molecule is stabilized primarily disulfide bonds, crosslinks between special parts of a polypeptide chain or between different polypeptide chains (quaternary structure). Most polar groups are on the surface and are hydrated, while the hydrophobic residues remain inside the molecule futher stabilizing the globular conformation. A special part of the enzyme molecule, where the catalyzed reaction occurs, usually located in the hydrophobic region, is the called the active site. Those parts of the molecule that bind the subsrate(s), in or very near to the active site, are called the substrate binding sites.[2]

Some enzymes are simple proteins, others are conjugated proteins, containing cofactors: prosthetic groups, coenzymes, or both. Prosthetic groups are tightly bound to the enzyme molecule. They are either metallic ions or other nonprotein structures, like the heme group in cytochrome C, covalently bound to the polypeptide chain of the enzyme. Coenzymes are complex small molecular weight organic molecules loosely attached to the enzyme molecule. The enzyme-cofactor complex is called a holoenzyme. When the cofactor is removed, the remaining protein, which is inactive, is called an apoenzyme. Cofactors usually serve as intermediate carriers of electrons, hydrogen atoms, or of specific groups (amino groups, methyl groups, acetyl groups, etc.), and aid their transfer from one substrate to another. Many coenzymes contain vitamins, organic nutrients required in trace amounts for normal cell function.[1]

The molecular weight of enzymes ranges from 10,000 to more than 1,000,000 Da. There is a good correlation between the molecular weight and size of enzyme molecules. These differences in molecular weight and size govern both their diffusion in tissues and their ability to pass physiological barriers: vessels walls, basal membranes, cytoplasmic membranes, etc.

Enzymes function as biological catalysts. They increase the rate of chemical reactions by 10^9 to 10^{20} times. An enzyme molecule is capable of transforming 10 to 1000 substrate molecules per second at body temperature.

At low substrate concentrations the enzymatic reaction is a first-order reaction, i.e., proportional to the concentration of the substrate. As substrate concentration is increased, a point is reached where the reaction rate does not increase and the rate becomes independent of substrate concentration. In this zone, the enzyme is saturated and the reaction is zero order with respect to substrate concentration. Each enzyme has a characteristic substrate concentration value (K_M, the Michaelis-Menten constant) at which the reaction velocity is one half maximal. The quantitative realtionships among K_M, the substrate concentration, and the maximum velocity of the enzyme reaction are given by the Michaelis-Menten equation, whose derivation is based on the assumption that an enzyme-substrate complex is formed as an essential step in catalysis.

The enzyme action can be inhibited by specific inhibitors. The inhibition can be achieved by two basically different mechanisms: competitive and noncompetitive inhibition. Competitive inhibitors of enzymes are those whose action can be reversed by increasing the substrate concentration. They usually have a structural resemblance to the substrate, with which they compete for the active site. They form reversible enzyme-inhibitor complexes with the enzyme. Noncompetitive inhibition cannot be reversed by raising the substrate concentration. It results from a reversible interaction of the inhibitor with some other essential group of the molecule. Irreversible inhibitors usually produce a permanent modification in the structure of the enzyme molecule. Kinetic tests are used to distinguish the various types of enzyme inhibition. Lineweaver-Burk and Eadie-Hofstee plots are especially useful in plotting and evaluating kinetic data.

Enzymes also have a pH optimum. They are usually assayed by measuring the initial reaction rate under conditions in which the enzyme is saturated with substrate and the pH is optimal. One unit of activity is the amount causing transformation of 1.0 μmol of substrate per minute at 25°C; specific activity is the number of units per milligram of protein.

Enzymes usually function in sequential multienzyme systems. Often such a system is organized into a complex that may be located in a membrane or in a cell organelle. Multienzyme systems usually have a rate-limiting step catalyzed by an allosteric, or regulatory, enzyme. The activity of the regulatory enzyme may be stimulated or inhibited by the binding of another molecule, often product of the sequence, at a second site on the enzyme molecule. Regulatory enzymes often possess subunits that can interact with each other.

From a clinical diagnostic point of view enzymes can be characterized by their specificity, selectivity, and efficiency.

Enzymes are present in different compartments or attached to different structures within the cell. Their localization is of utmost importance in clinical enzymology. Both membrane-bound and cytoplasmic enzymes can be released from the cells in various diseases. Quantitative and qualitative changes in their levels in body fluids can be used for diagnostic purposes, demonstrating the involvement of various tissues and organs in the pathological processes, enabling the detection of specific syndromes, and reflecting the severity of tissue destruction. The appearance of mitochondrial enzymes in serum, for instance, expresses pathological events of a more severe nature than do cytoplasmic enzymes. ALAT(GPT), a solely cytoplasmic enzyme, and ASAT(GOT), situated both in cytoplasm and the mitochondria, help both the differentiation between diseases and the evaluation of the severely pathological process. Various enzymes such as ALP, γ-GT, and 5′-NU are bound to the cytoplasmic membranes with different firmness and their sequential release may reflect the progress of the disease.

Specific cells in each tissue contain enzymes appropriate to its function, and they are largely determined by the nature of the chemical reactions occurring there. Enzymes therefore can be used as tissue markers. The organ specificity is not absolute. Certain enzymes such as ASAT(GOT) and LDH are present though in different amounts in various tissues: heart, liver, skeletal muscle, corneal and conjunctival epithelium, lacrimal gland, etc. In contrast ALAT(GPT) is a good liver marker, as is CK for heart and skeletal muscles. This lack of absolute specificity can be compensated by the study of isoenzymes such as LDH-1 and CK-MB (specific for myocardium), LDH-5 and CK-MM (specific for skeletal muscles), or the use of a combination of several enzyme assays providing an enzymatic profile.

The mechanism of enzyme release at the cellular level has been studied extensively. The first major hypothesis was based on the experimental work of Wilkinson and Robinson[3] showing a relationship between the energy status of the cell and the release of intracellular enzymes. By progressively reducing the ATP content of the lymphocyte culture medium, a parallel release of LDH from the lymphocytes was induced. Wilkinson's hypothesis was challenged and further developed by Friedel et al.[4] and Diederichs et al.[5] Based on the experimental data the following mechanism was postulated:

1. With a normal intracellular energy level, the cell volume is stable. APT-dependent pump mechanisms maintain this volume by accumulating potassium and expelling calcium, sodium, and water from the cell through the cytoplasmic membrane.
2. If intracellular level of ATP is relatively low, the ionic pump mechanisms are disturbed. This leads to the swelling of the cell. The intracellular calcium level was shown to increase with the distension of the cell. Calcium activates contractile filaments on the internal side of the cell membrane and activates other mechanisms so that channels and membrane pores are opened. Enzymes can leave the cell via these pores and channels. This is a reversible stage of cellular injury.
3. When the cell is totally devoid of ATP, membrane integrity can no longer be maintained. This is the irreversible stage of cell death. The dead cell soon empties all its contents into its surroundings.

Other authors link the increase of intracellular calcium levels to the activation of the membrane-bound enzyme phospholipase A_2. This enzyme converts structural membrane phospholipids (lecithin) into lysolecithin, leading to membrane disorganization, affecting membrane permeability, and causing the release of enzyme molecules into the extracellular space.

Classification and numbering of enzymes — The International Union of Biochemistry in 1961 and 1964 recommended a scheme for numbering enzymes, which was closely linked with new principles for classification.[6] The enzymes were divided into groups on the basis of the type of reaction they catalyze, and this, together with the name(s) of the substrate(s), provided a basis for naming individual enzymes. In this classification each enzyme number contains four elements, separated by points; the first number shows to which of the six main divisions of the enzyme list the particular enzyme belongs; the second and third numbers show the sub-class and the sub-sub-class, respectively, thus defining the type of reaction, and the fourth is the number of the enzyme within its sub-sub-class. The enzymes were divided into six main groups, as follows:

1. Oxidoreductases: enzymes which are concerned with biological oxidation and reduction, and therefore with respiration and fermentation processes. This class includes not only the dehydrogenases and oxidases, but also the peroxidases, which use H_2O_2 as the oxidant; the hydrolases, which introduce hydroxyl groups; and the oxygenases, which introduce molecular O_2 in place of a double bond in the substrate.
2. Transferases: enzymes which catalyze the transfer of one-carbon groups (methyl, formyl, carboxyl groups), aldehydic or ketonic residues, alkyl groups, nitrogenous groups, and phosphorus- and sulfur-containing groups.
3. Hydrolases: esterases, phosphatases, glycosidases, peptidases, etc.
4. Lyases: enzymes which remove groups from their substrates (not by hydrolysis), leaving double bonds, or which conversely add groups to double bonds. The class also includes decarboxylases, aldolases, dehydratases, etc.
5. Isomerases: racemases, epimerases, *cis-trans* isomerases, intramolecular oxidoreductases, and intramolecular transferases.
6. Ligases: enzymes which catalyze the joining together of two molecules coupled with the breakdown of a phosphate bond in ATP or similar triphosphate (also known as synthetases).

REFERENCES

1. **Lehninger, A. L.,** *Biochemistry. The Molecular Basis of Cell Structure and Function,* Worth, New York, 1972.
2. **Stryer, L.,** *Biochemistry,* W.H. Freeman, New York, 1988.
3. **Wilkinson, J. H. and Robinson, J. M.,** Factors Involved in the Release of Intracellular Enzymes, Proc. 6th Int. Symposium on Clinical Enzymology, 1979, 7.
4. **Friedel, R., Diederichs, F., and Lindena, J.,** Release and intracellular turnover of cellular enzymes, in *Advances in Clinical Enzymology,* Schmidt, E., Schmidt, W., Trautschold, I., and Friedel, R., Eds., S. Karger, Basel, 1979, 70.
5. **Diederichs, F., Mulhaus, K., Trautschold, I., and Friedel, R.,** On the mechanism of lactate dehydrogenase release from skeletal muscle in relation to the control of cell volume, *Enzyme,* 24, 404, 1979.
6. Recommendations of the International Union of Biochemistry on the Nomenclature and Classification of Enzymes, Enzyme Nomenclature, 1965.

Chapter 5

PRINCIPLES OF CLINICAL ENZYME DETERMINATIONS

I. INTRODUCTION

Determination of enzymes in various body fluids and secretions has become an indispensible aid in the diagnosis of a number of diseases. Clinical enzymology is one of the youngest but a very important and a quickly developing branch of clinical chemistry. Enzyme determinations can hardly be substituted by other methods in clinical chemical investigation.

Enzymes are proteins that are prone to inactivation and denaturation. Inactivation may be reversible but denaturation is always irreversible. It is therefore essential that the enzyme determination be performed as quickly as possible before the enzymes in the studied biological sample lose their activities. Enzymes are extremely sensitive to heavy metals, detergents, and acids. A common mistake in clinical enzymology is the use of insufficiently clean instruments, glassware, and reagents.[1]

Enzymes have catalytic activities, that is, they are able to accelerate chemical reactions, producing a 10^3- to 10^6-fold increase in the reaction rate. The enzyme does participate in the reaction but it remains qualitatively and quantitatively unchanged. Since the reaction usually can also occur spontaneously, though very slowly, it is always essential to use appropriate blank values. This usually means that with the omission of the enzyme all the other reagents have to be mixed, incubated, and measured to rule out or detect the effects of spontaneous reactions and other sources of error.[2]

Enzymes are proteins with specific catalytic functions. Their quantities in biological fluids can be determined either by measuring their activity, i.e., their capacity to convert a specific substrate, or they can be determined as proteins with specific (immunological or nonimmunological) methods. Both ways have their advantages and disadvantages. Activity measurements give information on the actual amounts of active enzymes. This is most important from a functional point of view. The disadvantage of activity determinations is the fact that these tests are greatly influenced by many factors: pH, temperature, ionic strength, relative concentration of enzyme, substrate, and product; inactive forms, proenzymes, or enzyme-inhibitor complexes are usuallly not detected with these methods. Enzyme-protein determinations on the other hand do not differentiate between the active and the inactive forms of the enzyme and are usually less specific due to cross-reactions or the inaccuracy of the nonimmunological methods.

Clinically these methods are more informative and technically make it much easier to determine the catalytic activity of the enzyme. Therefore, most procedures used in clinical enzymology are based on activity measurement

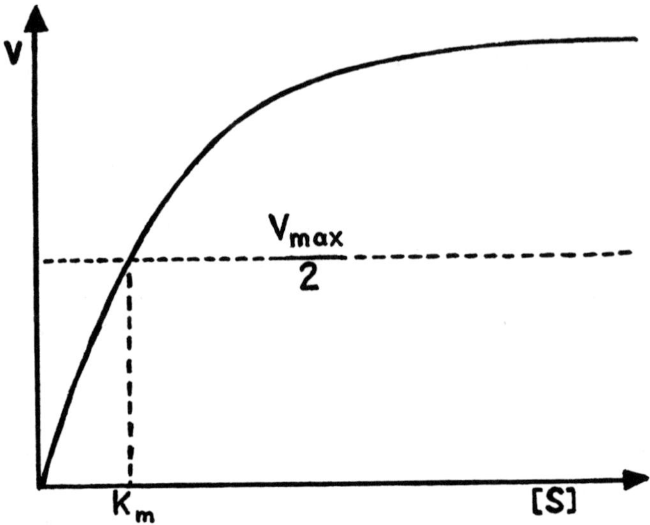

FIGURE 1. Dependence of the reaction velocity on the substrate concentration. v, velocity of the enzyme reaction; S, substrate concentration. (Modified and redrawn from Richterich, R. and Colombo, J. P., *Clinical Chemistry. Theory Practice and Interpretation,* John Wiley & Sons, New York, 1981.)

rather than enzyme (protein) concentration determinations. The enzyme activity is characterized either by the decrease in the amount of substrate(s) or by the increase in the amount of product(s) expressed per unit of time. The amount of substrate conversion in unit time catalyzed by an enzyme depends on many factors: substrate concentration, reaction temperature, hydrogen ion concentration, ionic strength, nature of the buffer, and if applicable the presence of cofactors, activators, and inhibitors.[2]

The velocity of the conversion of the substrate by the enzyme is largely dependent on the substrate conversion. The velocity curve when plotted as a function of concentration is shown in Figure 1. The first part of the curve is linear. In this region the enzyme is not yet saturated with the substrate and the reaction velocity is proportional to the substrate concentration. The reaction is a first-order reaction. This relationship is represented by the straight line on the semilogarithmic plot. The second part of the curve represents a situation when the enzyme is saturated with the substrate and the reaction velocity is dependent only on the enzyme concentration. This is a zero-order reaction. In clinical diagnostic enzyme activity determinations the conditions are designed to ensure a zero-order reaction. This means that the reaction velocity remains constant throughout the entire incubation period. At a constant period of time the relationship between the amount of enzyme and the extent of conversion is linear. Under otherwise identical conditions when the reaction is zero order, one can determine enzyme activities from substrate or product changes.[3]

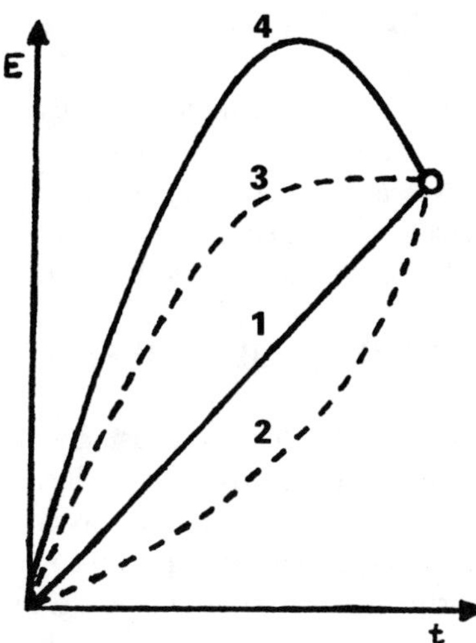

FIGURE 2. Drawbacks of the "two-point determination" method. Possible conversion/time curves. E, extent of conversion; t, time. (Modified and redrawn from Richterich, R. and Colombo, J. P., *Clinical Chemistry. Theory Practice and Interpretation,* John Wiley & Sons, New York, 1981.)

The enzyme activity may be determined using the so-called two-point method. This means two measurements: one activity value and one blank value. The difference between the two readings characterizes the activity of the enzyme. The drawbacks and uncertainties of the two-point method are demonstrated in Figure 2. The use of this approach is rather dubious since it does not reveal the kinetics of the enzymatic reaction. A few possible curves are demonstrated in Figure 2. Curve 1 demonstrates the desired linear relationship, but this is only one of the possibilities. The kinetics of the enzymatic conversion more often follows curves 2 and 3. At high enzyme activities a decrease in the amount of substrate or the interaction of the enzyme with end products (end-product inhibition) can result in declining enzyme activity. This possibility must be taken into consideration whenever the enzyme activity is high. It is therefore advisable to repeat the determination using half the quantity of enzyme or perform the measurements with half as long an incubation period. This frequently yields higher (and more correct) results than the first determination. The use of kinetic enzyme activity measurements often help to overcome these difficulties. Kinetic analysis can be performed either manually or using a continuously recording instrument. Kinetic methods are unquestionably preferable to the older two-point determinations.[3]

The best way is to carry out all enzyme determinations under optimum reaction conditions. Each condition and component of the incubation mixture should be chosen and standardized in such a way that the resulting activity cannot be increased by any further modifications.

II. INCUBATION PERIOD

The length of the incubation period should be chosen as short as possible, preferably not exceeding 1 h. With older techniques it was not uncommon to incubate for 12 or 24 h. Nowadays these long incubations are avoided with the result that there is less chance for the enzyme to suffer slow denaturation or for the reaction mixture to become contaminated by bacteria.

III. CHOICE OF SUBSTRATE

The substrate specificity of the enzyme is usually not absolute. The ideal solution of the problem would be the use of the physiological substrate, but very often this is not possible. In some cases the physiological substrate is not known or not easily obtainable, in other instances the reaction utilizes chromogenic or fluorogenic substrates or substrate analogues. Synthetic substrates are usually cheaper than their natural counterparts. There are no objections in principle to using nonphysiological substrates; the interpretation of the quantitative results, however, in these cases, as with other nonphysiological conditions, may be difficult. Most investigators accept the following general principles: the results must be reproducible, the determination should be kept as simple as possible, and the evaluation of the data has to involve considerations of possible nonphysiological conditions.

IV. SUBSTRATE CONCENTRATION

To use suitable substrate concentration in enzyme activity determinations is of utmost importance. If the substrate is in excess compared with the number of enzyme molecules, the measured activity is a function of the enzyme concentration. If the substrate concentration is relatively small the measured activity will be lower; if the enzyme is in excess the activity is a function of the subsrate concentration. The required constant velocity is met only at high substrate concentrations. Substrate concentration has to be chosen so as the maximum velocity of enzymatic conversion is attained. The knowledge of the K_M value can be of great help in practice. If the substrate concentration is chosen to be ten times as high as the K_M value, the reaction velocity will be at least 90% of the maximum velocity. Another recommendation is that not more than one fifth of the substrate introduced should react during the period of the measurement.

V. TEMPERATURE

Enzyme activity is highly influenced by temperature. When the temperature of the reaction mixture is higher, at least within a certain range, enzymes tend to work at a higher activity rate. Determinations carried out at different tem-

TABLE 1
The Influence of Temperature on the Activity of Some of the Enzymes Determined in Human Serum (Factors of Conversion)

	Conversion of enzyme activity at 25°C	
Enzyme	**To activity at 30°C**	**To activity at 37°C**
Acid phosphatase	1.49	2.08
Alanine aninotransferase	1.38	2.13
Alkaline phosphatase	1.19	1.45
Aminopeptidase	1.54	2.63
Aspartate aminotransferase	1.38	2.13
Cholinesterase	1.05	1.10
Creatine kinase	1.47	2.52
Glutamate dehydrogenase	1.23	1.61
Isocitrate dehydrogenase	1.49	2.78
Lactate dehydrogenase	1.33	1.67
Malate dehydrogenase	1.47	2.50

perature values (usually at 25, 30, or 37°C) give different results even if other factors of the reaction are kept constant. As the surface of the eye is somewhat cooler than the body temperature, tear enzymes might as well be tested at this temperature, 35°C. With the exception of the recommendation of the International Federation of Clinical Chemistry for enzyme activity determinations to be carried out at 30°C, there is no international agreement on the choice of temperature at which activity measurements should be performed. Table 1 shows the influence of temperature on the activity of some of the enzymes determined in human serum.[4]

VI. EFFECTS OF PH AND IONIC STRENGTH AND THE USE OF BUFFERS

All enzymes have their pH optimum, at which their activity is at its peak point. Their activities therefore ideally should be determined at their pH optimum. Another approach is to test enzymes under circumstances that are close to those found in the human body. It is important to remember that temperature, ionic strength, and the presence of activators and inhibitors may also have a considerable effect on the pH optimum of the enzymes. In general, these statements are valid for ionic strength of the reaction mixture, too. Certain enzymes, however, require the presence of special ions. Those enzymes that are active only in the presence of metallic ions are called metalloenzymes. Most buffers are ideal culture media for bacteria and fungi. Other buffers (e.g., phosphate buffers) may have a direct effect on the activity of various enzymes. Therefore, some investigators prefer the use of amine buffers based on TRIS, colloidine, ammediol, and triethanolamine in clinical enzyme activity measurements.[3]

VII. THE PRESENCE OF CELLS, CONTAMINATION WITH BLOOD, AND THE EFFECT OF HEMOLYSIS

Biological samples, body fluids, and secretions usually contain cells and are sometimes contaminated with blood. This is also true for tear samples. Cells contain intracellular enzymes. When the cells are disrupted enzymes of cellular origin may appear as free enzymes changing the results of activity measurements and enzyme profile determinations. One of the possible ways to overcome these difficulties is to centrifugate the samples and to use the supernatant for activity tests. The centrifugation should be performed right after sample collection as the lysis of cellular elements may soon occur. This is of utmost importance if the samples are frozen before activity determinations. Repeated freezing and thawing, besides causing the lysis of cells, may decrease the activity of free enzymes. Erythrocytes are extremely rich in enzymes. *In situ* or *in vitro* hemolysis of blood-contaminated samples disturbs enzyme activity measurements to such an extent that assays performed using hemolyzed samples (shown by the pink color of the otherwise clear solutions) do not yield reliable results.

VIII. ENZYME UNITS

Enzyme activity determinations are based on the idea that, the more substrate is converted per unit of time, the higher is the enzyme activity. As the basic aim of the measurements is to determine the concentration of enzymes in biological samples, the activity measurements have to be converted into concentration values by using enzyme standards, or by calculations based on specific activity values, or can be given simply in enzyme activity per volume units.

Since many enzymes have not yet been isolated in pure form, neither "the amount of substance" nor "the mass" can be used to express their quantities. The only possibility in these cases is to use catalytic activity expressed in basic or international units. According to the Committee for Biochemical Nomenclature[5] 1 μmol of substrate is changed by 1 enzyme unit (U) in 1 min, under standard conditions. The International Federation of Clinical Chemistry recommends the katal (kat) as basic unit of enzyme activity. 1 kat is defined as the amount of catalyst that catalyzes the conversion of 1 mol in 1 s under given conditions. 1 U/L corresponds to 16.67 nkat/L. The catalytic activity in body fluids is expressed as volume activity, usually referred to 1 L.[6]

IX. ENZYME VARIANTS

Enzymes catalyzing the same reaction may have slightly different molecular forms regarding their molecular mass and conformation. There are three types of molecular variants: isoenzymes, heteroenzymes, and alloenzymes. Isoenzymes, though they basically catalyze the same chemical reaction, differ in their affinity to subtrates and their reactivity to specific activators and inhibitors. Heteroenzymes differ in their molecular structure and can be distinguished by immunological methods. Alloenzymes are genetically determined variants of enzymes and isoenzymes. Enzymes variants may alter the results of enzyme activity and enzyme concentration measurements, and therefore have to be taken into consideration. This is especially true for body fluids that may contain enzymes originating from different parts and organs of the human body.[1,3]

X. ORGAN SPECIFICITY

Various organs and tissues have different enzyme patterns. The differences can be (1) qualitative: certain tissues or organs contain enzymes that others do not; (2) variational: various cells in different organs may use several isoenzymes or other enzyme variants to catalyze the same chemical reaction; (3) quantitative: different cells contain essentially the same enzymes but produce them in different quantities. Enzyme activities may vary by several orders of magnitude from organ to organ.

XI. INTRACELLULAR ENZYMES

Enzymes within the cell are localized in different compartments. Their localization and the firmness of their attachment to intracellular structures basically determine the extractability in experiments and the release by the cells of intracellular enzymes in various pathological states. The first group consists of enzymes located in the cytoplasm in soluble form, e.g., lactate dehydrogenase and alanine aminotransferase. These enzymes are easily released from the cells as a result of mild trauma, without the destruction of the cells. To the second group belong intracellular enzymes that become soluble only when the cell structure has been severely damaged. The mitochondrial enzyme glutamate dehydrogenase is a good example of this. Aspartate aminotransferase, malate dehydrogenase, and isocytrate dehydrogenase can be classified separately as the third group, since their cytosol isoenzymes have a high solubility while their mitochondrial isoenzymes are hardly soluble. Still another group of enzymes consists of those which, like the enzymes of the electron-transfer system, are so firmly a part of a structure that they lose their activity when attempts are made to bring them into solution.

XII. EXTRACELLULAR ENZYMES

From a diagnostic point of view enzymes in different body fluids and secretions are most important. These are either constant components of these fluids, e.g., circulating plasma enzymes, lysozyme in tears, or the enzyme components of the secretions of the gastrointestinal tract. Other enzymes are not normally present in secretions or body fluids and appear in these fluids only in various pathological states. They are considered mainly to originate from the

TABLE 2
Normal Ranges for Some of the Serum Enzymes
Used in Clinical Enzymological Diagnostics

Enzyme	Temperature (°C)	Range (IU/l)
Acid phosphatase	37	2–6
Alanine transaminase	25	5–14
Aldolase	37	2–6
Akaline phosphatase	37	13–45
Amylase	40	65–310
Aspartate transaminase	25	5–15
Creatinine kinase	30	0–150
Glutamyltransferase	25	5–25
Lactate dehydrogenase	25	30–120

parenchyma of various organs, from leukocytes, or from microorganisms. Changes in the activities of enzymes belonging to both groups, just like alterations in the ratio of two or more of them, and variations in isoenzyme patterns are widely used in clinical enzyme diagnostics.

For enzyme diagnostics it is advantageous to classify enzymes on the basis of their sites of action rather than on the reaction they catalyze or the role they play in metabolic processes. This classification is justified by the fact that it is the site of their action, the organ of their origin, that determines changes in their levels in blood plasma in various pathological states. A classification of some of the enzymes used in clinical enzymological diagnostics is given in Table 2. Elevated activities with an organ specificity are the result of acute or chronic tissue destruction occurring in a number of diseases with diverse etiologies.

Practically all intracellular enzymes used in clinical diagnostics can be found in normal plasma in very low activities. The constant level of these enzymes in blood is thought to be maintained by continuous production of enzymes and by maintaining the barrier between the two compartments (intracellular and extracellular) under normal circumstances. Various organs are known to contribute to the normal enzyme content of the plasma. An intensification of cellular metabolism such as that during intensive muscle work or cellular damage due to mechanical or chemical trauma results in increased plasma levels of intracellular enzymes. With the exception of a few genetically determined metabolic diseases such as hypophosphatemia, no reduced activity of cellular enzymes in plasma has been observed so far.[2,3] Tear enzyme levels are different from their plasma counterparts also in this respect. Decreased lysozyme concentrations in tears often occur and are used in the evaluation of the impaired secretory function of the lacrimal glands in Sjögren's syndrome and in other acute or chronic diseases of lacrimal apparatus.

REFERENCES

1. **Damm, H. C., Besch, P. K., and Goldwyn, A. J.,** *The Handbook of Biochemistry and Biophysics,* World, New York, 1966.
2. **Zimmerman, H. J. and Henry, J. B.,** Clinical enzymology, in *Clinical Diagnosis and Management by Laboratory Methods,* Henry, J. B., Ed., W.B. Saunders, Philadelphia, 1984, chap. 14.
3. **Richterich, R. and Colombo, J. P.,** *Clinical Chemistry. Theory Practice and Interpretation,* John Wiley & Sons, New York, 1981.
4. **Schmidt, F. W.,** Fundamentals of enzyme diagnostics, in *Physical Chemistry, Composition of Blood, Hematology, Somatometric Data. Geigy Scientific Tables,* Vol. 3, Lentner, C., Ed., CIBA-GEIGY, Basel, 1984, 166.
5. International Union of Pure and Applied Chemistry and International Union of Biochemistry, *Enzyme Nomenclature Recommendations 1972,* Elsevier, Amsterdam, 1973, 26.
6. **Bowers, G. N., Jr.,** The acceptability of the enzyme methods, in *Quality Control in Clinical Chemistry,* Anido, G. et al., Eds., Walter de Gruyter, Berlin, 1975, 441.

Chapter 6

LYSOZYME IN TEARS

Part A
**Structure, Characteristics, Mechanism of Enzyme Action,
Bacteriolytic Effect, and Biological Significance**

I. DISCOVERY AND GENERAL CHARACTERISTICS

Lysozyme is an antibacterial enzyme widely found in animal and human tissues, body fluids, secretions, and exudates. This enzyme hydrolyzes β-1,4-glucosidic linkages in the polysaccharide cell wall structure of a variety of microorganisms.[1] The enzyme is also called muraminidase or *N*-acetylmuraminide glycanohydrolase, an enzyme-digesting murein, the complex peptidoglycan structure of the cell wall of certain Gram-positive bacteria. The name murein comes from the Latin word "murus", meaning "wall". The Enzyme Commission number of lysozyme is 3.2.1.17, indicating that the enzyme was classified in the main group of hydrolases as a β-1,4-glucosido-hydrolase (Enzyme Nomenclature 1984).

Lysozyme was discovered by the English bacteriologist Alexander Fleming in 1922. He demonstrated in human nasal secretion the presence of an antibacterial substance, which he named lysozyme (lyso because of its capacity to lyse bacteria and zyme because it was an enzyme). He also discovered a bacterium that was particularly susceptible to lysozyme. He named it *Micrococcus lysodeikticus* because it displayed lysis ("deiktikos" means "able to show" in Greek). He found that human tears are rich in lysozyme, and used tears as a source of lysozyme for further experiments. Fleming was disappointed to find that lysozyme was not effective against the most harmful bacteria. He continued the experimental work and 7 years later discovered the first effective antibiotic, penicillin.[2]

Lysozyme was first isolated from egg white.[3] The molecular weight of egg white lysozyme was determined as 14,000,[4] its amino acid sequence by Jolles et al.[5] and Canfield,[6] and its tertiary structure by Phillips.[7] The extinction coefficient of egg white lysozyme $E_{281.5}^{1\%} = 26.4$,[8] the isoelectric point: pH 10.5 to 11,[3] the pH optimum: 9.2.[9] The enzyme is inhibited by surface-active reagents such as dodecyl sulfate, alcohols, and fatty acids.[10] The human enzyme has been crystallized from urine by Osserman.[11] Plant lysozyme was found in ficus and papaya latex, and was shown to be chemically distinct from egg white lysozyme.[12]

II. THE MOLECULAR STUCTURE OF LYSOZYME

Lysozyme is a relatively small enzyme; its molecular weight is 14.6 kDa. It consists of a single polypeptide chain that is built up of 129 amino acid residues. The conformation of the molecule is stabilized by four disulfide bridges. The three-dimensional structure of lysozyme was determined by Phillips and co-workers in 1965. Their high-resolution X-ray diffraction pattern was the first for an enzyme molecule. Lysozyme is a compact molecule, roughly ellipsoidal in shape, with dimensions $45 \times 30 \times 30$ Å. This form is due to complex folding of the main chain including hairpin turns at some places. There is less α-helical structure in the lysozyme molecule than in myoglobin or in hemoglobin. Long parts of the molecule are polypeptide regions in an extended β-sheet conformation. The interior part of the molecule is formed almost entirely by nonpolar groups held together by hydrophobic interactions.[2]

X-ray diffraction studies were performed on lysozyme-substrate complexes and compared with those of the free enzyme. The three-dimensional conformation of the molecule was significantly altered when it contained a bound substrate. This observation was the first experimental data confirming the hypothesis that substrate binding may cause temporary changes in the enzyme structure. These experiments also made possible the precise localization of the substrate binding site on the surface of the lysozyme molecule. High-resolution X-ray diffraction patterns even enabled the identification of the amino acid residues that are involved in bringing about the hydrolysis of the glycosidic linkage of the substrate (glutamic acid residue at position 35 and aspartic acid residue at position 52).[13]

III. MECHANISM OF ENZYME ACTION

The bacteriolytic action of lysozyme is based on its ability to lyse murein, the polysaccharide component of the cell wall of various Gram-positive bacteria. The exact enzymatic action involves the cleavage of the β-1-4-glycosidic bond between *N*-acetylglucosamine and *N*-acetylmuramic acid in the peptidoglycan structure of the bacterial cell wall. The enzyme binds a hexasaccharide segment of the peptidoglycan polymer. One of the monomers in the hexasaccharide, an *N*-acetylmuramic acid residue, is distorted by lysozyme from its usual chair conformation into a strained half-chair conformation. The enzyme promotes catalysis by inducing steric strain in the substrate. Bond cleavage occurs at the glycosidic bond of the strained half chair *N*-acetylmuramic acid residue.[14,15]

The rate of lysozyme catalysis depends on the pH of the medium. The pH-dependence curve of the rate of hydrolysis of murein caused by lysozyme is bell shaped with a maximum at pH 5.0 and inflections at pH 3.8 and 6.7 (Figure 1). X-ray diffraction data showed that the labile glycosidic bond was flanked by the carboxyl groups of aspartate 52 and glutamate 35. The pKas of these groups match the inflection points of the pH-rate profile. At pH 5.0, the pH optimum for lysozyme, aspartate 52 is unprotonated and glutamate 35 is protonated.[16]

The mechanism of action of lysozyme has been understood from a combination of data obtained from X-ray crystallographic and chemical studies. Bond breaking between the two sugar components gives a planar, resonance-stabilized acylium ion. The ion is further stabilized by electrostatic interaction

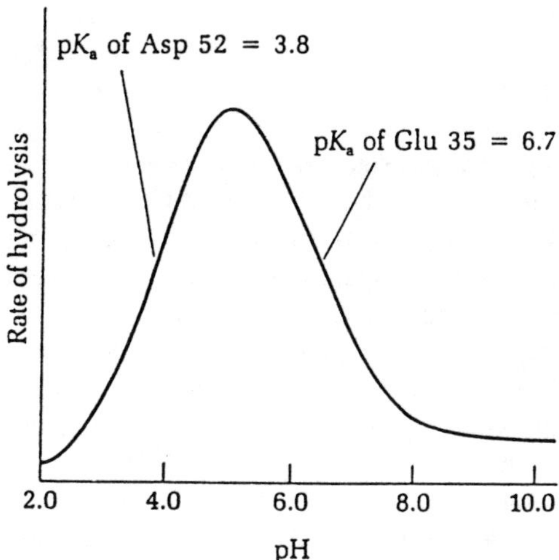

FIGURE 1. A pH-rate profile demonstrating that the rate of lysozyme catalysis depends upon the pH of the medium. The profile of poly-*N*-acetylglucosamine is bell shaped with a maximum at pH 5.0 and inflections at pH 3.8 and 6.7. (From Rawn, J. D., *Biochemistry*, Harper & Row, New York, 1983, 299. With permission.)

with the negatively charged carboxylate anion of aspartate 52. Theoretical calculations suggest that the increase in rate of the lysozyme catalyzed reaction relative to the noncatalyzed reaction is almost entirely due to stabilization of the acylium ion by electrostatic interactions with the negatively charged carboxylate anion of aspartate 52.[16]

The result of the enzymatic cleavage is the disintegration and the dissolution of the cell wall, usually resulting in the lysis of the bacterium. In experimental conditions the cell wall can be removed without destroying the cell when bacteria, susceptible to lysozyme action, are treated with the enzyme in the presence of an impermeant solute such as sucrose. This procedure yields a protoplast, a term applied to the "naked" bacterial cell surrounded only by its cytoplasmic membrane. Protoplasts remain viable as long as the culture fluid is isotonic to prevent swelling of the cell and rupture of the cell membrane.[13]

IV. BIOLOGICAL FUNCTION AND BACTERIOLYTIC EFFECT

Lysozyme is the most studied and best characterized enzyme and protein molecule. Since its discovery by Fleming in 1922 hundreds of biochemists and microbiologists have studied its molecular structure, enzyme action, and biological function. In spite of the large amount of information available about lysozyme, the most fundamental questions — why this rather specific and relatively ineffective antibacterial factor is so widespread in the living world, why various organisms produce and secrete it in large quantities and benefit seemingly so little from it — still remain unanswered.

Bacteria, unlike animal and human cells, are surrounded by a cell wall covering their cell membrane. Whether a bacterium is susceptible to lysozyme or not depends on the structure and chemical nature of its cell wall. Bacteria are classically divided into Gram-positive and -negative organisms, according to their reaction to the Gram stain, an empirical procedure in which the cells are succesively treated with the dye crystal violet, with iodine, and with safranine. Whereas the crystal violet-iodine complex is retained by Gram-positive bacteria washed by organic solvents, it is not retained when Gram-negative cells are decolorized with alcohol or acetone. The differences in their staining characteristics can be explained by the structure and composition of their cell walls. The cell wall of Gram-positive bacteria is a thick peptidoglycan layer composed of cross-linked polysaccharide chains interwoven with teichoic acid polymers and is in close connection with the cytoplasmic membrane containing structural and enzymatic proteins. The cell wall of Gram-negative bacteria is a multilayered structure. It consists of an outer membrane containing lipopolysaccharides separated by periplasmic space and lipoprotein "spacers" from a thin peptidoglycan layer immediately adjacent to the cytoplasmic membranes. The outer membrane and the lipoprotein components of the cell wall of Gram-positive bacteria prevent lysozyme molecules to reach the inner peptidoglycan layer. Besides, lipoproteins act as endotoxins and contribute to the development of the symptoms of Gram-negative infections causing fever, shock, or even disseminated intravascular coagulation in the infected organism. Together with other macromolecules in the cell walls of Gram-negative and -positive bacteria lipoproteins form the so-called somatic or O antigen structures on the surface the bacteria.[17]

The peptidoglycan (murein) framework constituting the cell wall of the Gram-positive and the innermost layer of the cell wall of the Gram-negative bacteria consist of polysaccharide chains connected by tetrapeptide bridges at each disaccharide unit forming a tight cross-linked structure. Besides polysaccharide and technoic acid polymers the cell walls of some of the Gram-positive bacteria contain polypetides and proteins that alter the structure and prevent the cleavage by lysozyme.[13] As a result, like all Gram-negative bacteria, some of the Gram positives are also resistant to the lytic action of lysozyme. This is why the protection provided by lysozyme contained in various secretions and exudates is very limited and not effective against the most harmful pathogenic microorganisms.

In spite of this lysozyme is ubiquitous and is present in almost all secretions and body fluids of the human body. Among others, Hankiewicz and Swierczek[18] using the agar-gel diffusion measured lysozyme activities in the blood serum of adults and children, umbilical cord blood serum, cerebrospinal fluid, amniotic fluid, urine, saliva, gastric juice, bile, tears, mother's milk, seminal fluid, and different exudates. Lysozyme was detected in all body fluids and secretions, though in different amounts. In serum the lysozyme levels, based on activity measurements, ranged from 7 to 13 mg/ml. The highest levels were found in tears, about 120 times, and in gastric juice, about 8 times higher than in serum. The lowest levels were detected in urine and cerebrospinal fluid. The lysozyme contents of other secretions were found to be in the same range as in serum. Besides the obvious but limited antibacterial action, other possible biological functions such as mucolytic action and interactions with other molecules have been proposed,[19,20] but none of these have been proved so far. The role and biological significance of lysozyme in tears is extremely difficult to determine, considering that the concentration of this enzyme is more than 100 times higher in tears than in blood and other secretions.

REFERENCES

1. **Hara, S. and Matsushima, Y.,** Studies on the substrate specificity of egg white lysozyme. I. The N-acyl substituents in the substrate mucopolysaccharides, *J. Biochem.,* 62, 118, 1967.
2. **Stryer, L.,** *Biochemistry,* W.H. Freeman, New York, 1988.
3. **Alderton, G., Ward, W. H., and Fevold, H. L.,** Isolation of lysozyme from egg white, *J. Biol. Chem.,* 157, 43, 1945.
4. **Sophianopoulos, A. J., Rhodes, C. K., Holcomb, D. N., and van Holde, K. E.,** Physical studies of lysozyme. I. Characterization, *J. Biol. Chem.,* 237, 1107, 1962.
5. **Jolles, J., Jauregui-Adell, J., Bernier, I., and Jolles, P.,** La structure chimique du lysozyme de blanc d'oeuf de poule: etude detaillee, *Biochim. Biophys. Acta,* 78, 668, 1963.
6. **Canfield, R. E.,** The aminoacid sequence of egg white lysozyme, *J. Biol. Chem.,* 238, 2698, 1963.
7. **Phillips, D. C.,** The hen egg-white lysozyme molecule, *Proc. Natl. Acad. Sci. U.S.A.,* 57, 484, 1967.
8. **Aune, K. C. and Tanford, C.,** Thermodynamics of the denaturation of lysozyme by guanidine hydrochloride. I. Dependence on pH at 25°C, *Biochemistry,* 8, 4579, 1969.
9. **Davies, A. C., Neuberger, A., and Wilson, B. M.,** The dependence of lysozyme activity on pH and ionic strength, *Biochim. Biophys. Acta,* 178, 294, 1969.
10. **Smith, G. N. and Stoker, C.,** Inhibition of crystalline lysozyme, *Arch. Biochem.,* 21, 383, 1949.
11. **Osserman, E. F.,** Crystallization of human lysozymes, *Science,* 155, 1536, 1967.
12. **Meyer, K., Hahnel, E., and Steinberg, A.,** Lysozyme of plant origin, *J. Biol. Chem.,* 163, 733, 1946.
13. **Lehninger, A. L.,** *Biochemistry. The Molecular Basis of Cell Structure and Function,* Worth, New York, 1970.
14. **Chipman, D. M. and Sharon, N.,** Mechanism of lysozyme action, *Science,* 165, 454, 1969.
15. **Jolles, P.,** Lysozyme, in *The Enzymes,* Vol. 4, Boyer, P. D., Lardy, H., and Myrback, K., Eds., Academic Press, New York, 1960, 431.
16. **Rawn, J. D.,** *Biochemistry,* Harper & Row, New York, 1983, 299.
17. **Rose, N. R. and Barron, A. L.,** *Microbiology. Basic Principles and Clinical Applications,* Macmillan, New York, 1983.
18. **Hankiewicz, J. and Swierczek, E.,** Lysozyme in human body fluids, *Clin. Chim. Acta,* 57, 205, 1974.
19. **Selinger, D. S., Selinger, R. C., and Reed, W. P.,** Resistance to infection of the external eye: the role of tears, *Surv. Ophthalmol.,* 24, 33, 1979.
20. **Schumacher, H.,** Neue Erkentnisse über die Biochemie des Lysozymes und ihre klinische Bedeutung für die Augenheilkunde, *Adv. Ophthalmol.,* 8, 142, 1958.

Chapter 6

LYSOZYME IN TEARS

Part B
**Origin, Special Characteristics of Tear Lysozyme —
Different Methods for Determination**

I. HUMAN TEAR LYSOZYME, CHARACTERISTICS, ORIGIN, COMPARISON WITH HEN EGG WHITE LYSOZYME

In 1922 Fleming[1] discoverd the bacteriolytic effect of lysozyme. He assayed the level of lysozyme in tears using a dilution technique. He showed that the concentration of lysozyme in tears was up to a thousandfold greater than in serum.[2,3] His findings were soon confirmed by others. The lysozyme titer in tears, the dilution still causing lysis of bacteria on culture plates, was found to range from 1:10,000 to 1:100,000 by different authors.[1,4,5]

Human tear lysozyme was isolated from normal stimulated tears. The specific activity and ultraviolet spectrum of human tear lysozyme were found to be different from those of hen egg-white lysozyme (HEL) though the two enzymes had similar electrophoretic mobility. Immunodiffusion and immuno-electrophoretic studies did not reveal any cross-reaction between HEL and human tear lysozyme with either antiserum to human tears or antiserum to HEL. The marked differences in antigenicity of the two lysozymes may reflect differences in their conformation, even though they catalyze the same enzymatic reaction.[6]

Human tear lysozyme was found to be very similar in structure and immunological characteristics to lysozyme found in serum and in different secretions.[7] Human serum lysozyme activity was shown to vary widely in a variety of hematological disorders,[8] mostly in relationship to the turnover of granulocytes and monocytes.[9] In further studies simultaneous tear and serum lysozyme activity measurements were performed in patients with various hematological disorders to find out whether or not the lysozyme activities of the two fluids were related. Lysozyme activities of serum and tears did not correlate with each other in the case of 39 patients with a wide range of serum lysozyme activities. On the basis of these findings it has been suggested that the lacrimal gland not only collects from the blood, concentrates, and excretes lysozyme into tears but also a *de novo* sythesis of lysozyme might take place in the lacrimal gland followed by an active secretion.[10]

TABLE 1
Methods Used for the Determination of Tear Lysozyme
Levels and Activities

I. Methods based on the measurement of enzyme activity
   Lysozyme titer determination
   Viscosimetry
   Agar diffusion test, "lysoplate" method
   Spectrophotometric and turbidimetric methods

II. Methods based on the determination of the enzyme as a protein
   1. Methods based on antigen-antibody reactions (immunological methods)
       Radial immunodiffusion
       Electroimmunodiffusion
       Radioimmunoassay (RIA)
       Laser nephelometry, immunoturbidimetry
   2. Methods based on protein-dye interactions
       Dye elution test
       Electrophoretic densitometric lysozyme determination

II. DIFFERENT METHODS DEVELOPED AND USED FOR THE DETERMINATION OF TEAR LYSOZYME LEVELS AND ACTIVITIES

Several methods were described and used for the determination of tear lysozyme levels, including turbidimetric, agar diffusion, spectrophotometric, electroimmunodiffusion, radial immunodiffusion, dye elution, and electrophoretic-densitometric tear lysozyme determinations (Table 1). These procedures can basically be classified in two different groups. The first is the group of enzymologic methods to which belong determinations that are based on the measurement of the enzyme action of lysozyme, such as the turbidimetric, the lysoplate, and the spectrophotometric methods. In the second group are classified the immunological methods that are based on specific antigen antibody reactions, such as the radial immunodiffusion, electroimmunodiffusion, and the laser nephelometric lysozyme determinations. Both ways have their advantages and disadvantages and usually produce different values that make the comparison of the results of different authors rather difficult.

The standardization of the determinations based on the measurement of enzyme activity is difficult due to the large differences in enzymatic action under different conditions, the presence of promoting and inhibiting factors in body fluids, and the lack of reliable standard enzyme and substrate sources.[11,12] Immunological methods that are based on immunoprecipitation are highly influenced by the relative concentration of antigens and antibodies, the circumstances under which the precipitation occurs, and to a great extent by the specificity of the antisera used in the reaction. The latter is very important because of the possible cross-reactions that may occur when incompletely purified enzymes or antisera are used in the determinations.

A third possibility for the determination of the concentration of enzyme proteins is the measurement of their dye-binding capacity. Enzymes like other proteins can be separated by different electrophoretic techniques and quantitatively stained by specific dyes. The determination of protein concentrations is accomplished either by the elution of the dye from the stained protein bands or by the densitometric evaluation of stained gels under well-defined conditions. Janssen and van Bijsterveld[13] used the dye elution technique for the determination of lysozyme and other tear proteins and discussed the advantages and disadvantages of the dye elution technique. We tried another possibility, studying the sensitivity and reliability of a tear lysozyme determination based on the densitometric evaluation of tear protein bands produced in a well-defined electrophoretic system.[14]

Ever since Fleming demonstrated that the bacterium *Micrococcus lysodeikticus* is a perfect test organism to study the action of lysozyme, this bacterium has been used in different methods developed for the determination of lysozyme activity. The first in the row of such procedures was the viscosimetric method, based on the measurement of changes caused by samples containing lysozyme in the viscosity of bacterial supensions.[15] Even though this method needs sample volumes of at least 0.5 ml, the first quantitative data concerning the lysozyme content of tears were obtained by this method. The range of lysozyme level in normal human tears was found to be between 1000 and 2000 units per cubic centimeters of tears.[16]

A quantitative method for the determination of tear lysozyme activities was published by Bonavida and Sapse in 1968. The method was based on the enzymatic action of lysozyme on the cell walls of *M. lysodeicticus*. The bacterial suspension was seeded in an agarose layer and tear samples collected with standard Schirmer strips were placed on the surface. The action of lysozyme resulted in a visible, clear zone of lysis depending on the concentration of lysozyme in the sample. Quantitative data could be obtained by the help of standards prepared from purified human tear lysozyme.[17]

van Bijsterveld used a modified version of the agar diffusion method with 15 ml meat infusion agar, containing 5×10^7 viable *M. lysodeicticus* per milliliter, cast in petri dishes 9 cm in diameter. Tear samples were collected by placing filter-paper disks (Whatman No. 3) with a diameter of 6 mm in the lower cul-de-sac, so that the disk was entirely covered by the lower lid. The disk was removed only after it was entirely wet; excess fluid was drained by blotting it lightly between Whatman No. 3 filter papers. The disks were put in sterile screwcap bottles and kept at 4°C until processed within 24 h. Disks were put on the surface of the culture media (four disks per dish) and incubated for 24 h at 37°C. The diameters of lysis zones were measured with calipers on the reverse side of the dish.[17]

Several authors tested the accuracy, reliability, and reproducibility of tear lysozyme determinations performed with the agar diffusion test and suggested several modifications. Because different culture media gave different results, van Bijsterveld suggested the use of standardized media, chemicals, and bacteria and expressed a view that the reagents should be obtained by all users from the same sources. He used diagnostic sensitivity test agar (Oxoid, Flow Laboratories, Inc., Rockville, MD), heart infusion broth culture (Difco), and a *M. lysodeicticus* strain deposited at the American Type Culture Collection under accession No. 27523.[18] Ensink and van Haeringen[11] emphasized the importance of the use of human tear lysozyme as a standard. With HEL used as a standard, they obtained three times higher values for tear lysozyme levels than with human tear lysozyme, as determined with the lysoplate method of Osserman and Lawlor.[19] The differences can be explained by the three-times higher specific activity of human tear lysozyme than that of HEL, and differences in their diffusion properties.[7] Possible differences in the measured lysozyme levels may be caused by different proteins and inorganic ions present in the tear samples.[11,12] Copeland et al.[20] demonstrated that the storage of tear samples at subzero temperatures lowered the lysozyme concentration. This change occurred in a predictable, linear manner in the case of samples frozen in glass capillaries, while proving to be nonlinear due to internal adsorption when samples were collected and frozen in Weck-Cel sponges. It was demonstrated that a power rather than the suggested exponential dependence between the lysozyme concentration and lysis diameter provides the most accurate fit and thus should be used for interpolation of the results of measurements performed on commercially available Quantiplates (Kallestad Laboratories, Inc., Chaska, MN).[20]

III. TEAR LYSOZYME LEVELS AT VARIOUS FLOW RATES, UNDER DIFFERENT CIRCUMSTANCES — LYSOZYME IN THE TEARS OF DIFFERENT SPECIES

Stuchell et al.[21] studied the lysozyme concentration in tear samples collected with standard Shirmer strips and with a much smaller paper strip (Periopaper). They called these latter samples "basal tears" arguing that the Periopaper is supposed to cause less irritation, consequently less reflex tearing than the common Shirmer paper. Tears were collected from normal controls, from keratoconjunctivits sicca (KCS) patients, and from contact lens (CL) wearers. Lysozyme concentrations were determined by electroimmune diffusion. The concentration of lysozyme in reflex tears was 160 ± 73 mg% in normals, 74 ± 41 mg% in KCS, and 186 ± 83 mg% in CL wearers. Lysozyme levels in the tear samples collected at low flow rates (less irritation) were significantly lower in all three groups: 65 ± 43 mg% in normals, 44 ± 26 mg% in KCS patients, and 81 ± 50 mg% in CL wearers.[21]

van Bijsterveld and Mansour investigated lysozyme levels in tears collected with filter paper strips without and with topical anesthesia. Lysozyme levels were determined with the agar diffusion method. HEL was used as standard and the concentration values were expressed in micrograms per milliliter HEL (hen egg white activity equivalents). The average lysozyme concentrations were 3274 µg/ml HEL in nonanesthetized and 3124.20 µg/ml HEL in anesthetized eyes. The difference was not statistically significant.[22]

The effect of collection technique of tear samples on the lysozyme content was studied by Stuchell et al.[23] in another series of experiments. Tears from the same individual were collected with Schirmer filter paper strip from the unanesthetized inferior conjunctival sac and were compared with tears collected by a capillary tube (without touching the conjunctiva) after stimulation of tear secretion by irritation of the nasal mucosa with ammonium vapor. Lysozyme levels in the tear samples were determined by electroimmune diffusion. The mean values of lysozyme concentrations measured in tear samples collected with Schirmer strips and with capillary tubes were 220 ± 79 and 213 ± 48 mg/ml, respectively. This difference was not significant. Similar results were obtained for lactoferrin levels, while serum albumin, transferrin, and immunoglobulin G levels were significantly lower in stimulated tears collected with capillary tubes than in samples eluted from Schirmer strips.[23]

Janssen and van Bijsterveld investigated the relations between tear fluid concentrations of lysozyme, tear-specific prealbumin, and lactoferrin. Tear samples collected from a heterogeneous group of 94 persons, ranging from normal persons with normal lacrimal gland function to patients with severe KCS, were analyzed by sodium dodecyl sulfate-polyacrylamide electrophoresis. The electrophoretically separated and Coomassie Blue stained proteins were quantified by spectrophoretic measurement of the eluted dye. A high degree of correlation was found between lysozyme concentration and that of lactoferrin as well as tear-specific prealbumin. They suggested that, besides lysozyme, lactoferrin and tear-specific prealbumin can serve as alternative parameters for tear gland function.[13]

We investigated the effects of flow rate, sampling, and changes in vascular permeability on the tear levels of lysozyme and other tear proteins.[14] Tears were obtained from healthy subjects and from patients suffering from various external eye diseases. Tears were collected with capillary tubes without stimulation or with the stimulation of the nasal mucosa. Tear flow rates were calculated by dividing the total volume of the tear sample by the time interval of the sample collection. Tear samples were assayed by polyacrylamide electrophoresis and the total protein content of the samples was determined spectrophotometrically. The concentrations of the electrophoretically separated tear proteins were determined by the densitometric evaluation of the Coomassie brillant blue stained bands using protein standards with known concentrations and were plotted against tear flow rate. The total protein concentration in normal tears was in the range of 5 to 10 µg/µl between flow rates of 10 and 50 µl/min and did not seem to change with flow rate within this interval. The protein levels in samples collected at low flow rates (<10 µl/min) were significantly higher (7 to 17 µg/µl), probably due to the effect of proteins deriving from the conjunctiva and the cornea. At very high flow rates there was a tendency for lower tear protein concentrations to be found. Tear lysozyme levels were found to be independent from the tear flow rate in the range of 2 to 50 µl/min, but lower lysozyme concentrations were detected in normal tears at very low and at very high flow rates, and in tears collected from patients with various inflammations of the conjunctiva. We suggested that tear flow rates and changes in the vascular pearmeability should be taken into consideration when tear lysozyme levels are evaluated. We pointed out that the changes in the tear fluid concentrations of various proteins secreted by the lacrimal gland are parallel not only because they are all produced by the lacrimal gland,[10,24] but also because several influences (the dilution in the conjunctival sac, changes in the secretion at extremely high flow rates, evaporation, transudation, etc.) affect their tear levels basically in the same way.[14]

REFERENCES

1. **Fleming, A.,** On a remarkable bacteriolytic element found in tissues and secretions, *Proc. R. Soc. London Ser. B,* 93, 306, 1922.
2. **Fleming, A. and Allison, V. D.,** Observations on a bacteriolytic substance ('lysozyme') found in secretions and tissues, *J. Exp. Pathol.,* 3, 353, 1922.
3. **Fleming, A.,** Further observations on a bacteriolytic element found in tissues and secretions, *Proc. R. Soc. London Ser. B,* 94, 142, 1923.
4. **James, W. M.,** The lysozyme content of tears, *Am. J. Ophthalmol.,* 18, 1109, 1934.
5. **von Hallauer, C.,** Klinische und experimentalle Untersuchungen über den Lysozymegehalt in der Bindehautsack und in der Tranenflüssigkeit, *Arch. Augenheilk.,* 103, 199, 1930.
6. **Bonavida, B., Sapse, A. T., and Sercarz, E. E.,** Human tear lysozyme. I. Purification, physicochemical, and immunochemical characterization, *J. Lab. Clin. Med.,* 70, 951, 1967.
7. **Jollés, J. and Jollés, P.,** Human tear and human milk lysozyme, *Biochemistry,* 6, 411, 1967.
8. **Perillie, P., Kaplan, S. S., Lefkowitz, E., and Finch, S. C.,** Muraminidase (lysozyme) activity in leukemia, *Blood,* 28, 1000, 1966.
9. **Perillie, P. E., Kaplan, S. S., and Finch, S. C.,** Significance of changes in serum muraminidase activity in megaloblastic anemia, *N. Engl. J. Med.,* 227, 10, 1967.
10. **Covey, W., Perille, P., and Finch, S. C.,** The origin of tear lysozyme, *Proc. Soc. Exp. Biol. Med.,* 137, 1362, 1971.
11. **Ensink, F. T. E. and van Haeringen, N. J.,** Pitfalls in the assay of lysozyme in human tear fluid, *Ophthal. Res.,* 9, 366, 1977.
12. **Huemer, P. and Rieger, G.,** Zur Problematik der Standardisierung des Lysozymgehaltes in der Tranenflüssigkeit, *Klin. Mbl. Augenheilk.,* 184, 42, 1984.
13. **Janssen, P. T. and van Bijsterveld, O. P.,** The relations between tear fluid concentrations of lysozyme, tear-specific prealbumin and lactoferrin, *Exp. Eye Res.,* 36, 773, 1983.
14. **Berta, A.,** Standardization of tear protein determinations. The effects of sampling, flow rate, and vascular permeability, in *The Preocular Tear Film in Health, Disease, and Contact Lens Wear,* Holly, F. J., Ed., Dry Eye Institute, Lubbock, TX, 1986, 418.
15. **Meyer, K. and Haknel, E.,** The estimation of lysozyme by a viscosimetric method, *J. Biol. Chem.,* 163, 723, 1946.
16. **Regan, E.,** The lysozyme content of tears, *Am. J. Ophthalmol.,* 33, 600, 1950.
17. **Bonavida, B. and Sapse, A. T.,** Human tear lysozyme. II. Quantitative determination with standard Schirmer strips, *Am. J. Ophthalmol.,* 66, 70, 1968.

18. **van Bijsterveld, O. P.,** Standardization of the lysozyme test for a commercially available medium, *Arch. Ophthalmol.,* 91, 432, 1974.
19. **Osserman, E. F. and Lawlor, D. P.,** Serum and urinary lysozyme (muraminidase) in monocytic and monomyelocytic leukemia, *J. Exp. Med.,* 124, 921, 1966.
20. **Copeland, J. R., Lamberts, D. W., and Holly, F. J.,** Investigation of the accuracy of tear lysozyme determination by the Quantiplate method, *Invest. Ophthalmol. Vis. Sci.,* 22, 103, 1982.
21. **Stuchell, R. N., Farris, R. L., and Mandel, I. D.,** Basal and reflex tear analysis. II. Chemical analysis: lactoferrin and lysozyme, *Ophthalmology,* 88, 858, 1981.
22. **van Bijsterveld, O. P. and Mansour, K. H.,** Lysozyme concentration in reflex and 'basic' secretion, *Graefe's Arch. Clin. Exp. Ophthalmol.,* 221, 130, 1983.
23. **Stuchell, R. N., Feldman, J. J., Farris, R. L., and Mandel, I.,** The effect of collection technique on tear composition, *Invest. Ophthalmol. Vis. Sci.,* 25, 374, 1984.
24. **Janssen, P. T. and van Bijsterveld, O. P.,** Origin and biosynthesis of human tear fluid proteins, *Invest. Ophthalmol. Vis. Sci.,* 24, 623, 1983.

Chapter 6

LYSOZYME IN TEARS

Part C
The Diagnostic Value of Tear Lysozyme Determinations, Normal Values, Tear Lysozyme Levels in Keratoconjunctivitis Sicca, in Sjögren's Syndrome, and in other Eye Diseases

I. NORMAL VALUES, THE EFFECTS OF THE USE OF DIFFERENT METHODS AND DIFFERENT STANDARDS

Various authors have reported different normal levels for the lysozyme in tears, to a large extent depending on the methods and standards they used (Table 1). Following the lysozyme titer determinations in the first part of this century, the first normal value for the lysozyme content of tears obtained by viscosimetric method was 1438 IU/ml.[1] The use of international units was soon replaced by that of concentration values (mg/ml). The normal value of tear lysozyme measured with the agar diffusion method was in the range of 0.8 to 2.2 mg/ml when human tear lysozyme was used as a standard.[2] van Bijsterveld and Mansour[3] gained a two times (3.274 mg/ml) and Avisar et al.[4] found three times higher (5.34 mg/ml) normal values with the agar diffusion (lysoplate, quantiplate, radial diffusion) method using hen egg white lysozyme (HEL) as a standard.

The spectrophotometric method based on the measurement of changes in the absorption (optical density) of solutions of *Micrococcus lysodeikticus* due to the effect of lysozyme contained in the sample has been also widely used. The normal values gained with this method were in the range of 3 and 7 mg/ml using HEL as a standard.[5-8] The differences in the normal values for methods based on the enzymatic action of lysozyme were shown to be determined by the specific activities of the used lysozyme standards. The specific activity of HEL was approximately three times lower than that of the human tear lysozyme, which well explained the higher normal values produced using egg lysozyme as a standard.[9] There are other values published in the literature concerning this difference; Osserman and Lawor[10] found 8 to 12 times higher activity for human lysozyme than for HEL.

The diameter of the lysis zone with the agar diffusion method was shown to depend on a number of variables including incubation time, temperature, concentration of the bacterium, and thickness of the agar.[2] van Bijsterveld[11] investigated different media for growing the *M. lysodeikticus* in order to find a medium on which the test can be standardized. It has been suggested that the zone of lysis obtained with lysozyme standard may be variable due to impurities.[12] Mackie and Seal[13] reported that the growth of the *M. lysodeikticus* on DST agar was never exactly the same, and direct comparison between zones of lysis was unreliable. They suggested that some of the difficulties can be

TABLE 1
Lysozyme Levels Found in Tears by Different Authors

Author(s)	Year	Sampling	Method	Value (mg/ml)
Aine and Mörsky[25]	1984	Capillary tube	IMM.TURB	1.67 ± 0.63
Avisar et al.[4]	1981	Filter paper	Agar diffusion	5.29 ± 2.13 (HEL)
Berta[26]	1986	Capillary tube	SDS-PAGE	2.0 ± 0.7
Bonavida and Sapse[2]	1968	Filter paper	Agar diffusion	0.8–2.0
deLuise and Tabbara[20]	1983	Capillary tube		2.0
Dougherty and McCulley[27]	1985	Capillary tube	Lysoplate	3.76 (HEL)
Ensink and Haeringen[5]	1977	Filter paper	SPEC Lysoplate	3.0–7.0
Gachon et al.[28]	1982	Capillary tube	SDS-PAGE	0.6–2.3
Grabner et al.[29]	1982	Filter paper	EID, RID, SPEC	1.5–3.9
Gupta et al.[30]	1986	Capillary	RID	1.4–1.6
Harada et al.[31]	1980	?		2.2–6.9
Hemmingsen et al.[32]	1986	Capillary		2.2
Horowitz et al.[33]	1978			0.6–2.2
Janssen and van Bijsterveld[23,34,35]	1982	Filter paper	Agar diffusion	1.89
	1983	Filter paper	SDS-PAGE Dye elution	0.2–3.8
	1986	Filter paper	SDS-PAGE	2.1 ± 0.6
Konig and van Bijsterveld[36]	1983	Filter paper	Agar diffusion	2.89
Mackie and Seal[12]	1976	Filter paper	SPEC	130 IU/ml
McGill et al.[37]	1984	Filter paper	SDS-PAGE	0.7–1.1
Regen[1]	1950	Pipettes	VISC	1438 IU/ml
Ronen et al.[8]	1975		SPEC	6.1 (HEL)
Scharf et al.[7]	1982	Filter paper	TURB	7.67 ± 1.74 (HEL)
Sen and Sarin[38]	1986	Capillary		0.8–1.6
Strassner and Grabner[39]	1982	Filter paper	EID	0.5–2.2
Stuchell et al.[40,41]	1981	Filter paper	EID	1.6
	1984	Capillary	EID	2.1
		Filter paper		2.2
van Bijsterveld and Mansour	1983	Filter paper	Agar diffusion	3.27 (HEL)
Vinding et al.[42]	1987	Capillary tube		2.2

Note: SPEC: spectrophotometric method, RID: radial immunodiffusion, EID: electro-immunodiffusion, VISC: viscosimetry, TURB: turbidimetry, IMM.TURB: immunoturbidimetry, HEL: hen egg white lysozyme standard, SDS-PAGE: sodium dodecyl sulfate-polyacrylamide gel electrophoresis.

overcome if the volume of the collected tear sample is taken into consideration, if a calibrated standard is used, and if the lysozyme content of tears is expressed in unit of activity per microliter.[13] They determined the lysozyme levels in the eyes of 54 normal volunteers (between the age of 18 and 86 years). The mean values of the lower levels of the normal range were found to decrease with age. The mean level roughly decreased by one unit of activity per microliter for each year. The lower limit was 70 units of activity per microliter at 20 years and 40 units of activity per microliter at 85 years. There was no correlation between the lysozyme concentration and the total volume of the tear sample (the quantitity of the tears absorbed by the filter paper disk). This was also true for tear flow rates. Some of the normal volunteers who produced small quantities of tear fluid had normal or even high lysozyme concentrations. There was a significant correlation between lysozyme levels measured in the left and in the right eye of the same person (correlation coefficient: 0.7).[13]

II. TEAR LYSOZYME LEVELS IN KERATOCONJUNCTIVITIS SICCA AND SJÖGREN'S SYNDROME. TEAR LYSOZYME LEVELS IN OTHER EXTERNAL EYE DISEASES

Meyer[14] was the first to show, by viscosimetric method, that the tear lysozyme concentration decreased in keratoconjunctivitis sicca (KCS). His observation was confirmed by Regen[1] and McEven and Kimura[15] by viscosimetric and electrophoretic methods. Thygeson and Kimura[16] demonstrated that a decrease in the concentration of lysozyme preceded the appearance of all other symptoms of KCS. Soon after the introduction of a simple and still reliable test, based on the lysis of a bacterial suspension of *M. lysodeikticus* seeded in an agarose layer,[2] tear lysozyme determination became a useful method for the laboratory diagnosis of KCS and Sjögren's syndrome.[17]

In an extensive study van Bijsterveld investigated the lysozyme levels, tear secretion with the Schirmer's test, and the Rose bengal staining of the cornea in 43 patients (86 eyes) with KCS, and in 550 healthy controls (1100 eyes). Tear lysozyme levels were determined with the agar diffusion test (the diameter of the filter paper disk was 6 mm, meat infusion agar contained 5×10^7 *M. lysodeikticus,* the lysis was measured after 24 h of incubation at 37°C). The diameters of the lysis zones ranged from 20 to 29 mm in the control group and from 10 to 22 mm in the KCS group. The more than 50% of the cases showed lysis diameter between 23 and 26 mm in the control group and between 14 and 17 mm in the KCS group. The difference was highly significant. van Bijsterveld calculated the probability of misclassification for the control group and for the patients with the three parameters. With a diameter limit of 21.5 mm of lysis under the above-described test conditions, only one out of every 100 diagnoses would have been wrong. For Rose bengal test, using a limit of score of staining intensity of 3.5, 1 out of every 20 to 25 times a diagnostic error could be expected. With the Schirmer's I test, using a limit of 5.5 mm wetting of the filter paper strip, one out of every six persons would have been misclassified.

The author concluded that tear lysozyme determination with the agar diffusion test was the most sensitive and most reliable from the three tests used for the diagnosis of the sicca syndrome.[17] Based on the recommendation in this publication and in other papers[3,11,18] of van Bijsterveld and co-workers, several authors used his method or its modified versions for the diagnosis of KCS and Sjögren's syndrome.

Mackie and Seal[13] compared the tear lysozyme activities of KCS patients with those of age-matched controls. Five out of the six patients with KCS had tear lysozyme activities (expressed in units of activity per microliter) well below the lowest normal concentration (critical limit) for their age. In other experiments[19] they studied lysozyme activities and Schirmer values of 30 patients with manifest KCS, and 35 patients with clinically "questionable dry eyes", and compared them with patients suffering from autoimmune diseases without dry eyes, with patients suffering from external eye diseases not associated with dry eyes, and age-matched normal controls. Each determined tear lysozyme concentration was divided by the critical lower limit value of the corresponding age yielding a tear lysozyme ratio (TLR). The mean values of TLR in the group of patients with autoimmune diseases and with external eye disease without dry eyes were 2.37 and 2.23, respectively. Patients with questionably dry eyes and with manifest dry eyes had mean values of TRL 1.22 and 0.55, respectively. In both groups the difference was significant ($p < 0.001$) as compared with normal controls. Within the group of questionably dry eyes those who demonstrated "particular matter" in the tear film produced even lower (0.77) TRL values. There was no significant correlation between Schirmer test values and tear lysozyme levels in the normal group, while there was good correlation between the two parameters in patients with undoubted KCS. They concluded that tear flow and tear lysozyme levels are not necessarily interrelated and that while the reduction in the Schirmer value is reliable only in manifest dry eyes the decrease in TLR is of diagnostic value both in borderline and in manifest cases of KCS.[19]

Ninety-seven rheumatoid arthritis patients receiving nonsteroid antiinflammatory drugs were checked for the prevalence and incidence of ocular problems in general, symptoms of KCS, and tear lysozyme levels by Pisko et al.[6] A prevalence of 31.9% and an incidence of 24.5%/year were found for all kinds of ocular pathology in these patients. Twenty-one of these patients were studied further for evidence of KCS. None of them had abnormal rose bengal staining. Over half of them had decreased tear production on both stimulated and unstimulated Schirmer's testing and many patients showed high concentrations of lysozyme in tears; none of the patients showed subnormal tear lysozyme levels. Increased tear lysozyme levels and decreased Schirmer's tests correlated with a rheumatoid activity index showing the severity and activity of the disease.[12]

Scharf et al.[7] compared the concentration of lysozyme in the tears of 53 patients with that of 34 healthy controls. Tear lysozyme level was significantly ($p < 0.0005$) lower in the tears of KCS patient than in the control group. This was especially true for patients with low Schirmer values. Even in rheumatoid arthritis patients with normal Schirmer test results, low concentrations of lysozyme in tears were found. It was suggested that tear lysozyme determinations may be used to detect subclinical involvement of lacrimal glands in collagen disease.

DeLuise et al.[20] used spectrophotometric methods to determine the lysozyme levels in the tears of patients with sicca syndrome and those of healthy individuals. The normal subjects were found to have lysozyme levels ranging from 0.5 to 4.0 mg/ml (mean 2.0 mg/ml). Subjects with dry eyes had tear lysozyme levels ranging from 0 to 3.7 mg/ml (mean 0.94 mg/ml). There was one person among the dry eye patients with paradoxically elevated tear lysozyme level (greater than 4.0 mg/ml) who was found to have underlying sarcoidosis. The reliability of the test to diagnose dry eye syndrome was found to be 83%.

Sapse et al.[20a] checked the lysozyme levels of subjects complaining of smog irritation. When tears were collected in smog-polluted areas from individuals who had symptoms of smog eye irritation, low levels of tear lysozyme were found, averaging 60% less than the mean normal levels. When subjects sensitive to smog were checked for their tear lysozyme content in pollution-free areas, their lysozyme levels were higher than the average mean found in normal subjects.

Ratnakar et al.[22] reported on lacrimal lysozyme alterations in experimental protein deficiency. Avisar et al.[4] estimated tear lysozyme levels in 130 eyes with external ocular infections during the active inflammation and after recovery and compared these findings to those for 286 normal eyes. They found an exact correlation between the rate of tear secretion in the eyes with external infection and the lysozyme content of tears. The greater the flow of tears, the greater was the decrease in the enzyme levels measured in tears.

Saari et al.[21] studied the lysozyme content of tears of patients' external eye infections, and compared them with that of 267 normal subjects. Patients with blepharitis, conjunctivitis, and keratitis showed normal tear lysozyme levels. The tears of patients with herpes simplex keratitis exhibited low levels of lysozyme in tears.

III. THE DIAGNOSTIC VALUE OF TEAR LYSOZYME DETERMINATIONS

Lysozyme is secreted by the main lacrimal gland. The lysozyme level in tears is known to decrease in a variety of diseases or pathological conditions that adversely affect the secretion of tears in the lacrimal gland such as Sjögren's syndrome, sarcoidosis, primary amyloidosis, Mikulitz's syndrome, various inflammatory, traumatic, or toxic lesions of the lacrimal gland as well as tumors, surgery, or irradiation of the lacrimal gland, certain systemic or neurological diseases, nutritional deficiencies, drug effects, and old age. The decreased secretion of lysozyme is usually, but not necessarily, accompanied by an aqueous tear deficiency. Low levels of lysozyme in the tears (in tear samples collected from the lower cul-de-sac or from the marginal tear menisci) were also demonstrated in eyes right after surgery, in bacterial conjunctivitis, in herpetic keratitis, in smog irritation, and in cases with non-Sjögren KCS. The decrease of the tear lysozyme level in these latter cases may be explained, to a large extent, by the dilution of the secretion of the lacrimal gland by a fluid, with very low lysozyme content, that gets into the conjunctival sac due to transudation through the walls of the conjunctival blood vessels. This diluting effect may contribute to that of the decreased lysozyme secretion also in cases when the primary cause is in the lacrimal gland but the dry eye is irritated or inflamed. The lower the rate of tear production and the lower the volume of tears in the eye, the greater the change in the tear level of lysozyme and of other tear proteins (lactoferrin, specific tear prealbumin) brought about by the diluting effect. A parallel determination of serum proteins could be helpful in determining the degree of transudation but the low volume of the tear samples obtained from dry eyes usually allows only one determination. In this respect the electrophoretic-densitometric lysozyme determination that we use for diagnostic purposes has an important advantage: it makes possible the determination of several proteins (including serum proteins) in one sample at the same time.

In spite of the general acceptance that tear lysozyme or tear lactoferrin levels are good indicators of impaired lacrimal gland function and decreased tear secretion, these test are not very popular among practicing ophthalmologists. According to a survey of the Dry Eye Institute, and also to my personal experiences, most ophthalmologists in the U.S., including dry eye specialists, do not routinely determine lysozyme levels in the tears of their dry eye patients.[23] The diagnosis of KCS and Sjögren's syndrome is still primarily based on clinical tests: Schirmer I test, breakup time (BUT) determinations, and vital staining of the cornea, in spite of the fact that the reliability of these clinical tests has also been strongly criticized.[17,19,24] Recently several recommendations were presented at an international symposium for the diagnostic criteria of Sjögren's syndrome and KCS.[9] None of them suggested that the diagnosis of these diseases should be based on tear protein determinations. They pointed out, however, that tear lysozyme determinations can be useful in the diagnosis of clinically atypical or borderline cases and agreed that in all cases such tests may confirm the clinical diagnosis and help in the follow-up of the progress of lacrimal gland destruction during the course of the disease.

REFERENCES

1. **Regan, E.,** The lysozyme content of tears, *Am. J. Ophthalmol.,* 33, 600, 1950.
2. **Bonavida, B. and Sapse, A. T.,** Human tear lysozyme. II. Quantitative determination with standard Schirmer strips, *Am. J. Ophthalmol.,* 66, 70, 1968.
3. **van Bijsterveld, O. P. and Mansour, K. H.,** Lysozyme concentration in reflex and 'basic' secretion, *Graefe's Arch. Clin. Exp. Ophthalmol.,* 221, 130, 1983.
4. **Avisar, R., Menaché, R., Shaked, P., and Savir, H.,** Lysozyme content of tears in some external eye infections, *Am. J. Ophthalmol.,* 92, 555, 1981.
5. **Ensink, F. T. E. and van Haeringen, N. J.,** Pitfalls in the assay of lysozyme in human tear fluid, *Ophthal. Res.,* 9, 366, 1977.
6. **Pisko, E. J., Turner, R. A., Yeatts, R. P., Burch, P. G., McKinley, P. H., and Collins, R. L.,** Ocular pathology, tear production, and tear lysozyme in rheumatoid arthritis, *J. Rheumatol.,* 9, 708, 1982.
7. **Scharf, J., Meshulam, T., Obedeanu, N., Nahir, M., Zonis, S., and Merzbach, D.,** Lysozyme concentration in tears of patients with sicca syndrome, *Ann. Ophthalmol.,* 1063, 1982.
8. **Ronen, D., Eylan, E., Romano, A., Stein, R., and Modan, M.,** A spectrophotometric method for quantitative determination of lysozyme in human tears, *Invest. Ophthalmol.,* 14, 479, 1975.
9. **Manthorpe, R. and Prause, J. U., Eds.,** Proceedings of the 1st international seminar on Sjögren's syndrome, *Scand. J. Rheumatol. Suppl.,* 61, 1986.
10. **Osserman, E. F. and Lawlor, D. P.,** Serum and urinary lysozyme (muraminidase) in monocytic and monomyelotic leukemia, *J. Exp. Med.,* 124, 921, 1966.
11. **van Bijsterveld, O. P.,** Standardization of the lysozyme test for a commercially available medium, *Arch. Ophthalmol.,* 91, 432, 1974.
12. **Pietsch, R. L. and Pearlman, M. E.,** Human tear lysozyme variables, *Arch. Ophthalmol.,* 90, 94, 1973.
13. **Mackie, I. A. and Seal, D. V.,** Quantitative tear lysozyme assay in units of activity per microlitre, *Br. J. Ophthalmol.,* 60, 70, 1976.
14. **Meyer, K.,** Mucopolysaccharides and mucoids of ocular tissues and their enzymatic hydrolysis, in *Modern Trends in Ophthalmology,* Vol. 2, Sorsby, A., Ed., Butterworth, London, 1948, 71.
15. **McEwen, W. K. and Kimura, S. J.,** Filter-paper electrophoresis of tears, *Am. J. Ophthalmol.,* 39, 200, 1955.
16. **Thygeson, P. and Kimura, S. J.,** Chronic conjunctivitis, *Trans. Am. Acad. Ophthalmol. Otolaryngol.,* 67, 494, 1963.
17. **van Bijsterveld, O. P.,** Diagnostic tests in the sicca syndrome, *Arch. Ophthalmol.,* 82, 10, 1969.
18. **van Bijsterveld, O. P.,** Queratoconjunctivitis sicca. Conceptos actuales, *Invest. Med. Int.,* 6, 125, 1979.
19. **Mackie, I. A. and Seal, D. V.,** The questionable dry eye, *Br. J. Ophthalmol.,* 65, 2, 1981.
20. **deLuise, V. P. and Tabbara, K. F.,** Quantitation of tear lysozyme levels in dry-eye disorders, *Arch. Ophthalmol.,* 101, 634, 1983.
20a. **Sapse, A. T., Bonavida, B., and Sone, W., Jr.,** Human tear lysozyme. III. Preliminary study on lysozyme levels in subjects with smog eye irritation, *Am. J. Ophthalmol.,* 66, 76, 1968.
21. **Saari, K. M., Aine, E., Posz, A., and Klockars, M.,** Lysozyme content of tears in normal subjects and in patients with external eye infections, *Graefe's Arch. Clin. Exp. Ophthalmol.,* 221, 86, 1983.
22. **Ratnakar, K. S., Kanta, R. C., Mehta, U., and Amoji, S. D.,** Lacrimal lysozyme alterations in experimental protein deficiency, *Ophthalmologica,* 181, 320, 1980.

23. **Janssen, P. T. and van Bijsterveld, O. P.,** The relations between tear fluid concentrations of lysozyme, tear-specific prealbumin and lactoferrin, *Exp. Eye Res.,* 36, 773, 1983.

24. **Schapiro, A. and Merin, S.,** Schirmer test and break up time of tear film in normal subjects, *Am. J. Ophthalmol.,* 88, 752, 1979.

25. **Aine, E. and Mörsky, P.,** Lysozyme concentration in tears, assessment of reference values in normal subjects, *Acta Ophthalmol.,* 62, 932, 1984.

26. **Berta, A.,** Standardization of tear protein determinations. The effects of sampling, flow rate, and vascular permeability, in *The Preocular Tear Film in Health, Disease, and Contact Lens Wear,* Holly, F. J., Ed., Dry Eye Institute, Lubbock, TX, 1986, 418.

27. **Dougherty, J. M. and McCulley, J. P.,** Tear lysozyme measurements in chronic blepharitis, *Ann. Ophthalmol.,* 17, 53, 1985.

28. **Gachon, A. M., Richard, J., and Dastugue, B.,** Human tears: normal protein pattern and individual protein determination in adults, *Curr. Eye Res.,* 2, 301, 1982.

29. **Grabner, G., Formanek, I., Dorda, W., and Luger, T.,** Human tear lysozyme. A comparison of electro-immunodiffusion, radial immunodiffusion and spectrophotometric assay, *Graefe's Arch. Clin. Exp. Ophthalmol.,* 218, 265, 1982.

30. **Gupta, A. K., Sarin, G. S., Lambda, P. A., and D'Souza, P.,** Immunoassay of tear lysozyme in acute adenovirus conjunctivitis, *Br. J. Ophthalmol.,* 70, 439, 1986.

31. **Harada, M., Miyata, M., and Ishikawa, S.,** Antibacterial substances in human tears, *Jpn. J. Ophthalmol.,* 24, 320, 1980.

32. **Hemmingsen, L., Holm, J., Ericksen, J., and Persson, B. E.,** Secretory immunoglobulin A, immunoglobulin G and lysozyme levels in tear fluid from subjects wearing contact lenses, *IRCS Med. Sci.,* 14, 950, 1986.

33. **Horowitz, B. L., Christensen, G. R., and Ritzman, S. R.,** Diurnal profiles of tear lysozyme and gamma A globulin, *Ann. Ophthalmol.,* 75, 1978.

34. **Janssen, P. T. and van Bijsterveld, O. P.,** Immunochemical determination of human tear lysozyme (muramidase) in keratoconjunctivitis sicca, *Clin. Chim. Acta,* 121, 251, 1982.

35. **Janssen, P. T. and van Bijsterveld, O. P.,** Tear fluid proteins in Sjögren's syndrome, *Scand. J. Rheumatol. Suppl.,* 61, 224, 1986.

36. **de Koning, E. W. J. and van Bijsterveld, O. P.,** Schirmer test values and lysozyme content of tears in acute dendritic keratitis, *Invest. Ophthalmol. Vis. Sci.,* 25, 55, 1984.

37. **McGill, J. I., Liakos, G. M., Goulding, N., and Seal, D. V.,** Normal tear protein profiles and age-related changes, *Br. J. Ophthalmol.,* 68, 316, 1984.

38. **Sen, D. K. and Sarin, G. S.,** Biological variations of lysozyme concentration in the tear fluid of healthy persons, *Br. J. Ophthalmol.,* 70, 246, 1986.

39. **Strasser, G. and Grabner, G.,** The influence of long-term treatment with Timolol on human tear lysozyme and albumin content, *Graefe's Arch. Clin. Exp. Ophthalmol.,* 218, 93, 1982.

40. **Stuchell, R. N., Farris, R. L., and Mandel, I. D.,** Basal and reflex tear analysis. II. Chemical analysis: lactoferrin and lysozyme, *Ophthalmology,* 88, 858, 1981.

41. **Stuchell, R. N., Feldman, J. J., Farris, R. L., and Mandel, I.,** The effect of collection technique on tear composition, *Invest. Ophthalmol. Vis. Sci.,* 25, 374, 1984.

42. **Vinding, J. T., Sindberg Ericksen, J., and Vesti Nielsen, N.,** The concentration of lysozyme and secretory IgA in tears from healthy persons with and without contact lens use, *Acta Ophthalmol.,* 65, 23, 1987.

LYSOZYME IN TEARS

Part D
**Lysozyme and Nonlysozyme Antibacterial Factors in Tears —
Lysozyme Therapy**

I. LYSOZYME AND NONLYSOZYME ANTIBACTERIAL FACTORS IN HUMAN TEARS

The integrity of the human body is protected from external influences by epithelial barriers such as the skin and mucosal surfaces. Most of these epithelial surfaces are in direct contact with the outside environment and developed a number of defense mechanisms to prevent the invasion of microorganisms. The defense mechanisms of the eye seem to be very effective as the epithelial covering of the eye, unlike skin and gut mucosa, is colonized only by a small number of bacteria and infrequently becomes infected. The defense mechanisms of the eye include the protective function of eyelids and eyelashes, the washing effect of tear flow, and the specific action of antibacterial factors present in tears. Antibacterial substances in tears include lysozyme, lactoferrin, tear-specific prealbumin, β-lysin, and immunological factors (secretory IgA, complement, IgG, etc.) (Table 1).[1-5]

It appears from studies of infections involving other epithelial surfaces that bacteria must first attach to the epithelial cells before they cause infection. Factors preventing bacterial adherence can play a protective role, similar in effect to those that inactivate or inhibit the growth of microorganisms.[6]

The most abundant antibacterial substance in tears is lysozyme, which accounts for approximately 40% of the protein content of human tears. This high level of lysozyme is in contrast to the small amounts of this enzyme found in other body fluids and secretions, suggesting a specific function of lysozyme in tears. Characteristics of the antibacterial action of lysozyme were first described by Fleming, who discovered this enzyme prior to the discovery of penicillin. He recognized that lysozyme acted mainly on harmless saprophytes and suggested that this was why they were harmless.[7] This suggestion can be doubted knowing that the normal bacterial flora consists of viable microorganisms while lysozyme causes lysis of Gram-positive bacteria by destroying their cell walls.[3]

Based on the assumption that nothing is without a definite reason in nature, one should really wonder why humans and some monkeys[8] keep such high lysozyme levels in tears. At first glance the antibacterial effect seems to be the most obvious explanation. That lysozyme inhibits only Gram-positive bacteria and is most effective on harmless saprophytes raises serious doubt whether this effect could be strong enough not to be lost in the selective processes of the evolution of species. Probably this is the reason why considerable energy is

TABLE 1
Antibacterial Substances in Tears

Substance	Effect
Lysozyme	Destroys the cell walls of Gram-positive bacteria
Lactoferrin	Binds iron needed in bacterial metabolism
β-Lysin	Ruptures bacterial cell membrane
Secretory IgA	Inhibits the adherence of bacteria to epithelial surfaces
IgG	Promotes phagocytosis and complement-mediated lysis
Complement	Causes lysis of various bacteria

devoted to find other important functions of lysozyme in tears and to study other tear components with antibacterial effects.

Thompson and Gallardo studied the antibacterial action of tears on staphylococci. They demonstrated the presence in pooled stimulated tears of an acid-heat labile antibacterial agent that inhibited the growth of certain strains of staphylococci.[39] Human tear lysozyme as measured spectrophotometrically by the clearing of a suspension of *Micrococcus lysodeikticus* was found to be heat resistant.[38] Friedland et al.[10] found a heat-labile, low molecular weight compound (5000 to 7500 Da) that had a bactericidal activity towards a wide range of microorganisms. They called this compound nonlysozyme antibacterial factor (NLAF). In their experiments this factor was more powerful than lysozyme, and was active against Gram-negative as well as Gram-positive bacteria including those known to cause ocular infections. They also suggested that the antibacterial effect, measured in the agar diffusion assay, had to be attributed mainly to the presence of NLAF and only for 0.5% to the action of lysozyme. Ford et al.[11] reported that this factor was identical with or very similar to β-lysin, an antibacterial compound with a molecular weight of 2000 Da. These two reports, however, did not explain two controversies. (1) NFLA did not cause the lysis of *M. lysodeikticus* though it seemed to have an antibacterial (bacteriostatic) effect on various bacteria including *M. lysodeikticus*, whereas the agar diffusion test is based primarily on the lytic action of lysozyme (the clearing of the originally opaque gel around the filter paper disk) and not on the prevention of bacterial growth. (2) Dialysis did not have an influence on the antibacterial activity of the studied tear samples. In case of an antibacterial factor with such a low molecular weight one would have expected that it should.

Others could not confirm the results concerning the presence of a low molecular weight NLAF in human tears. Selsted and Martinez reported the presence of three antibacterial proteins in the tear fluid: lysozyme, tear-specific prealbumin, and an antibacterial glycoprotein. They could not detect a low molecular weight factor with antibacterial activity.[4] Seal et al.[12] have shown that the antibacterial activity in tears disappeared completely after the addition of antilysozyme antiserum, and concluded that lysozyme was the only compound in tears that was active against *M. lysodeikticus*. Janssen et al.,[13] following the fractionation of pooled normal human tears, found that the antibacterial activity in the fractions invariably coincided with the presence of lysozyme. In the fractions that could have contained NLAF or β-lysin no antibacterial activity was observed. Based on their results they concluded that the antibacterial activity of tears demonstrated in spectrophotometric and agar diffusion assays has to be attributed to lysozyme and not to NLAF or β-lysin.

Other tear proteins such as lactoferrin and tear-specfic prealbumin were also found to have antibacterial activity, or at least play a role in modulating the effect of protective factors including that of tear lysozyme. Lactoferrin is an iron-binding protein with a molecular weight of 82,000 Da. Like transferrin it is capable to reversibly bind two Fe(III) ions, but its binding is 300 times stronger than that of transferrin. This characteristic feature is thought to be significant for the bacteriostatic activity of lactoferrin in tears so that it may act by depriving microorganisms of iron that is essential for their metabolism.[14] Lactoferrin may play a role as an adjunct to lysozyme in the antibacterial defense as some of the otherwise resistant Gram-negative bacteria may become susceptible to the lytic action of lysozyme in the presence of chelating agents.[15] Moreover, lactoferrin was shown to inhibit hemolytic complement activity in tears and therefore was suggested to play an important role in the modulation of complement-mediated reactions involved in the local immune response of the eye.[16]

Tear-specific prealbumin is a unique protein fraction in the tear fluid, also called specific tear prealbumin or anodal tear protein. Genetic polymorphism has been reported for tear-specific prealbumin.[17] Recent studies suggest the involvement of this protein as a carrier in the transport of retinol from the lacrimal gland to the corneal epithelial cells via the tear film.[5,9] An enhancement of lysozyme activity by tear-specific prealbumin has been reported by Josephson and Wald,[18] and a relatively weak bactericidal effect on *Bacillus subtilis* and *Listeria monocytogenes* was demonstrated by Selsted and Martinez.[4]

II. LYSOZYME THERAPY

The possibility of using lysozyme in eye drops for the treatment of different external eye diseases was first mentioned in the Russian ophthalmic literature. Tartovskaja[19] used lysozyme in the therapy of different types of keratitis and found this form of therapy effective in bacterial ulcers. Her observations were confirmed by Checkik-Kunina in 1945 who reported a higher healing rate of perforating corneal injuries when lysozyme eye drops were used as antibacterial agents.[20] To evaluate these results one has to keep in mind that no antibiotic eye drops were available at that time, and higher healing rates could have resulted from the less frequent or less severe bacterial complications. No evidence of a direct enhancing effect of lysozyme on corneal wound healing has been reported so far.

Uncontrolled studies have suggested beneficial effects of human tear lysozyme in corneal ulcers, trachoma, keratitis, corneal abrasions, acid and alkali burns, and conjunctivitis. Also, lysozyme from egg white has been used to treat conjunctivitis. With the development of better treatment modalities, however, the emphasis on the therapeutic application of lysozyme in infectious eye diseases has waned.[3]

The idea to treat keratoconjunctivitis sicca (KCS) by the replenishment of the "missing" lysozyme, using lysozyme-containing eye drops, dates back almost to time when the low levels of lysozyme in the tears of KCS patients were first detected. The rationale of the use of hen egg white lysozyme (HEL), serum proteins, and other natural substances (colostrum, saliva, autologous serum, etc.) in the treatment of dry eyes,[21-27] besides their antibacterial effect, is based on two assumptions. (1) Besides cases that can be characterized by the parallel decrease of the secretion of aqueous tears and the diminished production of tear proteins by the lacrimal gland, other conditions might exist in which the tear secretion is quantitatively normal (shown by normal Schirmer values) but the secretion of the lacrimal gland contains subnormal levels of proteins (lysozyme, lactoferrin, tear-specific tear prealbumin). This supposed qualitative tear deficiency is also called "lacrimal hyposecretory syndrome" in the Spanish literature. (2) The epithelial cells on the surface of the eye require a special protein milieau for their well being and normal function, and this milieau is disturbed when the lacrimal gland secretes quantitatively normal but too dilute tears, characterized by low levels of tear proteins. According to this view any form of tear substitution without sufficient protein content, including all presently used artificial tears, would exaggerate the local protein deficiency in the tear film (Suarez, J.C., personal communication). None of these assumptions have been proven so far, which, besides the technical difficulties of developing a stable, sterile, and reliable natural tear substitute, may account for the fact that no such preparation has been developed, approved, and marketed so far.

The first assumption may be supported by the observation that tear lysozyme and lactoferrin levels can be subnormal in the early phase of KCS in patients with normal Schirmer values.[28] The pathological role of decreased levels of tear proteins in the development of epithelial lesions, however, does not seem to be established as the epithelium is usually intact in the borderline cases of KCS (questionable dry eyes). Epithelial lesions develop in cases of manifest KCS, which, besides the low levels of lysozyme and other proteins of lacrimal gland origin, are also characterized by aqueous tear deficiency. Moreover, the total protein content of tears in severe KCS is usually above the normal level, showing that there is no absolute "tear protein deficiency" in these cases.[29,30] Low total protein and low lysozyme levels in tears are common, on the other hand, in asymptomatic elderly patients.[31-33]

The only clinical trial to treat KCS with lysozyme-containing eye drops I know of was performed by Kahán et al.,[34] who reported a beneficial effect of solutions of egg white lysozyme on the subjective and objective symptoms of KCS. Possible roles of lysozyme in lowering the surface tension of tears, enhancing the adhesion of epithelial cells, resolubilizing the aggregated mucus particles, stabilizing the mucous phase of the preocular tear film were suggested, but none of these possibilities have been confirmed so far (Holly, F. J., personal communication).

A novel lysozyme-containing contact lens solution was tested and found to be useful both for storing and applying contact lenses. This product, presently available only in Hungary, was designed to meet the special needs of patients sensitive to most of the preservatives and disinfectants used in the commercially available contact lens solutions.[35]

However logical the substitution of the missing lysozyme in tears may be, the problem seems to be more complicated. The simple replacement of a single missing component is clearly not enough to restore the original structure and function of the tear film. Several quantitative and qualitative factors should be taken into consideration, and a better understanding of the multiple functions of tear proteins including lysozyme is needed before lysozyme or other tear proteins can be used for therapeutic purposes in the future. Similar doubts rise concerning the use of lysozyme in contact lens solutions especially if we take into consideration that protein debris containing substantial amounts of precipitated lysozyme is often found on contact lenses.[36,37]

REFERENCES

1. **Chandler, J. W., Leder, R., Kaufman, H. E., and Caldwell, J. R.,** Quantitative determinations of complement components and immunoglobulins in tears and aqueous humor, *Invest. Ophthalmol.,* 13, 151, 1974.
2. **Jensen, R. S., Tew, J. G., and Donaldson, D. M.,** Extracellular beta-lysin and muraminidase in body fluids and in inflammatory exudates, *Proc. Soc. Exp. Med. Biol.,* 124, 545, 1967.
3. **Selinger, D. S., Selinger, R. C., and Reed, W. P.,** Resistance to infection of the external eye: the role of tears, *Surv. Ophthalmol.,* 24, 33, 1979.
4. **Selsted, M. E. and Martinez, R. J.,** Isolation and purification of bactericides from human tears, *Exp. Eye Res.,* 34, 305, 1982.
5. **Ubels, J. L., Foley, K. M., and Rismondo, V.,** Retinol secretion by the lacrimal gland, *Invest. Ophthalmol. Vis. Sci.,* 8, 1261, 1986.
6. **Reed, W. P. and Williams, R. C., Jr.,** Bacterial adherence: first step in the pathogenesis of certain infections, *J. Chronic Dis.,* 31, 67, 1978.
7. **Fleming, A. and Allison, V. D.,** Observations on a bacteriolytic substance ('lysozyme') found in secretions and tissues, *Br. J. Exp. Pathol.,* 3, 252, 1922.
8. **Erickson, O. F., Feeney, L., and McEwen, W. K.,** Filter-paper electrophoresis of tears, *Arch. Ophthalmol.,* 55, 800, 1956.
9. **Hong, B. S.,** Demonstration of retinol-binding activity of tear specific prealbumin, *Invest. Ophthamol. Vis. Sci. Suppl.,* 27, 129, 1986.
10. **Friedland, B. R., Anderson, D. R., and Forster, R. K.,** Non-lysozyme antibacterial factor in human tears, *Am. J. Ophthalmol.,* 74, 52, 1972.
11. **Ford, L. C., DeLange, R. J., and Petty, R. W.,** Identification of a non-lysozymal bactericidal factor (beta lysin) in human tears and aqueous humor, *Am. J. Ophthalmol.,* 81, 30, 1976.
12. **Seal, D. V., Mackie, I. A., Coakes, R. L., and Farooqi, B.,** Quantitative tear lysozyme assay: a new technique for transporting specimens, *Br. J. Ophthalmol.,* 64, 700, 1980.
13. **Janssen, P. T., Muytiens, H. L., and van Bijsterveld, O. P.,** Nonlysozyme antibacterial factor in human tears. Fact or fiction?, *Invest. Ophthalmol. Vis. Sci.,* 25, 1156, 1984.

14. **Broekhuyse, R. M.,** Tear lactoferrin: a bacteriostatic and complexing protein, *Invest. Ophthalmol.,* 13, 550, 1974.
15. **Repaske, R.,** Lysis of gram-negative bacteria by lysozyme, *Biochim. Biophys. Acta,* 22, 189, 1956.
16. **Veerhuis, R. and Kijlstra, A.,** Inhibition of hemolytic complement activity by lactoferrin in tears, *Exp. Eye Res.,* 34, 257, 1982.
17. **Azen, E. A.,** Genetic polymorphism of human anodal tear protein, *Biochem. Genet.,* 14, 225, 1976.
18. **Josephson, A. S. and Wald, A.,** Enhancement of lysozyme activity by anodal tear protein, *Proc. Soc. Exp. Biol. Med.,* 131, 677, 1969.
19. **Tartovskaja, S. I.,** Lysozyme in der Therapie von Hornhauterkrankungen, *Vestn. Oftalmol. Abstr. Zentralbl. Ophthalmol.,* 14, 19, 1939.
20. **Checkik-Kunina, E. A.,** Lysozyme in der Behandlumg von Verbrennungen und Perforationen des Auges, *Vestn. Oftalmol. Abstr. Zentralbl. Ophthalmol.,* 24, 42, 1945.
21. **Andrzejewska, W., Stolarska, H., Klera, C., and Kirylowicz, K.,** Artificial tears developed by the author in a case of keratoconjunctivitis sicca, *Klin. Oczna,* 42, 889, 1972.
22. **Bisantis, C.,** Emploi d'un collyre fait d'autosérum dans les affections conjonctivales liées á une sécrétion lacrymale altérée, *Ann. Ocul.,* 209, 759, 1976.
23. **Deschamps, F., Royer, J., Montard, M., Roth, A., and Andermann, C.,** Essai en double aveugle de larmes artificielles naturelles dans 33 cas de sécheresse oculaire, *Bull. Soc. Ophthalmol. Fr.,* 10, 827, 1981.
24. **Fox, R. I., Chan, R., Michelson, J. B., Belmont, J. B., and Michelson, P. E.,** Beneficial effect of artificial tears made with autologous serum in patients with keratoconjunctivitis sicca, *Arthritis Rheum.,* 27, 459, 1984.
25. **Kono, I., Matsumoto, Y., Kono, K., Ishibashi, Y., Narushima, K., Kabashima, T., Yamane, K., Sakurai, T., and Kashiwagi, H.,** Beneficial effect of topical fibronectin in patients with keratoconjunctivitis sicca of Sjögren's syndrome, *J. Rheumatol.,* 12, 487, 1985.
26. **Liotet, S. and Perrin, D.,** Amélioration spectaculaire d'un syndrome sec sévíere par instillation de colostrum, *Bull. Soc. Ophthalmol. Fr.,* 82, 15, 1982.
27. **Murube-Del-Castillo, J.,** Dacriologica Basica, Soc. Esp. Oftal., Las Palmas, 1981, 575.
28. **Mackie, I. A. and Seal D. V.,** The questionable dry eye, *Br. J. Ophthalmol.,* 65, 2, 1981.
29. **Berta, A.,** Standardization of tear protein determinations. The effects of sampling, flow rate, and vascular permeability, in *The Preocular Tear Film in Health, Disease, and Contact Lens Wear,* Holly, F. J., Ed., Dry Eye Institute, Lubbock, TX, 1986, 418.
30. **Berta, A. and Török, M.,** Tear glycoprotein determinations in the diagnosis and differential diagnosis of dry eyes, *Scand. J. Rheumatol. Suppl.,* 61, 228, 1986.
31. **deLuise, V. P. and Tabbara, K. F.,** Quantitation of tear lysozyme levels in dry-eye disorders, *Arch. Ophthalmol.,* 101, 634, 1983.
32. **Pietsch, R. L. and Pearlman, M. E.,** Human tear lysozyme variables, *Arch. Ophthalmol.,* 90, 94, 1973.
33. **van Bijsterveld, O. P.,** Diagnostic tests in the sicca syndrome, *Arch. Ophthalmol.,* 82, 10, 1969.
34. **Kahán, I. L. et al.,** First Experiences with Lysozyme Containing Eye Drops in the Treatment of KCS, paper presented at the International Congress of the Hungarian Ophthalmological Society, Szeged, 1989.
35. **Kahán, I. L.,** Lacrozym. A novel lysozyme containing contact lens solution, Biotechnica Rt., Budapest, 1989.
36. **Gachon, A. M., Bilbaut, T., and Dastuge, B.,** Adsorption of tear proteins on soft contact lenses, *Exp. Eye Res.,* 40, 105, 1985.
37. **Refojo, M. F. and Holly, F. J.,** Tear protein adsorption on hydrogels: a possible cause of contact lens allergy, *Contact Intraoc. Lens Med. J.,* 3, 23, 1977.
38. **van Haeringen, N. J.,** Clinical biochemistry of tears, *Surv. Ophthalmol.,* 26, 84, 1981.
39. **Thompson, R. and Gallardo, E.,** The antibacterial action of tears on staphylococci, *Am. J. Ophthalmol.,* 24, 635, 1941.

CORNEAL COLLAGENS AND TEAR COLLAGENASES

Part A
Collagen Structure — Collagenase Action — Different Types of Collagen and Collagenase, Activation, and Inhibition

I. STRUCTURE AND FUNCTION OF COLLAGEN

Collagen is a common macromolecule in animals and humans as an important component of the connective tissues. Various connective tissues have essentially the same basic structure consisting of fibrous meshwork of collagen and elastin embedded in and interacting with an amorphous polyanionic matrix of proteoglycans and glycoproteins.[1] The process of biological evolution has been accompanied by the parallel evolution of connective tissues leading to a great variety of connective tissues of differing structure and function.

Collagen is a ubiquitous structural protein and occurs in varying amounts in almost all tissues and organs of the human body. Collagen is able to fulfill its multiple structural role due to its ability to form insoluble fibers.[2] Collagen molecules are originally synthesized in a precursor form (procollagen) and are subsequently cleaved to a smaller molecule before aggregating to form collagen fibrils.[3] The procollagen molecule consists of three polypeptide chains (pro-α chains) coiled around each other in a right handed triple helical formation in the central core. There are substantial nonhelical extensions at the N- and C-terminals of the molecule with molecular weights of 15,000 to 20,000 and 33,000 to 35,000, respectively, the molecular weight of the whole chain being 145,000 to 180,000[4] (Figure 1).

The collagen in the extracellular space is made up of tropocollagen molecules produced by the claevage of the N- and C-terminal extensions of the procollagen molecule leaving only short nonhelical extensions (telopeptides). Seen under the electron microscope the collagen fibrils show a characteristic repeating pattern of 65 to 75 nm that results from the arrangement of tropocollagen molecules in a staggered array with a displacement of 223 to 234 residues. This structure is stabilized by covalent cross-links so that the collagen in the connective tissues is insoluble and resistant to the proteolytic effect of common proteases.[5]

The chemical analysis of collagen could not be performed for many years because of the insolubility of collagen fibers. A characterization was only possible after the recognition that immature collagen is not yet extensively cross linked. The absence of covalent cross-links made possible to extract the basic structural unit of collagen fibers, the tropocollagen. Tropocollagen has a molecular weight of 285 kDa, and consists of three polypeptide chains, each of which consists of about 1000 amino acid residues. Polypeptide chains found in the tropocollagen molecule are distinct in their amino acid composition and

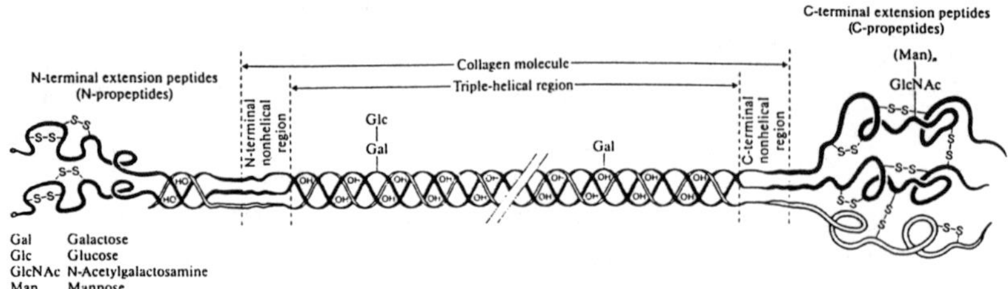

FIGURE 1. Schematic representation of a procollagen molecule. (From Prockop et. al., *N. Engl. J. Med.*, 301, 13, 1979. With permission.)

amino acid sequence. The proportion of glycine residues in all collagen molecules is nearly one third, which is unusually high in comparison with other proteins. Proline is also present in much higher proportion in collagen than in other proteins. Moreover, collagen contains two extremely rare amino acids, 4-hydroxyproline and 5-hydroxylysine. A special characteristic of the amino acid sequence is the remarkable regularity. Nearly every third residue is glycine, and the sequence glycine-proline-hydroxyproline occurs also very frequently. The special amino acid sequence plays an important role in the formation of the helical structure and of the interchain covalent links.[6]

At least six different types of collagen exist, which differ in the type, ratio, and primary structure of their constituent α-chains. The characteristics of collagen molecules are determined by several structural genes. The most common in mammalian connective tissues type I collagen consists of two α-1(I) and one α-2(I) chains. Other forms of type I collagen contain three α-1(I) chains. Types II, III and IV collagen consist of three identical α-1 chains differing in their primary structure designated as α-1(II), and α-1(III), respectively. Type IV collagen either consists of two different chains in the ratio of 1:2, or in case of some basement membrane collagens of small molecular weight contain three identical α-1(IV) chains. Type V collagen consists of two α-1(V) and one α-2(V) chains.[3] The recently discovered type VI collagen is a disulfide-linked fibrillar protein consisting of a short triple helix bounded by relatively large globular domains. This collagen was originally extracted from aorta and placenta, but since then it has been detected in a number of other organs and tissues including uterus, liver, kidney, skin, skeletal muscles, and cornea.[7] The composition and distribution in different organs and tissues of various types of collagen are summarized in Table 1.

The tissue specificity of collagens is based on the fact that various connective tissue cells synthesize different types of collagens as shown by cell culture experiments. Chondrocytes were shown to synthesize only type II, although some cartilages contain type I collagen, too. Osteocytes, tendon fibroblasts and corneal fibroblasts are known to synthesize type I collagen. Other cells such as smooth muscle cells seem to be capable of synthesizing four (I, III, IV, AB2) and corneal epithelial cells of three (I, III, IV) different types of collagen. There is a certain though not very strict correlation between the composition of the

TABLE 1
**The Composition and Tissue Distribution of Different Types of
Collagen**

Type	Composition	Distribution
I	$[\alpha\text{-}1(I)]_2$, $\alpha\text{-}2(I)$, or $[\alpha\text{-}1(I)]_3$	Skin, tendon, bone, cornea, dentin, fascia
II	$[\alpha\text{-}1(II)]_3$	Cartilage, vitreous body, invertebral disk
III	$[\alpha\text{-}1(III)]_3$	Blood vessels, uterus, fetal skin, reticular fibers
IV	$[\alpha\text{-}1(IV)]_2$, $\alpha\text{-}2(IV)$, or $[\alpha\text{-}1(IV)]_3$	Basement membranes, lens capsule, kidney glomeruli
V	$[\alpha\text{-}1(VI)]_2$, $\alpha\text{-}2(V)$	Placenta, skin, muscles, exoskeleton of fibroblasts
VI	Short triple helix and globular domains	Aorta, placenta, uterus, liver, kidney, skeletal muscles, cornea

collagen and the function of the connective tissue that contains it. The more rigid structures such as bone tendon and dentine contain predominantly type I collagen. The more elastic connective tissues contain type I and type III collagen in varying proportions. The corneal Descemet membrane and various basal membranes are the sites of type IV collagen. Tissues under continuous compression such as cartilagenous tissues and the vitreous humor of the eye contain type II collagen virtually exclusively. The intervertebral disk is an example of mixed function of different collagens with type I in the relatively rigid annulus and type II in the more elastic nucleus pulposa.[8]

II. THE ENZYMATIC CLEAVAGE OF COLLAGEN FIBERS

The catabolism of the insoluble cross-linked collagen fibrils is accomplished by a variety of collagenolytic enzymes. According to the current knowledge, three categories of collagen-degrading enzymes can be distinguished: (1) enzymes that depolymerize cross-linked fibrils, (2) enzymes that produce primary or specific collagenolysis, and (3) enzymes for further degradation of the products of the preceding steps. Enzymes that cause depolymerization of cross-linked collagen and further degrade primary collagenolysis products are called nonspecific collagenolytic proteases.[9]

A. Nonspecific Collagenolysis

Two routes of collagen catabolisms exist *in vivo:* the intra- and extracellular collagen digestion. Insoluble collagen fibers are depolymerized in different ways and possibly by different enzymes in the two compartments. Enzymes secreted by or released from cells into the extracellular matrix start the cleavage of collagen fibrils producing large fragments. These large particles are phagocytosed by appropriate cells, and the digestion continues within the secondary lysosomes. In general it is thought that enzymes acting at neutral pH are more important in the extracellular space, and those acting at acidic pH values are more involved in intracellular processes.[10] The presence of large particles in the extracellular space and very similar fragments in the phagolysosomes of cells have been detected by electron microscopy. Basic steps of the above-described process have been demonstrated in cases requiring rapid resorption of connective tissues, e.g., in wound healing, during the post-partum involution of uterus, and in other repair mechanisms. A candidate for intracellular collagen degradation is cathepsin B, and its action could be followed by a series of exopeptidases able to degrade peptides containing proline and hydroxyproline to amino acids.[11]

Enzymes involved in the cleavage of collagen fibers in the extracellular space either act on the telopeptide region of collagen molecules or are involved in the deprivation of the helical conformation and by this solubilize the otherwise insoluble collagen molecules. There are many enzymes including pepsin, trypsin, chymotrypsin, thermolysin, and elastase capable of performing cleavage on certain parts of the collagen molecule either in soluble or in insoluble forms. Most of these enzymes, with the exception of trypsin, are able to convert the intact tropocollagen molecule to α-chains by cleaving the cross-linked terminal peptides at the amino and at a lower rate also at the carboxyl end of the molecule. Trypsin is able to cleave the Lys9-Ser10 peptide bond, providing the Lys residue is not involved in a cross link. Most enzymes act only on soluble forms of collagen. Pepsin is particularly efficient in solubilization; therefore, it is a potent degradative of all collagen types. Cathepsin B is an effective agent against both soluble and insoluble collagen. The pH optimum of cathepsin B for collagen in solution is 4.5 to 5.0 and with insoluble collagen around pH 3.6. These are pH values that rarely if ever occur in the exracellular space, supporting the proposed involvement of cathepsin B in intracellular collagen digestion.[10,12]

Two neutral proteases, cathepsin G and lysosomal elastase, deserve special attention as they are also able to solubilize cross-linked collagen fibers by cleaving the telopeptide regions, and possibly some parts of the helical core, too. Lysosomal elastase was shown to degrade type I collagen, the major constituent of the corneal stroma, though at a slower rate compared with its action on type II collagen. Cathepsin G has little or no effect on type I collagen of skin or tendon origin. Both elastase and cathepsin G degrade type IV collagen of the basal membranes.[13,14]

B. Specific Collagenolysis

Specific collagenases are enzymes capable of cleaving across the native collagen triple helix under physiological conditions. The specific locus where vertebrate collagenases cut the triple helix is found at 75% of the length of the tropocollagen molecule starting from the amino terminal, and the released products correspond to three fourths and one fourth of the original molecular size.[15] The three-fourths and one-fourth fragments are called also TC-A and TC-B, respectively. It has been shown that mammalian and other vertebrate collagenases specifically cleave the α-1 chain of type I collagen at the Gly772-Ile773 peptide bond. The α-2 chain of type I collagen has been shown to be cut at a Glu-Leu position in a homologous region.[16] This feature is not unique to type I collagen; type II and III collagens are cleaved at the same locus. For the cleavage of the triple helix at a specific locus, the active centers of specific collagenases must contain binding sites able to recognize the primary, secondary, and tertiary structure of the substrate and catalytic centers specific for Gly-Ile and Gly-Leu bonds. The importance of secondary and tertiary structure is suggested by the fact that gelatin is generally a poor substrate for specific collagenases.[9]

Futher data on the enzymatic action of specific collagenases were obtained by gel electrophoretic studies and viscosimetric mesurements of solutions containing degradation products. Gel electrophoresis at acid pH revealed the presence of three faster-moving bands in addition to those two of intact α-chains and β-components. After dialysis against acetic acid and the addition of ATP, electron microscopy revealed fragments measuring three fourths and one fourth the length of native collagen molecules.[15] Isolation of the large and small fragments by differential ammonium sulfate precipitaion permitted the determination of the intactness of the helix in each fragment by optical rotatory dispersion. The molecular weight of these fragments was confirmed by equilibrium sedimentation and intrinsic viscosity measurements. Denaturation studies on the isolated intact helical fragments revealed a significant increase in solubility of the large and small pieces supporting the observation that both fragments contained nondenatured helical regions.[17] NH_2- and COOH-terminal amino acid analysis of the isolated fragments revealed two Ile and one Leu residues at the NH_2-terminal of the short (TC-B) and three Gly residues at the COOH-terminal of the larger (TC-A) fragment.[18]

Very little is known about the susbstrate specificity of collagenases and about the capability of highly purified mammalian collagenases in attacking noncollagen proteins. Highly purified human skin collagenase has been shown to be unable to cleave casein, bovine serum albumin, and hemoglobin.[19] The same observation was made with purified rheumatoid synovial collagenase with bovine serum albumin, transferrin, and fibrinogen,[20] while this latter enzyme seems to degrade gelatin at a slow rate. Ohlsson and Olsson[21] on the other hand reported the degradation of fibrinogen and proteoglycan molecules by purified granulocyte collagenase. Though the enzymatic action of collagenases seems to be rather specific, there is growing evidence about differences in substrate affinity and the rate of enzymatic action on different substrates of collagenases with different origin.

C. Different Collagenases

Clostridium histolyticum, a bacterium that causes gas gangrene, secretes several collagenases that slit each polypeptide chain in the collagen molecule at more than 200 sites. These collagenases contribute to the invasiveness of this highly pathogenic bacterium by destroying the connective tissue barriers of the host. The clostridia are unaffected by their secretory product because they are devoid of collagen, which is found only in multicellular eukaryocytes.[22] Tissue collagenases were found in amphibian and mammalian tissues undergoing growth or remodeling. Tissue collagenase was first demonstrated in the tail fin of tadpoles during their metamorphosis.[23] Collagenase was also detected in human granulocytes.[24] High collagenase activity was demonstrated in the uterus during the period of reorganization following pregnancy.[25] Human fibroblast collagenase, a 52-kDa protein, is formed by proteolytic removal of the amino-terminal segment of procollagenase, the inactive precursor.[26] The specificity of tadpole collagenase is remarkable. It cleaves tropocollagen across each of its three chains at a single site, between residues 775 and 776 in the sequence of 1056.[15] Vertebrate collagenases including corneal collagenases are metalloproteinases requiring Ca^{++} and Zn^{++} for their enzymatic action.[27,28]

Though collagenases isolated from different animal and human tissue sources cleave the native collagen molecule basically in the same way, there is emerging evidence on differences in the rate of action of the various collagenases on different types of collagen substrates. Horowitz et al.[29] have observed that the collagenase extracted from polymorphonuclear (PMN) cells attacked type I collagen about 15 times more effectively than type III collagen. The rate of cleavage of collagenases produced by fibroblasts of the human lung and that of collagenase produced by rabbit alveolar macrophages were the same on two different collagen substrates. Another observation of the same type was made by Liotta et al.,[30] showing that the collagenase, extracted from the highly invasive, rapidly growing mouse fibrosarcoma, failed to degrade type I, but was very effective in cleaving type IV collagen of basement membrane origin. In addition to slight differences in the binding and active sites of various collagenases, different collagen types are also known to exhibit unequal susceptibilities to collagenolytic effect. Type II collagen of human hyaline cartilage was found both in solution and fibril form to be more resistant to a variety of collagenases than type I or III collagen. The state of aggregation and cross linking, the occurrence of formaldehyde cross links, the greater extent of glycosylation might play a role in these differences.[2,30,31]

Further differences between various collagenases include different rates of action on various collagen types (type I through V collagenases are proposed names for enzymes that prefer type I through V collagens as their substrates); differences in the site of cleavage on the collagen molecule (collagen fragments produced by different collagenases may be identified on the basis of their molecular size by electron microscopy and by sodium dodecyl sulfate-polyacrylamide gel electrophoresis, or characterized by viscosity changes in the substrate solution); various collagenases may be differentiated by immunological methods; and possibly due to their different susceptibility to specific inhibitors.[32–34]

III. THE REGULATION OF COLLAGENASE ACTIVITY

The regulation of collagenase action can be divided into two areas: (1) the secretion and activation of latent collagenases and (2) the inhibition of the enzymes by specific inhibitors. Collagenases are present in different tissue extracts and secretions in inactive forms in large quantities. The inactive collagenases have been described as being either proenzymes or reversible complexes of inhibitors with the enzymes.[35]

The interpretation of latent collagenases as proenzymes was based on the observation that limited digestion with trypsin caused a marked activation and a decrease in molecular weight of 10,000 to 20,000 determined by gel filtration. The activation by cleavage of a small polypeptide from the inactive procollagenase was demonstrated in a number of tissues and secretions including mouse bone culture media, gingival fibroblast, alveolar macrophage, human skin fibroblast media, rheumotoid synovial fluid, human PMN extracts, etc. The enzymes capable to activate procollagenase include trypsin, plasmin, cathepsin B, kallikrein, lysosomal proteinases, mast cell serine protease, and endogenous serine proteinase of the involuting rat uterus.[36–43]

Several physiological inhibitors of collagenase were detected in serum, tissue extracts, and different culture media.[40,44] α-2-Macroglobulin is the major serum inhibitor of collagenases. It is known to form irreversible complexes with collagenases.[6] Limited reactivation by trypsin or plasmin treatment has been described. However, such complexes are unlikely to be a source of active enzyme *in vivo,* since the native inhibitor is very resistant to proteolytic breakdown.[45] Trypsin-activable latent collagenases do not bind to α-2-macroglobulin in the inactive form,[46] in good accordance with the assumption that collagenase activity is required for enzyme-inhibitor complex formation.[47]

Another serum inhibitor, β-1-anticollagenase, was described as specific for mammalian collagenases.[48] This inhibitor accounts for less than 10% of serum collagenase inhibitory activity in humans.[44] Its molecular weight is 40,000. Because of the low molecular weight, this inhibitor has far greater access to tissue sites than α-2-macroglobulin. β-1-Anticollagenase could act as a carrier for collagenase, passing it onto α-2-macroglobulin molecules, as has been described for α-1-proteinase inhibitor-enzyme complexes.[32]

Polycationic proteins like lysozyme and cationic proteins with different molecular weights ranging from 6000 to 40,000 isolated from different tissue sources were shown to have inhibitory effect on collagenases.[49–51] Small molecular weight collagenase inhibitors also have been extracted from connective tissues, but their exact role in the *in vivo* regulation of collagenase action still has to be elucidated.[50,52]

REFERENCES

1. **Módis, L.,** The molecular structure of the interfibrillar matrix in connective tissue, *Acta Biol. Acad. Sci. Hung.,* 29, 197, 1978.
2. **Leibovich, S. J. and Weiss, J. B.,** Failure of human rheumatoid synovial collagenase to degrade either normal or rheumatoid arthritic polymeric collagen, *Biochim. Biophys. Acta,* 251, 109, 1971.

3. **Jackson, D. S.,** The substrate collagen, in Collagenase in *Normal and Pathological Connective Tissues,* Woolley, D. E., Ed., John Wiley & Sons, Chichester, 1980, 1.
4. **Fessler, J. H. and Fessler, L. I.,** Biosynthesis of procollagen, *Annu. Rev. Biochem.,* 47, 129, 1978.
5. **Miller, A.,** in *Biochemistry of Collagen,* Ramachandran, G. N. and Reddi, A. H., Eds., Plenum Press, New York, 1976, 85.
6. **Stryer, L.,** *Biochemistry,* W.H. Freeman, New York, 1988, 261.
7. **Zimmermann, D. R., Trüeb, B., Winterhalter, K. H., Witmer, R., and Fischer, R. W.,** Type VI collagen is a major component of the human cornea, *FEBS Lett.,* 197, 55, 1986.
8. **Eyre, D. R.,** Biochemistry of the intervertebral disc, *Int. Rev. Connect. Tissue Res.,* 8, 227, 1979.
9. **Baici, A.,** Structure and breakdown of proteoglycans, collagen, and elastin, in *Proteinases and Tumor Invasion,* Strauli, P., Barrett, A. J., and Baici, A., Eds., Raven Press, New York, 1980, 17.
10. **Burleigh, M. C., Barrett, A. J., and Lazarus, G. S.,** Cathepsin B1. A lysozymal enzyme that degrades native collagen, *Biochem. J.,* 137, 387, 1974.
11. **McDonald, J. K. and Schwabe, C.,** Intracellular exopeptidases, in *Proteinases in Mammalian Cells and Tissues,* Barrett, A. J., Ed., North-Holland, Amsterdam, 1977, 311.
12. **Burleigh, M. C.,** Degradation of collagen by non-specific proteinases, in *Proteinases in Mammalian Cells and Tissues,* Barrett, A. J., Ed., North-Holland, Amsterdam, 1977, 285.
13. **Davies, M., Barrett, A. J., Travis, J., Snaders, E., and Coles, G. A.,** The degradation of human basement membrane with purified lysosomal proteinases: evidence for the pathogenetic role of the polymorphonuclear leukocyte glomeluronephritis, *Clin. Sci. Mol. Med.,* 54, 233, 1978.
14. **Starkey, P. M., Barrett, A. J., and Burleigh, M. C.,** The degradation of articular collagen by neutrophil proteinases, *Biochim. Biophys. Acta,* 483, 386, 1977.
15. **Gross, J. and Nagai, Y.,** Specific degradation of the collagen molecule by tadpole collagenolytic enzyme, *Proc. Natl. Acad. Sci. U.S.A.,* 54, 1197, 1965.
16. **Gross, J., Harper, E., Harris, E. D., Jr., McCroskery, P. A., Highberger, J. H., Corbett, C., and Kang, A. H.,** Animal collagenases: specificity of action and structure of the substrate cleavage site, *Biochem. Biophys. Res. Commun.,* 61, 605, 1974.
17. **Sakai, T. and Gross, J.,** Some properties of the products of reaction of tadpole collagenase with collagen, *Biochemistry,* 6, 518, 1967.
18. **Nagai, Y., Masui, Y., and Sakakibara, S.,** Substrate specificity of vertebrate collagenase, *Biochim. Biophys. Acta,* 445, 521, 1976.
19. **Woolley, D. E., Glanville, R. W., Roberts, D. R., and Evanson, J. M.,** Purification, characterization and inhibition of human skin collagenase, *Biochem. J.,* 169, 265, 1978.
20. **Woolley, D. E., Glanville, R. W., Crossley, M. J., and Evanson, J. M.,** Purification of rheumatoid synovial collagenase and its action on soluble and insoluble collagen, *Eur. J. Biochem.,* 54, 611, 1975.
21. **Ohlsson, K. and Olsson, I.,** The neutral proteases of human granulocytes. Isolation and partial characterization of two granulocyte collagenases, *Eur. J. Biochem.,* 36, 473, 1973.
22. **Mandl, I., MacLennan, J. D., and Howes, E. L.,** Isolation and characterization of proteinase and collagenase from *Cl. histolyticum, J. Clin. Invest.,* 32, 1323, 1953.
23. **Kang, A. H., Nagai, Y., Piez, K. A., and Gross, J.,** Studies on the structure of collagen utilizing a collagenolytic enzyme from tadpole, *Biochemistry,* 5, 509, 1966.
24. **Lazarus, G. S., Brown, R. S., Daniels, J. R., and Fullmer, H. M.,** Human granulocyte collagenase, *Science,* 159, 1483, 1968.
25. **Jeffrey, J. J. and Gross, J.,** Collagenase from rat uterus. Isolation and partial characterization, *Biochemistry,* 9, 268, 1970.
26. **Goldberg, G. I., Wilhelm, S. M., Kronberger, A., Baur, E. A., Grant, G. A., and Eizen, A. Z.,** Human fibroblast collagenase: complete primary structure and homology to an oncogene transformation-induced rat protein, *J. Biol. Chem.,* 261, 6600, 1986.

27. **Berman, M. B. and Manabe, R.,** Corneal collagenases: evidence for zinc metalloenzymes, *Ann. Ophthalmol.,* 5, 1193, 1973.

28. **Mandl, I.,** *Collagenases and Elastases. Advances in Enzymology,* Vol. 23, Interscience, New York, 1961.

29. **Horowitz, A. L., Hance, A. J., and Crystal, R. G.,** Granulocyte collagenase: selective digestion of type I relative to type III collagen, *Proc. Natl. Acad. Sci. U.S.A.,* 74, 897, 1977.

30. **Liotta, L. A., Abe, S., Robey, P. G., and Martin, G. R.,** Preferential digestion of basement membrane collagen by an enzyme derived from a metastatic murine tumor, *Proc. Natl. Acad. Sci. U.S.A.,* 76, 2268, 1979.

31. **Harris, E. D., Jr. and Farrell, M. E.,** Resistance to collagenase: a characteristic of collagen fibrils cross-linked by formaldehyde, *Biochim. Biophys. Acta,* 278, 133, 1972.

32. **Balldin, G., Laurell, C. B., and Ohlsson, K.,** *Hoppe-Seylers Z. Physiol. Chem.,* 359, 699, 1978.

33. **Berman, M., Dohlman, C. H., Gnadiger, M., and Davison, P.,** Characterization of collagenolytic activity in the ulcerating cornea, *Exp. Eye Res.,* 11, 255, 1971.

34. **Gross, J.,** The animal collagenases, in *Chemistry and Molecular Biology of the Intercellular Matrix,* Vol. 3, Balazs, E., Ed., Academic Press, London, 1970, 1623.

35. **Murphy, G. and Sellers, A.,** The extracellular regulation of collagenase activity, in *Collagenase in Normal and Pathological Connective Tissues,* Woolley, D. E. and Evanson, J. M., Eds., John Wiley & Sons, Chichester, 1980, 65.

36. **Birkedal-Hansen, H., Cobb, C. M., Taylor, R. E., and Fullmer, H. M.,** Activation of fibroblast collagenase by mast cell proteases, *Biochim. Biophys. Acta,* 438, 273, 1976.

37. **Dayer, J. M., Graham, R., Russell, R. G. G., and Krane, S. M.,** Collagenase production by rheumatoid synovial cells: stimulation by a human lymphocyte factor, *Science,* 195, 181, 1977.

38. **Eeckhout, Y. and Vaes, G.,** Further studies on the activation of procollagenase, the latent precursor of bone collagenase. Effects of lysosomal cathepsin B, plasmin and kallikrein, and spontaneous activation, *Biochem. J.,* 166, 21, 1977.

39. **Horovitz, A. L., Kelman, J. A., and Crystal, R. G.,** Activation of alveolar macrophage collagenase by neutral protease secreted by the same cell, *Nature,* 264, 772, 1976.

40. **Nagai, Y., Hori, H., Kawamoto, T., and Komiya, M.,** A regulation mechanism of collagenase activity *in vitro* and *in vivo,* in *Dynamics of Connective Tissue Macromolecules,* Burlegh, P. M. C. and Poole, A. R., Eds., North-Holland, Amsterdam, 1975, 171.

41. **Oronsky, A. L., Perper, R. J., and Schroder, H. C.,** Phagocytic release and activation of human leucocyte procollagenase, *Nature,* 246, 417, 1973.

42. **Stricklin, G. P., Bauer, E. A., Jeffrey, J. L., and Eisen, A. Z.,** Human skin fibroblast collagenase: chemical properties of precursor and active forms, *Biochemistry,* 17, 1331, 1978.

43. **Woessner, J. F.,** A latent collagenase in the involuting rat uterus and its activation by a serine proteinase, *Biochem. J.,* 161, 535, 1977.

44. **Woolley, D. E., Akroyd, C., Evanson, J. M., Soames, J. V., and Davies, R. M.,** Characterization and serum inhibition of neutral collagenase from cultured dog gingival tissue, *Biochim. Biophys. Acta,* 522, 205, 1978.

45. **Birkedal-Hansen, H., Cobb, C. M., Taylor, R. E., and Fullmer, H. M.,** Activation of latent bovine gingival collagenase, *Arch. Oral Biol.,* 20, 681, 1975.

46. **Vater, C. A., Mainardi, C. L., and Harris, E. D., Jr.,** Binding of latent rheumatoid synovial collagenase to collagen fibrils, *Biochim. Biophys. Acta,* 539, 238, 1978.

47. **Werb, Z., Burleigh, M. C., Barrett, A. J., and Starkey, P. M.,** The interaction of alpha-2-macroglobulin with proteinases. Binding and inhibition of mammalian collagenases and other metal proteinases, *Biochem. J.,* 139, 359, 1974.

48. **Woolley, D. E., Roberts, D. R., and Evanson, J. M.,** Small molecular-weight beta-1 serum protein which specifically inhibits human collagenases, *Nature,* 261, 325, 1976.

49. **Harris, E. D., Jr., Vater, C. A., Mainardi, C. L., and Werb, Z.,** Cellular control of collagen breakdown in rheumatoid arthritis, *Agents Actions,* 8, 36, 1978.
50. **Sakamoto, S., Sakamoto, M., Goldhaber, P., and Glimcher, M. J.,** The inhibition of mouse bone collagenase by lysozyme, *Calcif. Tissue Res.,* 14, 291, 1974.
51. **Shinkai, H., Kawamoto, T., Hori, H., and Nagai, Y.,** A complex of collagenase with low molecular weight inhibitors in the culture medium of embryonic chick skin explants, *J. Biochem.,* 81, 261, 1977.
52. **Kuettner, K. E., Hiti, J., Eisenstein, R., and Harper, E.,** Collagenase inhibition by cationic proteins derived from cartilage and aorta, *Biochem. Biophys. Res. Commun.,* 72, 40, 1976.

Chapter 7

CORNEAL COLLAGENS AND TEAR COLLAGENASES

Part B
Collagen Structures and Different Collagens in the Cornea

Collagen, the most abundant structural protein in connective tissues, has been characterized into at least five major isotypes, each being a distinct gene product.[1] Most connective tissues contain more than one collagen type, although one type usually predominates within a particular tissue.[2] The collagen composition of tissues is rather constant, slight differences between animal species in the ratio of different collagen types in the same tissue have been observed,[3] and alterations of tissue collagen content and composition in different diseases also occur.[4]

Of the dry weight of the corneal stroma, 71% is collagen.[5] Collagen is present in the cornea almost entirely in the form of fibrils, which are insoluble under physiological conditions of pH, temperature, and ionic strength. Under these circumstances the collagen molecules forming the fibrillar structures are remarkably resistant to the proteolytic action of common proteases. The triple helical form of collagen present in various structures of the cornea can be cleaved only by specific collagenases.[6]

Type I collagen has been found to be the primary collagen of the corneal stroma, forming collagen fibers arranged in a highly ordered lamellar array. The diameter of the collagen fibrils is 340 to 400 Å. The orderly layers, called lamellas, are attached at both ends at the limbus. Some of the layers are oriented parallel, others perpendicular to the corneal surface, still others are at various angles in between. The firmly arranged and to a great extent dehydrated structure of corneal stroma is maintained by barriers and by the function of the endothelial pump. Stromal collagen was found to be a very stable component of the cornea with little turnover in the course of a year.[7] It has been shown in cell culture that corneal fibroblasts secrete stromal collagen, which is very similar to type I collagen though more glycosylated than type I collagen found at other sites of the human body. Other collagen types also contribute to the collagen content of the stroma. These are type V collagen comprising up to 10% and type III forming 1 to 2% of the total collagen content of the corneal stroma. Embryonic corneas were found to contain type II collagen, too.[8]

The epithelial basement membrane is in close connection with the posterior surface of the basal epithelial cells on one side and with Bowman's layer on the other side. The basement membrane is 65 nm thick and was shown to be the product of epithelial cells. It is primarily built up of collagen type IV.[9] In addition laminin, heparan sulfate, and fibronectin were shown to be fairly constant components of the epithelial basement membrane.

The Bowman's layer is about 170 times the thickness of the epithelial basement membrane. By light microscopy it appears to be a distict layer, but electron microscopy demonstrated that Bowman's layer is actually a part of the anterior stroma. There are no cells in the Bowman's layer, and the fibrils form short fibers that are not arranged in layers, do not form lamellas, as they do in the deeper stroma.

The Descemet's membrane is the basement membrane of the endothelial layer that separates it from the posterior part of the corneal stroma. It is 10 μm thick and is known to be secreted by the endothelial cells. The anterior portion has a fibrous, banded structure, while the posterior part is formed of homogeneous, nonbanded, granular material. The Descemet's membrane is very elastic; once disrupted it tends to curl inward and does not regenerate. It can be replaced, in the defect area, by a new membrane-like material secreted by fibroblastic endothelial cells and laid on the surface of the bare stroma.[10,10a]

Epithelial basement membrane, Bowman's layer, and Descemet's membrane contain considerable amounts of different collagens. These layers are known to be built up of collagen fibers arranged in a less-oriented manner forming less compact structures than the stromal lamellas.[11] The exact composition of the Bowman's layer is not yet known, but its major collagen is thought to be type I. The Descemet's membrane was shown to contain collagens with high hydroxyproline and carbohydrate content, showing an amorphous appearance under electron microscope. Its biochemical character is not fully defined, but it is mainly composed of collagen type IV.[9,12–14]

The insolubility of collagen fibers and the differences in molecular weight, composition, and chemical characteristics of various collagen types as well as difficulties of the analysis account for the quantitative and qualitative differences found by different authors in the collagen content and composition of the corneal stroma. Polatnick et al.[15] found that only 1% of the total amount of collagen was acid soluble, and the alkali-soluble part was only 0.1%. Lewis et al.[16] extracted only 0.2% acid-soluble material. The solubilization of the otherwise insoluble collagen can be further promoted by enzymatic cleavage[17] and by thermal denaturation.[16] There are very few data available about the differences between the soluble and insoluble parts of the collagen of the cornea as usually the extracted fractions were chemically analyzed. It seems to be very probable, however, that such differences exist and, besides the difficulties in quantitative analysis, it has to be taken into consideration that certain types of collagen may be retained in the insoluble fraction.

Dische and Robert[18] have found that, when the supernatant from the extraction of the corneal stroma by 1 M $CaCl_2$ was dialysed, a precipitate appeared that contained collagen and glycosaminoglycans. When this collagen was extracted with 0.25 M TCA at 90°C for 30 min, its hexose content was much higher than that usually found in other mature collagens. When the minced corneal stroma was repeatedly washed, first with 0.15 M NaCl and then with 1 M $CaCl_2$ solution, different proportions of stromal collagen were solubilized by the consecutive washes. The amount of extracted collagen dropped sharply when NaCl solution was used several times, rose again when NaCl was replaced by $CaCl_2$, and dropped again when washing with $CaCl_2$ was repeated. These results suggest that the salt-soluble collagen of the corneal stroma is present in two different compartments or physical structures of the extracellular matrix of the corneal stroma that are affected in a different way by different salt solutions. The total amount of extractable collagen corresponded to about 4% of the total collagen. It has been suggested that this fraction might represent a freshly produced noncross-linked part of the collagen in the cornea, though glycine exchange measurements did not show any significant turnover of collagen in the adult cornea in animal experiments.[7]

To compare the hexose content of the extractable collagen with that of the nonextractable fraction the precipitates in the dialysates of the NaCl and $CaCl_2$ extracts were treated with collagenase for 16 h at room temperature, and the digest was dialysed against distilled water. The dialysate was then chemically analyzed. The use of two cysteine reactions for hexoses permitted the determination of galactose and glucose in these solutions, relative to the hydroxyproline and nitrogen contents. Hexosamine was completely absent indicating that hexoses were not present in form of glycosaminoglycans. The hexose content of the dialyzed collagen fragments that represented 90 to 93% of the total hydroxyproline of the degraded material was 2% in the NaCl extracts and 2.4% in the collagen of the $CaCl_2$ extracts. When the nonextractable collagen of the stroma was subjected to the same treatment as the salt-soluble collagen, the ratio of hexose in the dialysate was only 0.65%, in agreement with values found in other collagens. It has been suggested that the soluble collagen represents a precursor of the cross-linked nonsoluble collagen fibrils, which can be supported by the finding that the amount of hexose-rich soluble collagen was shown to be higher in growing calf corneas than in adult ones.[19]

The amino acid composition of corneal collagen is very similar to that of collagen from other tissues.[12] Collagen extracted from the corneal stroma, like collagens from other tissues, contain 33% of glycine, this amino acid occurring at every third site in the polypeptide chain, and a low percentage of tyrosine, which is concentrated at the ends of the molecules and is concerned with intermolecular links by which long fibers are formed. The ratio of tyrosine in corneal collagen is 1.5%.[15] Other amino acids of the collagen molecule include proline, which is also present in much higher proportion in collagen than in other proteins, and two extremely rare amino acids 4-hydroxyproline and 5-hydroxylysine. The ratio of sulfur amino acids is very low and tryptophan is typically absent from collagen molecules.[16]

Newsome et al.[17] studied the collagen composition of acetic acid-pepsin extracts of normal human corneal stroma by sodium dodecyl sulfate (SDS)-slab gel electrophoresis, enzyme sensitivity, cyanogen bromide peptide mapping, and carboxymethyl cellulose-column chromatography. Pepsin digestions of the minced corneas solubilized 41% of the native collagen as determined by quantitative amino acid analysis. Electrophoresis of collagen-containing fractions obtained by stepwise NaCl purification demonstrated bands comigrating with the bands of standard preparations of collagen types I, III, and V. Type I collagen was present in all fractions. The 1.7 M NaCl precipitate contained considerable amounts of type III collagen and the 4.5-M precipitate contained some of type V collagens [α1(V), α2(V)]. Type III before reduction with dithiothreitol did not penetrate the gel, while type V collagen entered the gel without reduction. Components of the three collagen types were detected by their migrating position following staining with Coomassie brillant blue; the quantitative analysis however due to the only partial solubilization and to differences in gel penetration was very difficult. Enzyme sensitivity assays confirmed that the detected bands were really collagen components, as the characteristic bands were not detected on the gels if the extracted fractions were pretreated with collagenase. Peptides produced by the cleavage of corneal extracts with cyanogen bromide were compared by slab gel electrophoresis to those obtained from collagen standards. In various fractions of the corneal stroma cleavage products characteristic of collagen types I, II, and V could be indentified. Carboxymethyl cellulose column chromatography of the corneal extracts yielded two prominent (α1 and α2) peaks and one small intermediate peak in the elution profiles. Radiolabeled type I collagen standard (rat skin collagen) produced similar peaks. Gel electrophoresis of pooled peak fractions both under reduced and nonreduced conditions revealed that the α1 peak contained predominantly α1 (I) chains and small amounts of type V collagen, and the α2 peak contained mostly α2 (I) material. The intermediate peak contained bands that comigrated with collagen standard type III.[17]

Radda and co-workers performed a quantitative assay of various collagen types in normal and in keratoconus corneas. Collagen types were separated by salt fractionation and thermal gelation in the pepsin-soluble fraction of lyophilized corneas. Collagen characterization was performed using SDS-polyacrylamide gel electrophoresis; collagen concentrations were determined using hydroxyproline measurements. Soluble collagens from normal corneas represented 85% collagen type I, 10% collagen type III, and 5% collagen type V. Soluble collagens from keratoconus corneas consited of 90% collagen type I, and not more than 5% of collagen type III and less than 5% of collagen type V.[20]

Zimmermann and co-workers demonstrated the presence of type VI collagen in normal human cornea, using a comparative peptide mapping of corneal extracts with type VI collagen isolated from placenta and immunoblot studies with antibodies specific for human type VI collagen. Densitometric evaluation of polyacrylamide gels indicated that type VI collagens comprise as much as one quarter of the dry weight of the cornea. Immunohistology using indirect immunofluorescence showed that this collagen is distributed throughout the corneal stroma. They suggested that based on these observations type VI collagen should be considered to be a major component of the extracellular matrix of the human cornea.[21]

Murata et al.[22] investigated the distribution of type VI collagen in the bovine cornea by indirect immunofluorescence microscopy and immunoelectron microscopy using a specific polyclonal antibody against type VI collagen. Type VI collagen was detectable throughout the stroma by immunofluorescence. In addition the uniform but rather weak staining, a laminar pattern similar to the arrangement of stromal lamellae was also detected. In agreement with the immunofluorescent pattern, immunoelectron micrographs showed strong linear reactions in the interlamellar spaces. Positive reaction was observed also in the pericellular region of keratocytes, while the epithelium, the endothelium, and the Descemet's membrane was negative.[22]

Using the immunogold technique combined with cryoultramicrotomy and London Resin white embedding, Mashall and colleagues studied the distribution of collagen types I to IV in the cornea of enucleated eyes. Type II collagen was not identified in any corneal layer. Type I and III collagen were found evenly distributed in the striated fibrils of the Bowman's layer and the stroma. Type IV collagen could be demonstrated only in the posterior nonbanded region of the Descemet's membrane. Laminin was detected in the subepithelial and subendothelial regions in accordance with its adhesive function.[23] They also studied the distribution of collagen types V and VI in the cornea of enucleated eyes using similar immunogold immunochemical techniques, cryoultramicrotomy, and London Resin white embedding. Type V collagen was demonstrated in the striated collagen fibrils in the Bowman's layer, the stroma, and a thin, nonbanded anterior zone of the Descemet's membrane. In contrast type VI collagen was detected in fine filaments in the interfibrillar matrix of the corneal stroma, the Bowman's layer, and the subepithelial basement-membrane complex. This implies a role of collagen type VI in the epithelial adhesion. Keratocytes were found to be surrounded by electrodens, amorphous deposits that reacted with antibodies against collagen types III, V, and VI. Long-spacing collagen was found in the corneal stroma that did not react with antibodies against any of the known collagen types.[24]

The presently available data suggest that all known collagen types, with the exception of type II, occur in the normal human cornea. Types I, III, V, and VI are distributed in the stroma, Descemet's membrane, and Bowman's layer and type IV collagen in the subepithelial basement membrane and in the subendothelial layer of the Descemet's membrane. Type III collagen being a constant component of the blood vessel walls may also appear in vascularized corneas under pathological conditions. A quantitative basis type I collagen is the predominant type in the cornea. There is no doubt concerning the local production and cellular origin of corneal collagen. Cultured stromal keratocytes were shown to synthetize and deposit collagen types I and III.[5,25] It has been suggested that corneal epithelial endothelial cells also contribute to the production of collagen types III, V, and VI. Based on the special location of collagen type IV the role of the epithelium and the endothelium in its production can also be suspected.

Disagreements concerning the collagen constituents of the cornea can be explained by differences in the specificity and sensitivity of the methods used for the detection, species variations, and age-related changes.[23] It is generally agreed that type I collagen is the major collagen type in the cornea in all mammalian species. However, there is a significant species variation. Type I collagen constitutes 85% of the total collagen content of the normal human cornea,[17] 94% in the calf,[26] and nondefined "substantial portion" in the rabbit.[9]

There is some controversy concerning the presence or absence of type I collagen in the Bowman's layer and Descemet's membrane.[5,23,27] Different results can be explained by the use of various methods for detection. Type II collagen has been reported to be absent from human[17,27,28] and bovine[9] cornea but was detected in murine[29] and avian[30,31] cornea. In the latter species it seems to be present in the stroma of the developing cornea, but probably persists in later life only in the Bowman's layer and Descemet's membrane.[32] Conflicting data were produced by various methods concerning collagen type III in the cornea of different species. Biochemical studies indicated the absence of type III collagen from the bovine[9] and rabbit[26] cornea while immunohistological studies revealed its presence in the Descemet's membrane.[33] The presence of type III collagen in the stroma appears to be a unique characteristic of the human cornea.[23] Immunofluorescent studies demonstrated the presence of type IV collagen in the subepithelial basement membrane;[28,34] others detected this collagen type only under the endothelium.[23] It has been suggested that such differences may be the result of aging. Type IV collagen has been reported to be associated with nerve in the avian corneal stroma[35] and with microfibrils in the stroma of mice.[36] It has been suggested that successful identification of collagen type V requires fibril disruption before labeling.[37,38] Others have shown that fibril disruption is not essential for the localization of this collagen type in the corneal stroma.[24] Several studies demonstrated the presence of type VI collagen in the interfibrillar matrix and in the subepithelial and subendothelial anchoring plaques of the human cornea.[21,22,24]

The function of collagen fibrils in the cornea is basically mechanical. The special arrangement of stromal lamellae is thought to be responsible for transparency of the cornea, a rather unique quality among tissues in the human body. Collagen types IV and VI may play a role in anchoring epithelial and endothelial cells assisting the role of laminin, fibronectin, and other attachment proteins. Recent studies revealed the presence of special long-spacing collagen, the stromal lamellae of normal corneas.[24] Such collagen was previously reported to be present in the Descemet's membrane of normal[22] and Fuchs' dystrophic corneas.[39] The significance of long-spacing collagen is not yet known. It has been suggested that it may appear in various corneal layers as a part of the aging process.[24]

REFERENCES

1. **Chung, E., Rhodes, K., and Miller, E. J.,** Isolation of three collagenous components of probable basement membrane origin from several tissues, *Biochem. Biophys. Res. Commun.,* 71, 1167, 1976.
2. **Miller, E. J.,** Biochemical characteristics and biological significance of the genetically distinct collagens, *Mol. Cell. Biochem.,* 13, 165, 1976.

3. **Hong, B.-S., Cannon, D. J., and Davison, P. F.,** Isolation and identification of a distinct collagen from several different tissues, *Fed. Proc.,* 37, 1528, 1978.

4. **Yue, B., Baum, J. L., and Smith, B. D.,** Collagen synthesis by cultures of stromal cells from normal human and keratoconus corneas, *Biochem. Biophys. Res. Commun.,* 86, 465, 1979.

5. **Newsome, D. A., Foidart, J. M., Hassell, J. R., Krachmer, J. H., Rodrigues, M. M., and Katz, S. I.,** Detection of specific collagen types in normal and keratoconus corneas, *Invest. Ophthalmol. Vis. Sci.,* 20, 738, 1981.

6. **Slansky, H. H. and Dohlman, C. H.,** Collagenase and the cornea, *Surv. Ophthalmol.,* 14, 402, 1970.

7. **Smelser, G. K., Pollock, F. M., and Ozaniks, V.,** Persistence of donor collagen in corneal transplants, *Exp. Eye Res.,* 4, 349, 1965.

8. **Schmut, O.,** The identification of type II collagen in calf and bovine corneas and sclera, *Exp. Eye Res.,* 25, 505, 1977.

9. **Davidson, P. F., Hong, B.-S., and Cannon, D. J.,** Quantitative analysis of collagens in the bovine cornea, *Exp. Eye Res.,* 29, 97, 1979.

10. **Hamano, H. and Kaufman, H. E.,** *The Physiology of the Cornea and Contact Lens Applications,* Churchill Livingstone, New York, 1987.

10a. **Jakus, M. A.,** Studies on the cornea. II. The fine structure of Descemet's membrane, *J. Biophys. Biochem. Cytol.,* 2, 243, 1956.

11. **Kephalides, N.,** Structure and biosynthesis of basement membranes, *Int. Rev. Connect. Tissue Res.,* 6, 63, 1973.

12. **Freeman, I. L.,** Collagen polymorphism in mature rabbit cornea, *Invest. Ophthalmol. Vis. Sci.,* 17, 171, 1978.

13. **Lee, R. E. and Davison, P. F.,** Collagen composition and turnover in ocular tissues, *Exp. Eye Res.,* 32, 737, 1981.

14. **Linsenmayer, T. F.,** Collagen, in *Cell Biology of the Extracellular Matrix,* Hay, E. D., Ed., Plenum Press, New York, 1981.

15. **Polatnick, J., La Tessa, A. J., and Katzin, H. M.,** *Biochim. Biophys. Acta,* 26, 365, 1975.

16. **Lewis, M. S., Dubin, M. W., and Aandahl, V. A.,** *Exp. Eye Res.,* 6, 57, 1967.

17. **Newsome, D. A., Gross, J., and Hassell, J. R.,** Human corneal stroma contains three distinct collagens, *Invest. Ophthalmol. Vis. Sci.,* 22, 376, 1982.

18. **Dische, Z. and Robert, J.,** 1962; citation by **Dische, Z.,** Biochemistry of connective tissue of the vertebrate eye, in *International Review of Connective Tissue Research,* Hall, D. A. and Jackson, D. S., Eds., Academic Press, New York, 1970.

19. **Dische, Z.,** Biochemistry of connective tissue of the vertebrate eye, in *International Review of Connective Tissue Research,* Hall, D. A. and Jackson, D. S., Eds., Academic Press, New York, 1970.

20. **Radda, T. M., Menzel, E. J., Freyler, H., and Gnad, H. D.,** Collagen types in keratoconus, *Graefe's Arch. Clin. Exp. Ophthalmol.,* 218, 262, 1982.

21. **Zimmermann, D. R., Trüeb, B., Winterhalter, K. H., Witmer, R., and Fischer, R. W.,** Type VI collagen is a major component of the human cornea, *FEBS Lett.,* 197, 55, 1986.

22. **Murata, Y., Yoshioka, H., Iyama, K.-I., and Usuka, G.,** Distribution of type VI collagen in the bovine cornea, *Ophthalmic Res.,* 21, 67, 1989.

23. **Marshall, G. E., Konstas, A. G., and Lee, W. R.,** Immunogold fine structural localization of extracellular matrix components in aged human cornea. I. Types I–IV collagen and laminin, *Graefe's Arch. Clin. Exp. Ophthalmol.,* 229, 157, 1991.

24. **Marshall, G. E., Konstas, A. G., and Lee, W. R.,** Immunogold fine structural localization of extracellular matrix components in aged human cornea. II. Types V and VI collagen, *Graefe's Arch. Clin. Exp. Ophthalmol.,* 229, 164, 1991.

25. **Stoesser, T. R., Church, R. L., and Brown, S. I.,** Partial characterization of human collagen and procollagen secreted by human corneal stromal fibroblasts in cell culture, *Invest. Ophthalmol. Vis. Sci.,* 17, 264, 1978.

26. **Welsh, C., Gay, S., Rhodes, R. K., Pfister, R., and Miller, E. J.,** Collagen heterogeneity in normal rabbit cornea. I. Isolation and biochemical characterization of the genetically-distinct collagens, *Biochim. Biophys. Acta,* 625, 78, 1980.

27. **Ben-Zvi, A., Rodrigues, M. M., Krachmer, J. H., and Fujikawa, L. S.,** Immunohistochemical characterization of extracellular matrix in the developing human cornea, *Curr. Eye Res.,* 5, 105, 1986.

28. **Nakayasu, K., Tanaka, M., Konomi, H., and Hayashi, T.,** Distribution of types I, II, III, IV and V collagen in normal and keratoconus corneas, *Ophthalmic Res.,* 18, 1, 1986.

29. **Harnisch, J.-P., Buchen, R., Sinha, P. K., and Barrach, H. J.,** Ultrastructural identification of type I and II collagen in the cornea of the mouse by means of enzyme labeled antibodies, *Graefe's Arch. Clin. Exp. Ophthalmol.,* 208, 9, 1978.

30. **Hayashi, M., Ninomiya, Y., Hayashi, K., Linsenmayer, T. F., Olsen, B. R., and Trelstad, R. L.,** Secretion of collagen types I and II by epithelial and endothelial cells in the developing chick cornea demonstrated by *in situ* hybridization and immunohistochemistry, *Development,* 103, 27, 1988.

31. **Hendrix, M. J. C., Hay, E. D., Von der Mark, K., and Linsenmayer, T. F.,** Immunohistochemical localization of collagen types I and II in the developing chick cornea and tibia by electron microscopy, *Invest. Ophthalmol. Vis. Sci.,* 22, 359, 1982.

32. **Von der Mark, K., Von der Mark, H., Timpl, R., and Trelstad, R. L.,** Immunofluorescent localization of collagen types I, II and III in the embryonic chick eye, *Dev. Biol.,* 59, 75, 1977.

33. **Cintron, C., Hong, B.-S., Covington, H. I., and Macarak, E. J.,** Heterogeneity of collagens in rabbit cornea: type III collagen, *Invest. Ophthalmol. Vis. Sci.,* 29, 767, 1988.

34. **Konomi, F., Hayashi, T., Nakayasu, K., and Arima, M.,** Localization of type V collagen and type IV collagen in human cornea, lung and skin. Immunohistochemical evidence by anti-collagen antibodies characterized by immunoelectroblotting, *Am. J. Pathol.,* 116, 417, 1984.

35. **Bee, J. A., Kuhl, U., Edgar, D., and Von der Mark, K.,** Avian corneal nerves: co-distribution with collagen type IV and acquisition of substance P immunoreactivity, *Invest. Ophthalmol. Vis. Sci.,* 29, 101, 1988.

36. **Pratt, B. M. and Madri, J. A.,** Immunolocalization of type IV collagen and laminin in nonbasement membrane structures of murine corneal stroma: a light and electron microscope study, *Lab. Invest.,* 52, 650, 1985.

37. **Linsenmayer, T. F., Fitch, J. M., and Mayne, R.,** Extracellular matrices in the developing avian eye: type V collagen in corneal and noncorneal tissue, *Invest. Ophthalmol. Vis. Sci.,* 25, 41, 1984.

38. **Von der Mark, K. and Ocalan, M.,** Immunofluorescent localization of type V collagen in the chick embryo with monoclonal antibodies, *Collagen Relat. Res.,* 2, 541, 1982.

39. **Iwamoto, T. and DeVoe, A. G.,** Electron microscopic studies on Fuchs' combined dystrophy. I. Posterior portion of the cornea, *Invest. Ophthalmol. Vis. Sci.,* 10, 9, 1971.

CORNEAL COLLAGENS AND TEAR COLLAGENASES

Part C
Collagenase and the Cornea

Collagen fibrils present in the corneal stroma are remakably resistant to degradation by the usual proteolytic enzymes under physiological conditions. A collagenolytic enzyme able to digest both reconstituted collagen and collagen of the corneal stroma was detected in human epithelium by Slansky et al.,[1] Brown et al.,[2] and Itoi et al.,[3] and in animal corneal epithelium by Slasnky et al.[4] This collagenolytic enzyme has been characterized as tissue collagenase. Its presence has been demonstrated both in normal and in alkali-burned corneas. The activity of this collagenase was found to be very high in the marginal epithelium of corneal ulcerations in animal[2,5] and in ulcerating human corneas.[2,4]

It had been known from clinical experience that various infectious and traumatic corneal diseases just like bacterial complications of corneal surgery can involve the destruction of stromal collagen leading to the formation of corneal ulcers. The possibility that proteolytic enzymes, released from the corneal epithelium, may be causative factors in the ulcerating process was first proposed by Slansky et al.[1,4] and Itoi.[6] They measured collagenase activity in homogenized epithelial cell suspensions using a synthetic amino acid substrate, an assay originally designed to study bacterial collagenase. Collagenase activity was detected at neutral pH, and could be inhibited by EDTA.[4] Collagenolytic activity of animal corneal epithelial cells was also demonstrated using a method in which explants of corneal epithelium were laid on the surface of reconstituted collagen gels in tissue culture.[3]

Full-thickness explants of ulcerating rabbit corneas, following experimental alkali burns, were also shown to cause significant lysis on reconstituted collagen gels.[7] Some collagenolytic activity was demonstrated also in areas in the corneal tissue as far as 3 mm from the site of the ulcer. The corneal tissue origin of collagenases in alkali-burned ulcerating corneas was debated at that time realizing that the activity may have been caused by invading granulocytes in the alkali-burned cornea. Important data to answer this question were provided by Itoi et al.,[3] who detected collagenolytic activity in epithelial explants from nonulcerating rabbit corneas. The activity was shown to be inhibited by EDTA and by sera from different species. No such activity was detected in corneal stromal and endothelial explants. These experiments were carried out using reconstituted collagen, and it has not been proven that the detected enzyme would also digest intact collagen fibrils of the corneal stroma.

In further experiments the same authors used sections of rabbit and calf corneal stroma as substrates. The stromal sections were placed adjacent to epithelial explants from normal rabbit corneas in tissue culture. After 5 d of incubation a disappearance of the fibrous structure of the substrate stromal sections was observed.[3] These results were confirmed by experiments revealing a loss of Van Gieson staining in the digested area.[5] It has to be taken into consideration, however, that the circumstances in tissue cultures are not identical to those in the intact eye. Moreover, the preparation of an epithelial explant involves non-negligable trauma to the epithelial cells, and this trauma may have induced the secretion of collagenases or activation of their proenzymes under experimental conditions. At last but not least there are striking differences among different species concerning the secretion and characteristics of the action of enzymes produced by corneal epithelial cells.

In contrast to earlier findings in rabbit cornea,[3] Slansky and co-workers[4] found no detectable collagenolytic activity of the corneal epithelium of nonulcerating corneas. They suggested that *de novo* synthesis of the collagenolytic enzymes might occur only after epithelial injury as Weimar and Haraguchi[8] had demostrated with other epithelial enzymes. They did however find high collagenolytic activity in the epithelium-bordering corneal ulcers with different etiologies (chemical burns, bacterial, herpetic, metaherpetic, rosacea, and radiation keratitis, just like corneal ulcers associated with scleroderma, Sjögren's syndrome, ocular pemphigoid, and Stevens-Johnson syndrome).[1,4]

In 1969, which proved to be a very prosperous year in the study of corneal collagenolysis, Gnadinger et al.[5] demonstrated that corneal stroma from ulcerating corneas was more susceptible to the collagenolytic action of alkali-burned epithelium than normal corneal stroma. It has been suggested that more rapid digestion was related to the loss of mucopolysaccharides (glycosaminoglycans) that may have occurred when the cornea was subjected to alkali. Further evidence concerning the possible protective role of mucopolysaccharides against the enzymatic digestion of corneal collagen fibrils was provided by experiments with hyaluronidase treatment of corneas. Pretreatment of normal corneal stroma with hyaluronidase made the stromal substrate more susceptible to epithelial collagenase digestion.[5]

Though the secretion of active collagenase by intact corneal epithelium *in vivo* seems to be rather improbable, there is enough evidence to accept that corneal epithelial cells are able to produce collagenolytic enzymes as a result of mild physical and chemical trauma and in early phases of inflammation. It has been suggested that this is a characteristic feature that the corneal epithelium does not share with the corneal stroma and the endothelium and that this secretion may also occur in the absence of leukocytic infiltration.[9]

Berman et al.[10] provided data concerning the characteristics of the collagenolytic enzyme found in the ulcerating rabbit cornea. Tissue collagenase according to Gross[11] cleaves tropocollagen in a limited manner, resulting in one-fourth and three-fourths fragments of the tropocollagen molecule. The specific collagenolytic action can be demonstrated by a well-defined reduction in viscosity of an undenatured collagen solution, by the demonstration of the produced one-fourth and three-fourths fragments of the tropocollagen molecule by electron microscopy, and by the identitification of the molecular fragments with appropriate molecular weights by sodium dodecyl sulfate (SDS)-polyacrylamide gel electrophoresis. Ulcerating rabbit corneas were shown to contain a collagenolytic enzyme capable of digesting collagen in a manner of tissue collagenase. The extracted crude corneal enzyme was shown to digest collagen when placed on the surface of reconstituted collagen gels, to reduce the viscosity of nondenatured collagen solution in a limited manner, and to produce fragments with one fourth and three fourths length of the tropocollagen molecule when studied by electron microscope.[10]

Berman et al.[12,13] detected collagenase in the culture media from normal eye bank cornea and in the media from removed corneal buttons of patients with keratoplasty with no history of corneal ulceration. Enzymes from both sources were shown to cleave tropocollagen at the same site as tadpole and rabbit corneal collagenases. Because of the restricted cleavage of collagen, the collagenolytic enzyme found in the culture media of nonulcerating human corneas was suggested to be tissue-type collagenase. Human corneal collagenase was also shown to be able to digest reconstituted collagen gel, and its activity was found to be inhibited by calcium-EDTA and N-acetyl-L-cysteine, both of which had previously been shown to inhibit collagenases produced by the ulcerating rabbit cornea. The inhibiting effect of EDTA and its salts on collagenases is based on the fact that tissue collagenases are metalloproteinases requiring Ca^{++} and Zn^{++} ions to perform their enzymatic action. Calcium-EDTA is able to stop the enzyme action by forming complexes with Zn^{++} and EDTA by forming complexes with both Ca^{++} and Zn^{++}. Evidence for the metalloproteinase nature of corneal collagenases was provided by Berman and Manabe in 1973;[14] their findings stimulated futher studies concerning the use of metal-binding agents in the therapy of corneal ulcers.

On the basis of clinical observations it has been suggested that human plasma inhibits corneal ulceration. The beneficial effect of autologous serum on ulcerating alkali burned corneas both in the form of subconjunctival injections[15] and in the form of topically administered eye drops[16,17] has been reported. In an experimental study Berman et al.[12] demostrated that the serum antiprotease α-1-antitrypsin inhibited collagenase from the ulcerating rabbit cornea, while it had no effect on human corneal collagenase obtained from human keratoplasty tissues. α-2-Macroglobulin on the other hand was found to inhibit both rabbit and human corneal collagenases by forming high molecular weight enzyme-inhibitor complexes.[12,18]

Very little information is available on the mechanism and regulation of the secretion of collagenase by the cells of the cornea. Most proteolytic enzymes are secreted in the form of a zymogen. Zymogen is a proenzyme, with a molecular weight somewhat higher than that of the active enzyme. The proenzyme is turned into the active form in a process called limited proteolysis usually carried out by another protease. A "latent" form of collagenase has been detected in the culture media of ulcerating and nonulcerating rabbit corneas. This latent collagenase could be activated by brief trypsin treatment of the crude protein extract obtained from tissue culture media of both ulcerating and nonulcerating corneas. The generated active collagenase was shown to cleave reconstituted collagen fibrils and soluble collagen into three-fourths and one-fourth length fragments in the manner of vertebrate collagenases. The molecular weight of latent collagenase was determined by column chromatography and SDS-polyacrylamide gel electrophoresis, and was found to be higher than that of the active enzyme.[19]

Besides corneal epithelial cells other sources of collagenases were found in ulcerating and regenerating corneas. It has been shown that collagenases acting in the alkali-burned rabbit cornea are produced not only by the regenerating epithelium but also by cells in the underlying stroma.[2,20] A large number of studies employing cells derived from skin,[21] gingiva,[22] synovium,[23,24] and cornea[25] have shown that fibroblasts are capable of producing collagenase. Hook et al.[25] demonstrated a procollagenase with a molecular weight of 65,000 in the culture fluid of human and rabbit corneal fibroblasts. The inactive enzyme could be activated by trypsin and polymorphonuclear leukocyte (PMN) lysosomal lysate. Corneal fibroblasts were shown to produce active collagenase when dexamethasone was included in the culture media. This observation is in good accordance with the clinical observation that corticosteroid therapy may worsen the condition of corneal ulcer patients and may lead to corneal perforation. These data together with others indicating that corticosteroids suppress host defense mechanisms and enhance the invasion of microorganisms led to the conclusion that corticosteroid therapy is not indicated in case of corneal ulcers.[25,63]

Gordon and co-workers[27] studied the degree of immunologic similarity between skin fibroblast collagenase and the collagenase isolated from organ cultures of ulcerating human corneas. The two collagenases reacted with antiserum against skin collagenase in immunodiffusion plates in a manner indicating that these two enzymes were antigenically closely related. Further evidence proving the immunologic similarity of the two collagenases was provided by the comparison of standard radioimmunoassay curves. The authors suggested that immune serum prepared against human skin collagenase can be used to assay human corneal collagenase in both immunological and immunohistological studies. They studied the distribution of collagenase using immunofluorescent staining of the sections of ulcerating and nonulcerating corneas removed by keratoplasty. Collagenase was localized only in the stromal portion of ulcerating human corneas (both in the ulcer bed and in stromal regions covered by seemingly intact epithelium). No staining was observed in or around the corneal epithelium. Collagenase was demonstrated also in corneas that were ulcerating but showed no infiltrate of PMN leukocytes. No tissue collagenase was detected by immunohistological methods in normal corneas or in corneas with various nonulcerating pathological conditions including keratoconus, aphakic bullous keratopathy, Fuchs' corneal dystrophy, and lattice corneal dystrophy. Tissue collagenase was also detected by immunological methods in the harvest of tissue cultures of minced human corneas.

Based on their findings Gordon and co-workers suggested that epithelial cells might not be the primary source of collagenase in the ulcerating cornea; rather, cellular elements (i.e., fibroblasts) in the corneal stroma produce collagenase that is responsible for the enzymatic degradation of collagen in the ulcerating cornea.[27]

Ulcerating corneas have long been known to be invaded by PMN leukocytes.[28,29] Brown and Hook[30] suggested that leukocytes invading the cornea during ulceration may be the major sources of collagenase. PMNs are known to have several enzymes capable of collagen degradation.[31,32] Among these PMN leukocyte neutral collagenase has been characterized by Ohlsson and Ohlsson,[33] and was extensively studied by Prause in ulcerating corneas and in the tears of patients with corneal ulcers.[34–36] Macrophages and PMNs were shown to secrete factors that stimulate collagenase production by cultured fibroblasts.[37] Corneal epithelial cells, when co-cultured with stromal fibroblasts, were found to have similar effects on collagenase production.[38,39] Collagenases, active at neutral pH, have been localized within the specific granules of neutrophyl leukocytes[33,40] and were shown to be immunologically and biochemically distinct from fibroblast collagenase.[41,42] Monoclonal antibodies produced against collagenase of leukocyte origin (isolated from rabbit uterus) also reacted with collagenase isolated from the culture media of alkali-burned corneas, strongly suggesting that leukocyte collagenases are present in ulcerating alkali-burned corneas. PMN leukocytes in the superficial layer of alkali-burned corneas showed indirect immunofluorescent staining while stromal cells and PMNs within the stroma were negative.[43] The authors suggested that PMNs in the alkali-burned corneas secrete all or most of their collagenases by degranulation at the anterior surface of the cornea.

The seemingly conflicting data concerning the origin of collagenase(s) in corneal ulceration can easily be explained if we accept that various cell types (PMNs, fibroblasts, epithelial cells, and certain microorganisms) are capable of producing collagenolytic enzymes including true collagenases. The specificity of the methods applied in most of the studies enabled the identification of only one type of collagenase or detected the combined action of various collagenolytic enzymes. That antibodies produced against leukocyte collagenase reacted with an enzyme in the culture media of ulcerating corneas does not rule out the possibility of the presence of fibroblast collagenase in the the same culture fluid or in the same cornea. Similarly, the detection of collagenase by antiskin fibroblast collagenase antiserum in the stroma of ulcerating corneas left the question unanswered whether with other specific antisera also leukocyte collagenase may have been demonstrated in the same ulcerating corneas. It can also be easily accepted however that, in certain types of ulceration, some of the possible sources are missing (ulceration without leukocyte infiltration) or all sources are present but one of them is very active and produces the overwhelming majority of the detected collagenase.

Plasminogen activator was shown to be released from ulcerating corneas partly from the destroyed and affected epithelial cells and from the ulcerating stroma.[44] Evidence proving that the plasminogen activator-plasmin system plays an important role in the inflammatory processes of the cornea, especially corneal ulceration, was also provided by Berman et al.,[26,45] demonstrating that plasmin is able to activate latent collagenase isolated from organ cultures of ulcerating rabbit corneas and from fibroblast cultures of such corneas. The activation process of latent collagenase caused by plasmin was shown to be very similar to that caused by trypsin, resulting in the conversion of a high molecular weight latent collagenase to the active form with a molecular weight of 23,000 Da.[19,45] The role of the plasminogen activator-plasmin system in different diseases of the cornea is discussed in detail in Chapters 8 and 10.

Similar observations concerning the presence of collagenases in the cornea were made in different species, corneal ulcers with diverse etiologies, and a few nonulcerating corneal diseases. High collagenase activity was demonstrated in the culture media of ulcerating rabbit,[2] human,[4] and rat[46] corneas with various etiologies. The role of collagenases in the pathogenesis of corneal ulcers in corneas following alkali burns,[20,47] thermal burns,[48,49] herpetic keratitis,[50] Pseudomonas infection of the cornea,[51] and in keratomalatia[52,53] has been proved in animal experiments. Corneal buttons removed from eyes with advanced keratoconus have also been shown to produce collagenase in tissue culture experiments. It has been suggested that corneal collagenase may play a role in the development of corneal ectatic dystrophies.[54]

Vitamin A deficiency, causing xerophthalmia and keratomalatia in severe cases, is a condition with increased risk for corneal ulceration. Pirie et al.[55] were the first to demonstrate the relase in culture of collagenolytic enzymes from vitamin A-deficient rat corneas. In a series of experiments Seng et al.[56] found increased collagenolytic activity in the culture media of thermal-burned corneas of vitamin A-deficient rats as compared with that in the media of corneas from pair-fed control rats with comparable corneal burns. The elevated collagenolytic activity measured in the culture media was shown to be proportional to the severity of vitamin A deficiency.[56] Besides the suggested increased production and activation of collagenases in the vitamin A-deficient cornea, other sources of collagenolytic enzymes, e.g., the large number of infiltrating leukocytes and the superinfection with bacteria, may also contribute to corneal ulceration, which occurs more often and is more severe in vitamin A deficiency than with normal nutrition.[57] Besides collagenases, other proteases (cathepsin B and a cathepsin D-like protease) have been found to play a role in the degradation process of the cornea in vitamin A deficiency.[53] Plasminogen activators, with a suggested role in the activation process of procollagenase (through the action of plasmin), were also demonstrated in the wounded epithelium of vitamin A-deficient rat corneas with a plasminogen-containing fibrin film overlay technique by Hayashi et al. in 1988.[58] It has been suggested that both excessive and inadequate levels of plasminogen activator activity may result in impaired epithelial wound healing in vitamin A deficiency.

Different levels of corneal collagenase production and action were suggested to be possible sites where inhibition could be performed and the effect of active collagenase might be prevented. The multiple levels of suggested inhibition are (1) biosynthesis, (2) secretion, (3) the process of activation, and (4) the actual level of enzymatic action. (1) No effective way to stop the biosynthesis of corneal collagenase without destroying the cells has been reported so far. (2) Dibutyryl cyclic AMP (cAMP) and theophyllin have been shown to partially inhibit the appearance of collagenase activity in the culture media when added to cultures of ulcerating rabbit corneas. It has been proposed that these drugs prevent the secretion of corneal collagenase by raising endogenous cAMP in corneal cells through stimulation of the adenyl cyclase system.[59] (3) Important data concerning the role of the plasminogen activator-plasmin system in corneal ulceration were provided by Berman et al. in 1980.[26] Their findings opened new perspectives for the possibility of using plasmin inhibitors and possibly plasminogen activator inhibitors to prevent the activation of procollagenase into collagenase. (4) There are basically two ways to inhibit active corneal collagenase. Calcium and zinc ions inevitable for the enzyme action can be bound by metal-binding agents (EDTA, Ca-EDTA, L-cysteine, N-acetyl-cysteine, etc.). The second possibility is the inhibition by serum antiproteases. α-1-Antritrypsin and α-2-macroglobulin have been shown to inhibit rabbit corneal collagenase. In case of human corneal collagenase only α-2-macroglobulin was found to be an effective inhibitor. It has been demonstrated that only the active enzyme and not the proenzyme is being bound by the serum antiprotease α-2-macroglobulin.[19] Aprotinin was suggested to be effective in the treatment of nonhealing ulcers by limiting the ulceration process[60] and also in the prevention of ulcer formation in various ocular surface diseases.[61] The effect of aprotinin is based on the inhibition of plasmin and possibly other serine proteases, of which beside their wide spectrum proteolytic action can also activate procollagenase into active collagenase.

A very important observation concerning the role of collagenases and collagenase inhibitors in corneal ulceration was made by Francois and colleagues in 1973.[62] Based on histological studies of ulcerating corneas with and without treatment with inhibitors they demonstrated that the use of collagenase inhibitors resulted in excessive scar formation probably due to the lack of debridment of necrotic tissue in the ulcer area. This observation, besides revealing an important function of collagenases and other hydrolytic enzymes in the formation and healing of corneal ulcers, also provided important data concerning the therapeutic use of inhibitors. The inhibition of the enzyme action may be beneficial by limiting the ulceration and preventing corneal perforation but such an intervention also affects other processes like the debridment of necrotic tissues, and various events in corneal wound healing. Excessive scar formation may be unfavorable for the visual outcome, but by saving the eye may be a price worth paying, especially if the scar is later removed by keratoplasty. The unfavorable effects of inhibitors in the healing stage of the corneal ulcers, on the other hand, suggest that inhibitors should be used only in the active stage of the ulceration when collagenase and other proteolytic enzymes are acting in the cornea and can also be detected in the tears of the patients.

REFERENCES

1. **Slansky, H. H., Freeman, M. I., and Itoi, M.,** Collagenolytic activity in bovine corneal epithelium, *Arch. Ophthalmol.,* 80, 496, 1968.
2. **Brown, S. I., Weller, C. A., and Wassermann, H. E.,** Collagenolytic activity of alkali burned corneas, *Arch. Ophthalmol.,* 81, 370, 1969.
3. **Itoi, M., Gnadinger, M. C., Slansky, H. H., Freeman, M. I., and Dohlman, C. H.,** Collagenase in the cornea, *Exp. Eye. Res.,* 8, 369, 1969.
4. **Slansky, H. H., Gnadinger, M. C., Itoi, M., and Dohlman, C. H.,** Collagenase and corneal ulcerations, *Arch. Ophthalmol.,* 82, 108, 1969.
5. **Gnadinger, M. C., Itoi, M., Slansky, H. H., and Dohlman, C. H.,** The role of collagenase in the alkali-burned cornea, *Am. J. Ophthalmol.,* 68, 478, 1969.
6. **Itoi, M.,** Discussion on the collagenolytic activity in the cornea and its role in normal and pathological conditions, *Arch. Ophthalmol.,* 81, 147, 1969.
7. **Brown, S. I., Akiya, S., and Weller, C. A.,** Prevention of the ulcers of the alkali-burned cornea. Preliminary studies with collagenase inhibitors, *Arch. Ophthalmol.,* 82, 95, 1969.
8. **Weimar, V.L. and Haraguchi, K. H.,** The lag phase of wound repair. A comparison of leucocytic and corneal connective tissue enzyme activities, *Exp. Eye Res.,* 6, 283, 1967.
9. **Slansky, H. H. and Dohlman, C. H.,** Collagenase and the cornea, *Surv. Ophthalmol.,* 14, 402, 1970.
10. **Berman, M., Dohlman, C. H., Gnadinger, M., and Davison, P.,** Characterization of collagenolytic activity in the ulcerating cornea, *Exp. Eye Res.,* 11, 255, 1971.
11. **Gross, J.,** The animal collagenases, in *Chemistry and Molecular Biology of the Intercellular Matrix. Structural Organization and Function of the Matrix,* Vol. 3, Balazs, E. A., Ed., Academic Press, New York, 1970.
12. **Berman, M. B., Barber, J. C., Talamo, R. C., and Langley, C. E.,** Corneal ulceration and serum antiproteases. I. Alpha-1-antitrypsin, *Invest. Ophthalmol.,* 12, 759, 1973.
13. **Berman, M. B., Kerza-Kwiatecki, A. P., and Davison, P. F.,** Characterization of human corneal collagenase, *Exp. Eye Res.,* 15, 367, 1973.
14. **Berman, M. B. and Manabe, R.,** Corneal collagenases: evidence for zinc metalloenzymes, *Ann. Ophthalmol.,* 5, 1193, 1973.
15. **Sallai, S., Valu, L., Fehér, J., and Podhorányi, G.,** On subconjunctival autochemotherapy of experimental lime burns, *Dtsch. Gesundheitswes.,* 22, 1994, 1967.
16. **Lebekov, P. I.,** The use of new blood preparation (Biological Ointment) in complex treatment of eye burns, *Vestn. Oftalmol.,* 77, 56, 1964.
17. **Németh, B., Bencsik, R., and Opauszki, A.,** Therapeutische Erfahrungen mit einem Blutextrakt bei Kornealasionen, *Klin. Mbl. Augenheilk.,* 178, 390, 1981.
18. **Berman, M., Gordon, J., Garcia, L. A., and Gage, L.,** Corneal ulceration and the serum antiproteases. II. Complexes of corneal collagenases and alpha-macroglobulins, *Exp. Eye Res.,* 20, 231, 1975.
19. **Berman, M., Leary, R., and Gage, J.,** Latent collagenase in the ulcerating rabbit cornea, *Exp. Eye Res.,* 25, 435, 1977.
20. **Brown, S. I., Weller, S., and Akiya, S.,** The pathogenesis of ulcers of the alkali burned cornea, *Arch. Ophthalmol.,* 83, 205, 1970.
21. **Bauer, E. A., Stricklin, G. P., Jeffrey, J. J., and Eisen, A. Z.,** Collagenase production by human skin fibroblasts, *Biochem. Biophys. Res. Commun.,* 64, 232, 1975.
22. **Birkedal-Hansen, H. M.,** Synthesis and release of procollagenase by cultured fibroblasts, *J. Biol. Chem.,* 251, 3162, 1975.
23. **Dayer, J.-M., Krane, S. M., Russell, R. G. G., and Robinson, D. R.,** Production of collagenase and prostaglandins by isolated adherent rheumatoid synovial cells, *Proc. Natl. Acad. Sci. U.S.A.,* 73, 945, 1976.
24. **Werb, Z. and Burleigh, M. C.,** A specific collagenase from rabbit fibroblasts in monolayer culture, *Biochem. J.,* 137, 373, 1974.

25. **Hook, R. M., Hook, C. W., and Brown, S. I.,** Fibroblast collagenase partial purification and characterization, *Invest. Ophthalmol.,* 12, 771, 1973.

26. **Berman, M., Leary, R., and Gage, J.,** Evidence for a role of the plasminogen activator-plasmin system in corneal ulceration, *Invest. Ophthalmol. Vis. Sci.,* 19, 1204, 1980.

27. **Gordon, J. M., Bauer, E. A., and Eisen, A. Z.,** Collagenase in human cornea. Immunologic localization, *Arch. Ophthalmol.,* 98, 341, 1980.

28. **Prause, J. U. and Jensen, O. A.,** PAS-positive PMN leucocytes in corneal ulcers, *Acta Ophthalmol.,* 58, 556, 1980.

29. **Weimar, V.,** Polymorphonuclear invasion of wounded corneas, *J. Exp. Med.,* 105, 141, 1957.

30. **Brown, S. I. and Hook, C. W.,** Isolation of stromal collagenase in corneal inflammation, *Am. J. Ophthalmol.,* 72, 1139, 1971.

31. **Lazarus, G. S., Daniels, J. R., Brown, R. S., Bladen, H. A., and Fullmer H. M.,** Degradation of collagen by a human granulocyte collagenolytic system, *J. Clin. Invest.,* 47, 2622, 1968.

32. **Yamada, M., Kishida, K., Yuasa, T., Mimura, Y., and Manabe, R.,** Neutrophil collagenolytic activity in patients with Behcet disease, *Curr. Eye Res.,* 3, 779, 1984.

33. **Ohlsson, K. and Ohlsson, I.,** The neutral proteases of human granulocytes. Isolation and partial characterization of two granulocyte collagenases, *Eur. J. Biochem.,* 36, 473, 1973.

34. **Prause, J. U.,** Serum albumin, serum antiproteases and polymorphonuclear leucocyte neutral collagenolytic protease in the tear fluid of normal healthy persons, *Acta Ophthalmol.,* 61, 261, 1983.

35. **Prause, J. U.,** Serum albumin, serum antiproteases and polymorphonuclear leucocyte neutral collagenolytic protease in the tear fluid of patients with corneal ulcers, *Acta Ophthalmol.,* 61, 272, 1983.

36. **Prause, J. U.,** Cellular and biochemical mechanisms involved in the degradation and healing of the cornea, *Acta Ophthalmol.,* 168(suppl.), 1, 1984.

37. **Newsome, D. A. and Gross, J.,** Regulation of corneal collagenase production: stimulation of serially passaged stromal cells by blood mononuclear cells, *Cell,* 16, 895, 1979.

38. **Johnson-Muller, B. and Gross, J.,** Regulation of corneal collagenase production: epithelial-stromal cell interactions, *Proc. Natl. Acad. Sci. U.S.A.,* 75, 4417, 1978.

39. **Johnson-Wint, B.,** Regulation of stromal cell collagenase production in the adult rabbit cornea: *in vitro* stimulation and inhibition by epithelial cell products, *Proc. Natl. Acad. Sci. U.S.A.,* 77, 5331, 1980.

40. **Murphy, G., Reynolds, J. J., Bretz, U., and Baggolini, M.,** Collagenase is a component of the specific granules of human neutrophil leukocytes, *Biochem. J.,* 162, 195, 1977.

41. **Hasty, K. A., Hibbs, M. S., Kang, A. H., and Mainardi, C. L.,** Heterogeneity among human collagenases demonstrated by monoclonal antibody that selectively recognizes and inhibits human neutrophil collagenase, *J. Exp. Med.,* 159, 1455, 1984.

42. **Macartney, H. W. and Tschesche, H.,** The collagenase inhibitor from human polymorphonuclear leukocytes. Isolation, purification and characterization, *Eur. J. Biochem.,* 130, 79, 1983.

43. **Kao, W. W.-Y., Ebert, J., Kao, C. W.-C., Covington, H., and Cintron, C.,** Dvelopment of monoclonal antibodies recognizing collagenase from rabbit PMN; the presence of this enzyme in ulcerating corneas, *Curr. Eye Res.,* 5, 801, 1986.

44. **Berman, M.,** Regulation of collagenase. Therapeutic considerations, *Trans. Opthamol. Soc. U.K.,* 98, 397, 1978.

45. **Berman, M., Leary, R., and Gage, J.,** Collagenase from corneal cell cultures and its regulation by phagocytosis, *Invest. Ophthalmol. Vis. Sci.,* 18, 588, 1979.

46. **Badenoch, P. R., Finlay-Jones, J. J., and Coster, D. J.,** Enzymatic disaggregation of the infected rat cornea, *Invest. Ophthalmol. Vis. Sci.,* 24, 253, 1983.

47. **Anderson, R. E., Kuns, M. D., and Dresden, M. H.,** Collagenase activity in the alkali-burned cornea, *Ann. Ophthalmol.,* 3, 619, 1971.

48. **Francois, J. and Fehér, J.,** Collagenolysis and regeneration in corneal burnings, *Ophthalmologica,* 165, 137, 1972.

49. **Kenyon, K. R., Berman, M., Conn, H., Henriques, A., Gage, J., and Rose, J.,** Ulceration and repair in rabbit cornea following thermal injury, *Invest. Ophthalmol. Vis. Sci. Suppl.,* 17, 252, 1977.

50. **Campbell, R., Pavan-Langston, D., Lass, J., Berman, M., Gage, J., and Albert, D.,** Collagenase levels in a new model of experimental herpetic interstitial keratitis, *Arch. Ophthalmol.,* 98, 919, 1980.

51. **Kessler, E., Mondino, B. J., and Brown, S. I.,** The corneal response to *Pseudomonas aeruginosa:* histopathological and enzymatic characterization, *Invest. Ophthalmol. Vis. Sci.,* 16, 116, 1977.

52. **Leonard, M. C., Maddison, L. K., and Pirie, A.,** A comparison between the enzymes in the cornea of the vitamin-A deficient rat and those of rat leukocytes, *Exp. Eye Res.,* 33, 479, 1981.

53. **Twining, S. S., Hatchell, D. L., Hyndiuk, R. A., and Nassif, K. F.,** Acid proteases and histologic correlations in experimental ulceration in vitamin A deficient rabbit corneas, *Invest. Ophthalmol. Vis. Sci.,* 26, 31, 1985.

54. **Rehany, U., Lahav, M., and Shoshan, S.,** Collagenolytic activity in keratoconus, *Ann. Ophthalmol.,* 14, 751, 1982.

55. **Pirie, A., Werb, Z., and Burleigh, M. C.,** Collagenase and other proteinases in cornea of retinol-deficient rat, *Br. J. Nutr.* 34, 297, 1975.

56. **Seng, W. L., Glogowski, J. A., Wolf, G., Berman, M. B., Kenyon, K. R., and Kiorpes, T. C.,** The effect of thermal burns on the release of collagenase from corneas of vitamin A-deficient and controls rats, *Invest. Ophthalmol. Vis. Sci.,* 19, 1461, 1980.

57. **Seng, W. L., Kenyon, K. R., and Wolf, G.,** Studies on the source and release of collagenase in thermally burned corneas of vitamin A-deficient and control rats, *Invest. Ophthalmol. Vis. Sci.,* 22, 62, 1982.

58. **Hayashi, K., Frangieh, G., Kenyon, K. R., Berman, M., and Wolf, G.,** Plasminogen activator activity in vitamin A-deficient rat corneas, *Invest. Ophthalmol. Vis. Sci.,* 29, 1810, 1988.

59. **Berman, M. B., Cavanagh, H. D., and Gage, J.,** Regulation of collagenase activity in the ulcerating cornea by cyclic-AMP, *Exp. Eye Res.,* 22, 209, 1976.

60. **Salonen, E.-M., Tervo, T., Törma, E., Tarkkanen, A., and Vaheri, A.,** Plasmin in tear fluid of patients with corneal ulcers: basis for new therapy, *Acta Ophthalmol.,* 65, 3, 1987.

61. **Berta, A., Holly, F. J., and Tözsér, J.,** Proteinase-inhibitors in the diagnosis and treatment of ocular surface diseases, in *Proc. Symp. Pharmacotherapy of the Eye,* Springer-Verlag, Heidelberg, in press.

62. **Francois, J., Cambie, E., Fehér, J., and van den Eeckhout, E.,** Collagenase inhibitors (penicillamine), *Ann. Ophthalmol.,* 5, 391, 1973.

63. **Berman, M. B.,** Collagenase and corneal ulceration, in *Collagenase in Normal and Pathological Connective Tissues,* Woolley, D. E. and Evanson, J. M., Eds., John Wiley & Sons, New York, 1980, 141.

CORNEAL COLLAGENS AND TEAR COLLAGENASES

Part D
Collagenases and Their Physiological Inhibitors in the Tears

In 1969 Slansky and colleagues[1,2] and Brown and colleagues[3] provided data concerning the possible role in corneal ulceration of collagenolytic enzymes capable of cleaving the collagen structure of the corneal stroma. The idea of the enzymatic degradation of the cornea in the ulceration process gave an enormous impulse to the research of collagenolytic proteases responsible for the degradation of the cornea. Over 100 publications deal with the origin, characteristics, enzyme action, and possible inhibition of corneal collagenases, detected in the culture media or in the frozen sections of the removed animal and human corneas.

Surprisingly few data are available in the literature on the collagenolytic activity and the collagenase content of the tear fluid. The first attempts to detect collagenase activity in the tears were unsuccessful, even in the case of patients with "central melting ulcers", probably because of the limited sensitivity of the applied methods.[4,5]

Prause et al.[5] have developed a micromodification of the electroimmunoassay that proved to be suitable for the immunological determination of different tear proteins[5] and granulocyte collagenase and serum antiproteases[6] present in tears in very small quantities. The test was designed to assay small samples of very dilute and complex solutions of different proteins like the tear fluid. The sensitivity of the method for granulocyte collagenase was shown to be 0.1 ng applied in 25 µl. The immunoglobulin requirement of the procedure was found to be 100 times less than that of the quantitative rocket immunoelectrophoresis.[4]

Burnett and colleagues[7] reported at the ARVO Meeting in 1981 to have measured, with the microimmunoelectrophoresis technique of Prause,[5] collagenase levels in the tears of patients with "melting corneas" in the range 5 to 20 OSU/µl, whereas controls measured less than 3 OSU/µl.[5] Their results, to the best knowledge of the author, have not been published in detail, so the nature of the used antibody, the nature of the detected collagenase, and clinical data concerning the melting corneas and the controls are not available.

Prause[8] reported a study concerning the serum albumin, serum antiprotease, and granulocyte collagenase content of the tears of 53 healthy persons of different ages. In none of the tear samples could he detect by microimmunoelectrophoretic technique polymorphonuclear (PMN) leucocyte neutral collagenolytic protease. In further experiments[6,9] he measured the level of the same enzyme in the tears of patients suffering from corneal ulcers with diverse origin. All samples taken from patients at the time of "maximun corneal ulcer" (grade +++), graded clinically on a semiquantitative scale, were shown to contain granulocyte collagenase, though in very different concentrations ranging from 0.2 to 76.8 µg/ml. The highest levels of the enzyme were found in the tears of patients with corneal ulcers who were suffering from Stevens-Johnson syndrome and rheumatoid arthritis. Moderately high levels of granulocyte collagenase were found in the tears of patients whose corneal ulcers had developed on the basis of herpetic virus infections, alkali burns, and neuroparalytic and rosacea keratitis. The only case with a true bacterial infection also fell in this range. Very low but still detectable levels of the enzyme were found in both eyes of the only patient with Sjögren's syndrome, and in some cases with already-mentioned etiologies. The number of cases in this study was not high enough to decide whether these differences have something to do with the specific causes of the corneal ulcers. It has been demonstrated, however, that the levels of granulocyte collagenase in tears had diminished during the healing of the ulcers. The decrease in collagenase levels parallelled the fall in number of PMNs counted in the pellet of the same tear samples; the number of PMNs, however, was still increased at the time when PMN collagenase could no longer be detected.[4] When the healing of the corneal ulcer was promoted with cyanoacrylate glue the level of granulocyte collagenase was found to be low or completely missing from the tears of the patients.[9] This may have been the result of mechanical barrier formed by the glue in the ulcerating area preventing the invasion of PMNs, or some unknown chemical effect blocking the liberation or action of chemotactic factors and/or that of proteolytic enzymes.

Using the method of Berman et al.,[10] we detected collagenolytic activity in the tears of patients suffering from Mooren's ulcer. The tear levels showed good correlation with the severity of the disease, and were suggested to be used to differentiate between the two clinical types, the active and inactive stages of the disease, as well as to determine the optimal time for surgical interventions to be carried out on patients suffering from Mooren's ulcer.[11,12]

In 1988 we developed a microtiter plate method to measure collagenase levels in tears. The procedure was based on the measurement of the absorption (optical density) of Coommassie brillant blue stained collagen gels reconstituted in the wells of microtiter plates.[13,14] Acid-soluble type IV collagen was gelled and dried in the wells of microtiter plates and was used as substrate in the collagenolytic activity measurements. After incubation with tear samples, the digested water-soluble protein fraction was washed away and the remaining, undigested substrate was stained with Coomassie brillant blue. Absorbancy values were determined at 550 nm by an automatic microtiter plate reader. Increasing amounts of collagen substrate were used to produce a calibration curve (Figure 1). Of 100 µl/ml bovine trypsin, 10 µl was used to activate inactive collagenase. Molar amount of soybean trypsin inhibitor (SBTI) was used to inhibit the added trypsin and other seril proteases that were also present in some of the tear samples.

The absorbance at 550 nm proved to be proportional to the amounts of collagen coated at the bottom of the wells over the range of 4 to 24 µg collagen

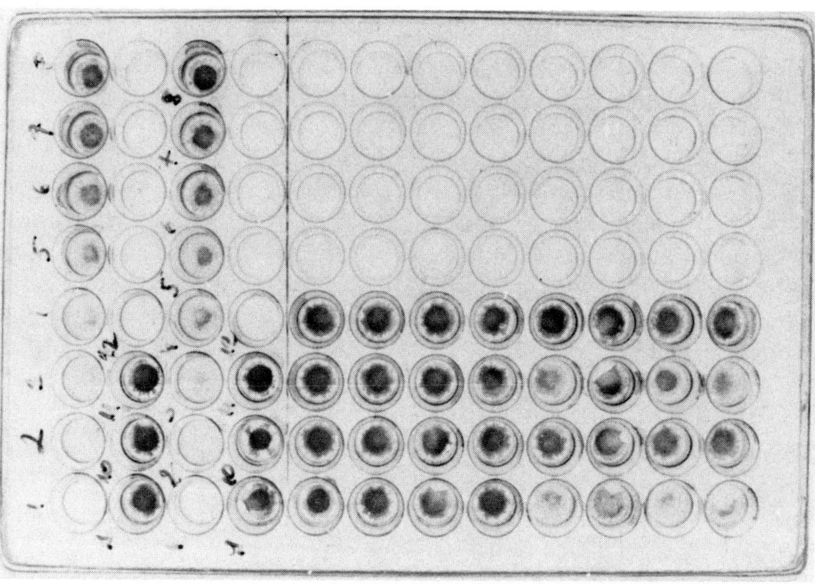

FIGURE 1. Reconstituted collagen in the wells of a microtiter plate stained with Coomassie brillant blue. Left: calibration; right: samples. (From Berta, A., Punyiczki, M., and Tözsér, J., in *Ophthalmology Today*, Ferres de Oliviera, L. N., Ed., Elsevier Science, Amsterdam, 1988, 503. With permission.)

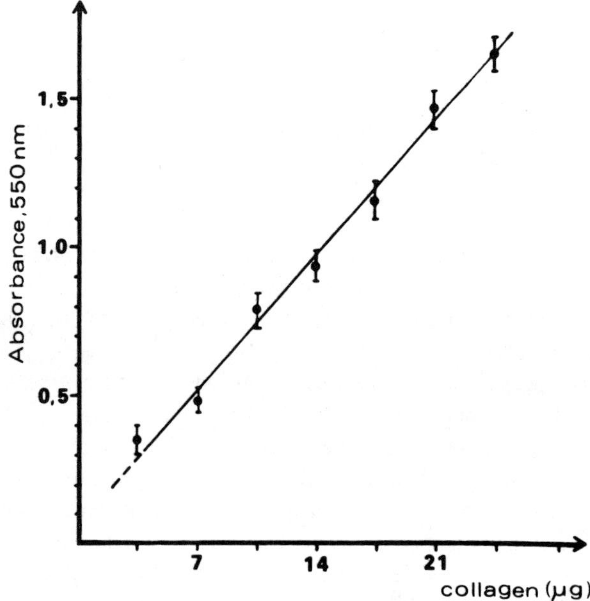

FIGURE 2. Standard curve for the determination of undigested collagen on the bottom of the wells of microtiter plates. (From Berta, A., Punyiczki, M., and Tözsér, J., in *Ophthalmology Today*, Ferres de Oliviera, L. N., Ed., Elsevier Science, Amsterdam, 1988, 503. With permission.)

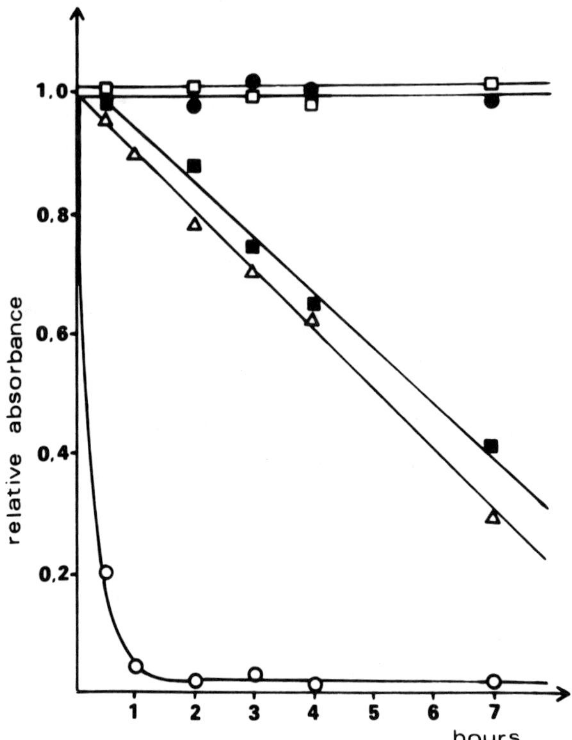

FIGURE 3. Digestion of coated collagen by different collagenases with and without EDTA.; ■, 0.3 U bacterial collagenase; □, 0.3 U bacterial collagenase and 20 mM EDTA; △, 1 μg human PMN granulocyte collagenase; ○, 10 μg human PMN granulocyte collagenase; ●, 10 μg human PMN granulocyte collagenase and 20 mM EDTA. (Modified and redrawn from Berta, A., Punyiczki, M., and Tözsér, J., in *Ophthalmology Today*, Ferres de Oliviera, L. N., Ed., Elsevier Science, Amsterdam, 1988, 503. With permission.)

(Figure 2). The time dependence of collagen digestion caused by bacterial and human granulocyte collagenase is demonstrated in Figure 3. The same enzymes failed to change the relative absorbance values in the presence of 20 mM EDTA. Ten times higher amounts of collagenase caused total digestion of the substrate within 1 h (Figure 3).

Beside the specific collagenase action a number of other proteases may act in a collagenolytic manner. This is especially true for soluble substrates and for partly denatured collagen fibrils that have partly lost their helical structure. Within 1 h, 25 μg bovine trypsin digested more than 50% of the coated collagen substrate. The rest of the substrate was unaffected during the 6 h of the assay. The trypsin digestion was completely inhibited by molar amount of SBTI (Figure 4). Collagenases being the only metalloproteinases to be found in tears are specifically inhibited by EDTA. Therefore, parallel measurements were made with and without EDTA. The absorbancy difference between the two wells served as a measure of collagenase action.

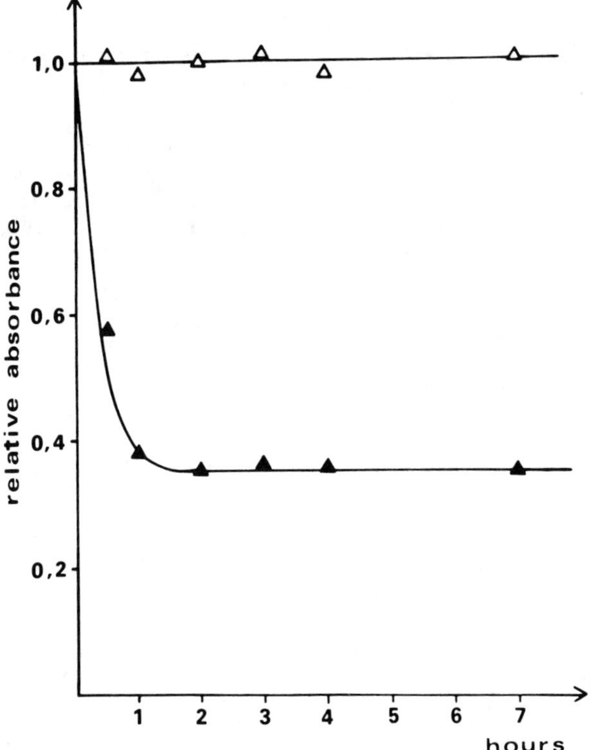

FIGURE 4. The effect of 25 μg bovine trypsin with (Δ) and without (▲) 125 μg SBTI. (Modified and redrawn from Berta, A., Punyiczki, M., and Tözsér, J., in *Ophthalmology Today*, Ferres de Oliviera, L. N., Ed., Elsevier Science, Amsterdam, 1988, 503. With permission.)

Collagenolytic activity values were calculated from absorbance changes due to the digestion, the total amount of the collagen substrate, the absorbance of the undigested substrate, the length of the digestion, and the total volume of the tear samples. Activity values were given in micrograms per hour × milliliter units. High activity levels of active and inactive collagenase were found in the tears of patients suffering from herpetic keratitis, alkali burns, and corneal ulcers especially those with bacterial origin. Active collagenase activity levels showed good correlation with the severity of the disease. Surprisingly high levels of inactive collagenase were found in bacterial conjunctivitis and keratoconjunctivitis sicca (Table 1). Significantly lower but still measurable inactive collagenase levels were demonstrated in 78% of the studied normal tear samples. Inactive collagenase was present in relatively higher concentration in normal tears collected at low flow rates (Table 1). It was lower but still detectable at very high flow rates when the tear samples practically represented the secretion of the lacrimal gland.

TABLE 1
Collagenase Levels in Human Tears Determined by the Microtiter Plate Method

Patient number	Diagnosis	Collagenase activities (μg collagen/h $\times$ ml tears)	
		Active collagenase	Inactive collagenase
1	Herpetic keratitis	197.86	93.32
2	Herpetic keratitis	230.74	83.96
3	Herpetic keratitis	150.80	123.26
4	Herpetic keratitis	171.00	175.60
5	Herpetic keratitis	181.80	737.46
6	Herpetic keratitis	159.36	332.08
7	Corneal ulcer (nonbacterial)	231.00	482.80
8	Corneal ulcer (bacterial)	537.60	473.00
9	Corneal ulcer (bacterial)	643.10	514.00
10	Corneal ulcer (bacterial)	845.34	388.18
11	Alkali burn (fresh case)	813.72	742.07
12	Alkali burn (healing stage)	132.48	184.08
13	Keratoconjunctivitis sicca	286.60	1069.40
14	Bacterial conjunctivitis	210.43	315.88
Normal tears (16)		0	Mean: 34.97

REFERENCES

1. **Itoi, M., Gnadinger, M. C., Slansky, H. H., Freeman, M. I., and Dohlman, C. H.,** Collagenase in the cornea, *Exp. Eye. Res.,* 8, 369, 1969.
2. **Slansky, H. H., Gnadinger, M. C., Itoi, M., and Dohlman C. H.,** Collagenase and corneal ulcerations, *Arch. Ophthalmol.,* 82, 108, 1969.
3. **Brown, S. I., Weller, C. A., and Wassermann, H. E.,** Collagenolytic activity of alkali burned corneas, *Arch. Ophthalmol.,* 81, 370, 1969.
4. **Prause, J. U.,** Cellular and biochemical mechanisms involved in the degradation and healing of the cornea, *Acta Ophthalmol.,* 168(suppl.), 7, 1984.
5. **Prause, J. U., Sjöström, H., Norén, O., and Josefsson, L.,** A micromodification of the electroimmunoassay, *Anal. Biochem,* 85, 564, 1978.
6. **Prause, J. U.,** Serum albumin, serum antiproteases and polymorphonuclear leucocyte neutral collagenolytic protease in the tear fluid of patients with corneal ulcers, *Acta Ophthalmol.,* 61, 272, 1983.
7. **Burnett, J. M., Smith, L. E., Prause, J. U., and Kenyon, K. R.,** Acute inflammatory cells and collagenase in tears of human melting corneas, *Invest. Ophthalmol. Vis. Sci. Suppl.,* 20, 173, 1981.
8. **Prause, J. U.,** Serum albumin, serum antiproteases and polymorphonuclear leucocyte neutral collagenolytic protease in the tear fluid of normal healthy persons, *Acta Ophthalmol.,* 61, 261, 1983.

9. **Prause, J. U.,** Serum antiproteases and polymorphonuclear leucocyte neutral collagenolytic protease in the tear fluid of patients with corneal ulcers treated with N-butylcyano acrylate glue, *Acta Ophthalmol.,* 61, 283, 1983.

10. **Bermen, M., Manabe, R., and Davison, P. F.,** Tissue collagenase: a simplified, semiquantitative enzyme assay, *Anal. Biochem.,* 54, 522, 1973.

11. **Berta, A., Zajácz, M., and Jaross, N.,** Tear protein tests in the diagnosis of Mooren's ulcer, *Szemeszet,* 124, 220, 1987.

12. **Zajácz, M. and Berta, A.,** Proteolytische Enzymaktivitat und chirurgische eingriffe bei ulcus Mooren, *Klin. Mbl. Augenheilk.,* 187, 401, 1985.

13. **Berta, A., Punyiczki, M., and Tözsér, J.,** Collagenase in normal and pathological human tears determined by microtiter plate method, in *Ophthalmology Today,* Ferras de Oliviera, L. N., Ed., Elsevier Science, Amsterdam, 1988, 503.

14. **Punyiczki, M., Berta, A., and Tözsér, J.,** Determination of collagenolytic activity in human tears by microtiter method, *Clin. Chem. Enzym. Comns.,* accepted.

THE PLASMINOGEN ACTIVATOR-PLASMIN SYSTEM
OF THE TEARS

Part A
Basic Facts about the Plasminogen Activator-Plasmin System

The human body possesses a complex enzyme system capable of dissolving fibrin deposits by the formation of a fibrinolytic protease, plasmin at the sites of fibrin deposition. The basic componets of the fibrinolytic system are plasminogen, plasmin, and plasminogen activators. Plasminogen, an inactive circulating zymogen, is converted to plasmin, the fibrinolytically active enzyme through the action of other proteases called plasminogen activators.[1]

Human plasminogen is a single-chain glycoprotein containing 790 amino acids, with a molecular weight of 92,000 and a carbohydrate content of around 2%.[2-4] Plasminogen is susceptible to specific proteolytic cleavage with removal of 76 residues from the amino-terminal portion of the molecule resulting in the conversion of the native form Glu- to Lys-plasminogen. These two forms have conformational differences and differ in their rate of activation to plasmin.[1] Plasminogen binds ω-aminocarboxylic acids such as lysine and its analogues (EACA, AMCA). Native plasminogen contains at least one lysine-binding site with strong affinity for EACA and AMCA.[5,6] A conformational change takes place in the plasminogen molecule on the binding of the ω-amino acids to plasminogen that is similar to that seen on the conversion of Glu- into Lys-plasminogen.[7]

Plasminogen readily adsorbs on to fibrin.[8] The interaction between plasminogen and fibrin is mediated through the lysine-binding sites.[9] Glu-plasminogen has a weaker Lys-plasminogen and a stronger affinity for fibrin.[10] The proteolytic cleavage of the so-called preactivation peptide from the amino-terminal part of the plasminogen molecule and the accompanying conformational change are required for exposure of the strong binding site. Lys-plasminogen showes a high affinity for both cross-linked and non-cross-linked fibrin.[11]

The activation of plasminogen starts with the specific cleavage of the Arg(560)-Val(561) bond in the carboxy-terminal portion of the Glu-plasminogen molecule brought about by the proteolytic action of plasminogen activators. This initial step results in the formation of catalytic quantities of Glu-plasmin, which in turn attacks the susceptible bonds in the amino-terminal parts of the remaining Glu-plasminogen molecules, leading to the removal of the preactivation peptide and the formation of Lys-plasminogen. The conformational change accompanying this cleavage increases the rate of Arg(560)-Val(561) bond cleavage by the plasminogen activator with Lys-plasmin formation, the final form found after activation.[12]

The fibrinogen molecule exists as a dimer, each of the identical halves consisting of three polypeptide chains (α, β, γ) connected by disulfide bonds. The two half-molecules are joined by further disulfide bridges. Thrombin converts fibrinogen into fibrin by cleaving fibrinopeptides A and B from the amino-terminal ends of the α and β chains. This is followed by end-to-end and lateral aggregation of fibrin molecules. Finally the fibrin network is stabilized through the formation of covalent bonds between the adjacent γ and α chains, promoted by factor XIII that is activated by thrombin in the presence of calcium ions. Such stabilized fibrin is more resistant to digestion by plasmin than noncross-linked fibrin.[1]

Plasmin is able to digest both fibrinogen and fibrin. The initial attack by plasmin takes place on the α chains removing peptides of around 40,000 Da from the carboxy ends. This is followed by the removal of peptides from the amino terminal of β chains and the asymmetric cleavage of the three chains at one side of the partially degraded dimer with the release of a fragment termed D. Subsequent lysis of the α, β, and γ chain remnants generates a second fragment D leaving a core fragment, a dimer of molecular mass 52,000 that contains parts of all three fibrinogen chains. During the lysis of fibrin the cross-links near the carboxy ends of the γ chains are retained. The major difference between the end products in digestion of fibrinogen and fibrin is the presence of D-dimer, the cross-linked complex of D fragments.[13]

In addition to its natural substrate fibrin, plasmin has the ability to cleave a variety of proteins including clotting factors V, VIII, and IX.[14,15] It may act as a kininogenase, releasing kinin directly from kininogen,[16] and is able to activate some of the complement components (C1, C1s, C4). It has the ability to cleave C3a from the C3 molecule, which in turn is able to release histamin from mast cells and basophils, contracting smooth muscles and increasing capillary permeability.[17] Plasmin is capable of degrading both the α and β chains of laminin,[18] the glycoprotein present in basement membranes, and fibronectin, another attachment protein. Plasmin probably plays an important role in the degradation of attachment proteins during tissue remodeling and repair.[19]

Plasminogen activators are proteases capable of turning plasminogen into plasmin by specific action of limited proteolysis. Plasminogen activators are basically classified into two groups: tissue-type (t-PA) and urokinase-type (u-PA) plasminogen activators, with distinctive molecular masses and enzymatic and immunological properties. t-PA activates plasminogen to plasmin on the surface of the fibrin network, and produces fibrin-bound plasmin with an action restricted to fibrin digestion. u-PA on the other hand activates plasminogen to plasmin in secretions and body fluids, and produces free plasmin with a wide proteolytic activity to be performed in various extracellular spaces.[8,20–21a] Other agents such as factor XII-dependent activator, prekallikrein, high molecular mass kininogen, and an activator formed through the complement system may play a role in the activation of plasminogen.[22,23]

Most normal tissues contain plasminogen activators and plasminogen activator activity with the exception of the liver. The level in different organs varies widely with high concentrations in the uterus, adrenals, and thyroid and low concentrations in the testis and spleen.[24] It has been shown that the fibrinolytic activity of tissues is directly related to their vascularity. Immunohistological studies showed that t-PA is concentrated in the endothelium of blood vessels. The activator is released from the vessel wall in response to stimuli such as physical excercise or venous occlusion.[25] Tissue activator has been isolated from a variety of organs including mammalian heart, uterus, ovaries, and from melanoma cells in tissue culture.[26,27] Most preparations of t-PA have had a molecular mass close to 70,000; some have had higher masses around 100,000 Da, but these are probably complexes of the activator with inhibitors. t-PA can be isolated as either a single- or a double-chain molecule. The conversion into the two-chain form results from the proteolytic cleavage of a bond in the central part of the native one-chain form. t-PA has been produced by recombinant DNA techniques: the heavy chain had a molecular mass of almost 31,000 and it is derived from the amino-terminal part of the molecule, while the light chain with a molecular mass of around 28,000 Da comprises the carboxy terminal and contains the catalytic site. The fibrinolytic activity and affinity for fibrin are very similar in both single- and two-chain forms of t-PA.[28]

Urokinase (u-PA) is a term given to the plasminogen activator found in urine.[29] It has been established that urokinase is produced and released from cells in the glomeruli of the kidney.[30] While the renal source of urinary u-PA is not disputed, it has become apparent in recent years that many other tissues can produce an activator with the properties and immunological characteristics of urokinase. u-PA exists in urine in two molecular forms: a 54,000-Da high molecular mass form and a 31,000-Da low molecular mass form.[31] It has been shown that the low molecular mass form can be derived from the high molecular mass form by limited proteolysis through the action of trypsin and plasmin.[32] The native high molecular mass form of u-PA is composed of two chains connected by a single disulfide bond. The heavy chain has been found to be functionally active and to have a molecular weight, specific activity, and amino acid composition identical to that found for low molecular mass u-PA.[33] The basic role of u-PA is the activation of plasminogen to plasmin by the specific cleavage of the Arg(560)-Val(561) bond in the plasminogen molecule. Besides this it has esterolytic and amidolytic properties that have been utilized for its assay. u-PA differs from t-PA in many respects. They have different substrate requirements towards peptide amides,[34] they are immunologically distinct,[20] and whereas t-PA is bound avidly to fibrin u-PA is not significantly adsorbed.[27] It has been shown that u-PA can be produced by cultured human cells of various origin. In the cultures of most of these cells t-PA was also detected.[35] u-PA produced by cultures of kidney cells appears to be in proenzyme form.[36] Anti-uPA antibodies were shown to quench a part of the plasma fibrinolytic activity. An activator with properties very similar to those of u-PA has been isolated from plasma.[37] It has been suggested that the u-PA-type activator found in plasma is in precursor form, and the conversion into the active form may be achieved by factor XII or kallikrein.[38,39]

A number of plasma proteins have the ability to neutralize the proteolytic activity of plasmin. Such physiologic plasmin inhibitors comprise α-2-antiplasmin, α-2-macroglobulin, α-1-antitrypsin, inter-α-trypsin inhibitor, C1 inactivator, and antithrombin III.[1]

α-2-Antiplasmin is the major plasma inhibitor of plasmin. It is a single-chain glycoprotein with a molecular weight of around 70,000 Da and a carbohydrate content of about 14%.[40] The concentration of α-2-antiplasmin in normal plasma is around 1 μM.[41] The formation of an enzymatically inactive complex results from the interaction between the light chain of plasmin and α-2-antiplasmin.[42] The lysin-binding sites play an important role in this interaction. The reaction takes place in two steps. The first step is a very rapid reversible reaction that is followed by a slower first-order irreversible transition.[43] Two molecular forms of α-2-antiplasmin exist in plasma. They differ in their reactivity towards plasmin and their ability to bind to fibrin.[44] α-2-Antiplasmin has an inhibitory action on fibrinolysis by interfering with the absorption of plasminogen onto fibrin.[45] Besides, it is capable of inhibiting urokinase and to a less extent t-PA.[46] It has also been shown to inhibit the activated forms of a number of clotting factors in purified systems.[47] Patients with the congenital deficiency of this inhibitor show hemorrhages demonstrating the importance of α-2-antiplasmin in the regulation of fibrinolysis.[48] Following the activation of plasminogen into plasmin the active enzyme is preferentially bound to α-2-antiplasmin. Only after this inhibitor is saturated is the excess plasmin neutralized by α-2-macroglobulin.[49] α-2-Antiplasmin is the only inhibitor showing a significant decrease in plasma concentration in patients with disseminated intravascular coagulation and fibrinolysis.[50]

α-2-Macroglobulin is one of the major plasma proteins with a capacity to inhibit a wide range of proteolytic enzymes including trypsin, chymotrypsin, papain, collagenase, elastase, thrombin, plasmin, kallikrein, etc. α-2-Macroglobulin is a glycoprotein with a molecular weight of 725,000 that is composed of four identical subunit chains. Each subunit consists of 1450 amino acid residues and has a molecular weight of 185,000. The polypeptide chains are held together in pairs by disulfide bonds.[51] The two pairs are held together by noncovalent bonds forming a dimeric structure. Each subunit chain is capable to form a complex with the light chain of plasmin. Such complexes have negligable proteolytic activity, and retain only 0.1% of their original fibrinolytic activity.[52] α-2-Macroglobulin does not bind to the active site of plasmin so its catalytic activity toward small molecules is preserved to a great extent. The enzyme in spite of the complex formation remains esterolytically and amidolytically active. When plasminogen is activated in plasma the plasmin formed is preferentially bound to α-2-antiplasmin. Only when this inhibitor is saturated is the excess plasmin neutralized by α-2-macroglobulin.[1] In contrast to the hemorrhagic state associated with α-2-antiplasmin, patients with congenital deficiency of α-2-macroglobulin do not suffer from abnormal bleeding.[53]

α-1-Antitrypsin is a single-chain glycoprotein consisting of 394 amino acid residues and three cabohydrate side chains. Its molecular weight is 51,000.[54] It has the ability to inhibit a number of proteases including trypsin, chymotrypsin, elastase, and plasmin. The inhibitor forms complexes with these proteases in an irreversible reaction.[55] In purified systems α-1-antitrypsin inhibits plasmin, but the association rate constant of α-1-antitrypsin with plasmin is very low; that is why it probably has no significant role as a natural plasmin inhibitor.[56] It has been shown that the primary target of α-1-antitrypsin is leukocyte elastase. Emphysema of the lung is likely to be produced by uninhibited elastase activity.[57]

Inter-α-trypsin inhibitor is a glycoprotein with a molecular weight of 160,000 Da. It was isolated from human serum in 1965.[58] It has been shown to be the precursor of the acid-stable trypsin inhibitor present in human serum and urine with a molecular weight of 30,000 Da.[59] This latter inhibitor is the result of limited proteolysis just like further degradation products with molecular weights of 14,000 and 8000 Da.[60] The enzymatic cleavage of inter-α-trypsin inhibitor can be brought about by granulocytic elastase, plasmin, trypsin, and kallikrein. The rate of cleavage by granulocytic elastase is much higher than that by the other proteases suggesting that elastase is probably the physiologic enzyme to produce the small molecular weight forms of trypsin inhibitor.[61] Human plasmin was found to be only weakly inhibited by inter-α-trypsin inhibitor and its cleavage products, supporting the opinion that inter-α-trypsin inhibitor and its derivates probably have no physiological role as antiplasmins.[62]

C1 inactivator was identified originally as an inhibitor of the esterase activity of the first component of the complement cascade. This is probably its principal physiological role. It has, however, inhibitory effects on the enzymes of the intrinsic pathway of the coagulatory system including plasma kallikrein and similar effect on plasmin.[63] C1 inactivator is a single-chain glycoprotein with a molecular weight of around 100,000 and a carbohydrate content of 35%.[64] The molecule contains three disulfide bridges, two of which can be reduced without loss of activity while the third seems to be essential in maintaining the conformation around the active site of the inhibitor. C1 inhibitor is a substrate of plasmin. Plasmin decreases the inhibitor activity of C1 inactivator in a time-dependent reaction, leading to degradation products, one of which retains the capacity to form a complex with plasmin.[52] The plasmin-inhibitor complex does not dissociate in the presence of fibrin. Plasmin generated in the presence of both fibrin and C1 inactivator, on the other hand, is not neutralized by the inhibitor because of the greater affinity of plasmin for fibrin.[65] There is accumulating evidence that C1 inactivator makes only a very minor contribution to the inhibition of plasmin formed in the circulation when α-2-antiplasmin and -macroglobulin are present in the plasma at usual concentrations.

Antithrombin III is a single-chain glycoprotein with a molecular weight of some 60,000.[66] It is known to form complexes with all serine proteases. As indicated by its name, it is the major physiological inhibitor of thrombin. It forms a 1:1 stoichiometric complex with plasmin in a purified system. Heparin markedly accelerates the rate at which antithrombin III reacts with proteinases including plasmin.[67] The lack of antithrombin III from plasma does not alter the rate or extent of inhibition of plasmin supposing that α-2-antiplasmin and -macroglobulin are present in normal quantities.[68]

Plasminogen activator inhibitors (PAIs) are specific inhibitors of plasminogen activators (PAs) which fulfill their regulatory function through complex formation of PA and PAI. Recent investigations have shown the presence of three immunologically different PAIs. These are: the endothelial cell type PAI (PAI-1), the placental or macrophage type (PAI-2), and the protease nexin-1. PAI-1 is produced by a variety of cells in culture, including human and bovine endothelial cells, hepatocytes, and certain tumor cell lines. It seems to be primarily involved in pathophysiological thrombotic and thrombolytic regulation. PAI-1 is present in plasma and in platelets, too. High plasma levels were detected in deep venous thrombosis and myocardial infarction. PAI-1 is probably secreted in an inactive form, but PAI in plasma is present mainly in active form. The inactive form is immunologically indistinguishable from the active one.

REFERENCES

1. **Ogston, D.,** *Antifibrinolytic Drugs. Chemistry, Pharmacology and Clinical Usage,* John Wiley & Sons, Chichester, 1984.
2. **Sottrup-Jensen, L., Claeys, H., Zajdel, M., Petersen, T. E., and Magnusson, S.,** The primary structure of human plasminogen: isolation of two lysine-binding fragments and one mini-plasminogen (MW 38,000) by elastse-catalysed-specific limited proteolysis, in *Progress in Chemical Fibrinolysis and Thrombolysis,* Vol. 3, Davidson, J. F., Rowan, R. M., Samama, M. M., and Desnoyers, P. C., Eds., Raven Press, New York, 1978, 191.
3. **Wiman, B.,** Primary structure of peptides released during activation of human plasminogen by urokinase, *Eur. J. Biochem.,* 39, 1, 1973.
4. **Wiman, B. and Wallén, P.,** On the primary structure of human plasminogen and plasmin. Purification and characterization of cyanogenbromide fragments, *Eur. J. Biochem.,* 57, 387, 1975.
5. **Markus, G., DePasquale, J. L., and Wissler, F. C.,** Quantitaitve determination of the binding of epsilon-aminocaproic acid to native plasminogen, *J. Biol. Chem.,* 253, 727, 1978.
6. **Markus, G., Priore, R. L., and Wissler, F. C.,** The binding of tranexamic acid to native (glu) and modified (lys) human plasminogen and its effect on conformation, *J. Biol. Chem.,* 254, 1211, 1979.
7. **Wiman, B. and Wallén, P.,** Structural relationship between glutamic acid and lysine forms of human plasminogen and their interaction with the NH_2-terminal activation peptide as studied by affinity chromatography, *Eur. J. Biochem.,* 50, 489, 1975.
8. **Thorsen, S.,** Difference in the binding to fibrin of native plasminogen modified by proteolytic degradation. Influence of ω-aminocarboxylic acids, *Biochim. Biophys. Acta,* 393, 55, 1975.
9. **Wiman, B. and Wallén, P.,** The specific interaction between plasminogen and fibrin. A physiological role of the lysine binding site in plasminogen, *Thromb. Res.,* 10, 213, 1977.
10. **Suenson, E. and Thorsen, S.,** Secondary-site binding of Glu-plasmin, Lys-plasmin and miniplasmin to fibrin, *Biochem. J.,* 197, 619, 1981.

11. **Lucas, M. A., Fretto, L. J., and McKee, P. A.,** The binding of human plasminogen to fibrin and fibrinogen, *J. Biol. Chem.,* 258, 4249, 1983.
12. **Violand, B. N. and Castellino, F. J.,** Mechanism of the urokinase-catalysed activation of human plasminogen, *J. Biol. Chem.,* 251, 3906, 1976.
13. **Gaffney, P. J., Joe, F., and Mahmoud, M.,** Giant fibrin fragments derived from crosslinked fibrin: structure and clinical implication, *Thromb. Res.,* 20, 647, 1980.
14. **Donaldson, V. H.,** Effect of plasmin *in vitro* on the clotting factors in plasma, *J. Lab. Clin. Med.,* 56, 644, 1960.
15. **Pasquini, R. and Hershgold, E. J.,** Effects of plasmin on human factor VIII (AHF), *Blood,* 41, 105, 1973.
16. **Donaldson, V. H. and Kleiniewski, J.,** The role of plasmin in kinin-release by preparations of human thrombin, *Thromb. Res.,* 16, 401, 1979.
17. **Bokisch, V. A., Müller-Eberhard, J., and Cochrane, C. G.,** Isolation of a fragment (C3a) of the third component of human complement containing anaphylatoxin and chemotactic activity and description of an anaphylatoxin inactivator of human serum, *J. Exp. Med.,* 129, 1109, 1969.
18. **Liotta, L. A., Goldfarb, R. H., and Terranova, V. P.,** Cleavage of laminin by thrombin and plasmin: alpha thrombin selectively cleaves the beta chain of laminin, *Thromb. Res.,* 21, 663, 1981.
19. **Kurkinen, M., Vartio, T., and Vaheri, A.,** Polypeptides of human plasma fibronectin are similar but not identical, *Biochim. Biophys. Acta,* 624, 490, 1980.
20. **Rijken, D. C., Wijngaards, G., and Welbergen, J.,** Immunological characterization of plasminogen activator activities in human tissues and body fluids, *J. Lab. Clin. Med.,* 97, 477, 1981.
21a. **Thorsen, S., Glas-Greenwalt, P., and Astrup, T.,** Differences in the binding to fibrin of urokinase and tissue plasminogen activator, *Thromb. Diath. Haemorrh.,* 28, 65, 1972.
21. **Thorsen, S.,** The inhibition of tissue plasminogen activator and urokinase induced fibrinolysis by some natural proteinase inhibitors and by plasma and by serum from normal and from pregnant subjects, *Scand. J. Clin. Lab. Invest.,* 31, 51, 1973.
22. **Schreiber, A.D. and Austen, K. F.,** Hageman fator-independent fibrinolytic pathway, *Clin. Exp. Immunol.,* 17, 587, 1974.
23. **Taylor, F. B., Nilsson, U. R., Creech, R. H., Carroll, E. T., and Beisswenger, J. G.,** Coagulysis: mechanism of formation and lysis of dilute whole blood clots and application of this assay into study of certain hypercoagulable states, *Ser. Haematol.,* 4, 529, 1973.
24. **Albrechtsen, O. K.,** The fibrinolytic activity of human tissues, *Br. J. Haematol.,* 3, 284, 1957.
25. **Rijken, D. C., Wijngaards, G., and Welbergen, J.,** Relationship between tissue plasminogen activator and activators in blood and vascular wall, *Thromb. Res.,* 18, 815, 1980.
26. **Binder, B. R., Reissert, G., and Beckmann, R.,** Isolation and characterization of plasminogen activator (PA) from human myocardial tissue, *Thromb. Haemostas.,* 46, 11, 1981.
27. **Rijken, D. C. and Collen, D.,** Purification and characterization of the plasminogen activator secreted by human melanoma cells in culture, *J. Biol. Chem.,* 256, 7035, 1981.
28. **Rijken, D. C., Hoylaerts, M., and Collen, D.,** On the fibrinolytic properties of single-chain and two-chain human tissue plasminogen activator, *Thromb. Haemostas.,* 46, 12, 1981.
29. **Sobel, G. W., Mohler, S. R., Jones, N. W., Dowdy, A. B. C., and Guest, M. M.,** Urokinase: an activator of plasma profibrinolysin extracted from urine, *Am. J. Physiol.,* 171, 768, 1952.
30. **Bernik, M. B. and Kwaan, H. C.,** Origin of fibrinolytic activity in cultures of the human kidney, *J. Lab. Clin. Med.,* 70, 650, 1967.
31. **White, W. F., Barlow, G. H., and Mozen, M. M.,** The isolation and characterization of plasminogen activators (urokinase) from human urine, *Biochemistry,* 5, 2160, 1966.
32. **Lesuk, A., Terminiello, L., Traver, J. H., and Groff, J. L.,** Biochemical and biophysical studies of human urokinase, *Thromb. Diath. Haemorrh.,* 18, 293, 1967.

33. **Sumi, H. and Robbins, K. C.,** A functionally active heavy chain derived from human high molecular weight urokinase, *J. Biol. Chem.,* 258, 8014, 1983.

34. **Nieuwenhuizen, W., Wijngaards, G., and Groeneveld, E.,** Synthetic substrates and the discrimination between urokinase and tissue plasminogen activator activity, *Thromb. Res.,* 11, 87, 1977.

35. **Vetterlein, D., Bell, T. E., Young, P.L., and Roblin, R.,** Immunological quantitation and immunoabsorption of urokinase-like plasminogen activators secreted by human cells, *J. Biol. Chem.,* 255, 3665, 1980.

36. **Nolan, C., Hall, L. S., Barlow, G. H., and Tribby, I. I. E.,** Plasminogen activator from human embryonic kidney cell cultures, *Biochim. Biophys. Acta,* 496, 384, 1977.

37. **Wijngaards, G., Kluft, C., and Groeneveld, E.,** Demonstration of urokinase-related fibrinolytic activators from human urine, *Haematology,* 51, 165, 1982.

38. **Kluft, C., Wijngaards, G., and Jie, A. F. H.,** The factor XII-independent plasminogen proactivator system of plasma includes urokinase-related activity, *Thromb. Haemostas.,* 46, 343, 1981.

39. **Ogston, D., Ogston, C. M., Ratnoff, O. D., and Forbes, C. D.,** Studies on a complex mechanism for the activation of plasminogen by kaolin and by chloroform: the participation of Hageman factor and additional cofactors, *J. Clin. Invest.,* 48, 1786, 1969.

40. **Wiman, B. and Collen, D.,** Purification and characterization of human antiplasmin, the fast acting plasma inhibitor in plasma, *Eur. J. Biochem.,* 78, 19, 1977.

41. **Wiman, B. and Collen, D.,** On the mechanism of the reaction between human antiplasmin and plasmin, *J. Biol. Chem.,* 254, 9291, 1979.

42. **Wiman, B. and Collen, D.,** On the kinetics of the reaction between human antiplasmin and plasmin, *Eur. J. Biochem.,* 84, 573, 1978.

43. **Christensen, U. and Clemmensen, I.,** Kinetic properties of the primary inhibitor of plasmin from human plasma, *Biochem. J.,* 163, 389, 1977.

44. **Kluft, C., Los, P., and Jie, A. F. H.,** Assay and occurrence of two molecular forms of alpha-2-antiplasmin in plasma, *Thromb. Haemostas.,* 46, 281, 1981.

45. **Aoki, N., Moroi, M., and Tachiya, K.,** Effects of alpha-2-plasmin inhibitor on fibrin clot lysis. Its comparison with alpha-2-macroglobulin, *Thromb. Haemostas.,* 39, 22, 1978.

46. **Korninger, C. and Collen, D.,** Inhibition of human tissue plasminogen activator by human plasma: no evidence for a specific antiactivator, *Thromb. Haemostas.,* 46, 280, 1981.

47. **Moroi, M. and Aoki, N.,** Inhibition of proteases in coagulation, kinin-forming and complement systems by alpha-2-plasmin inhibitor, *J. Biochem.,* 82, 969, 1977.

48. **Aoki, N., Saito, N., Kamiya, T., Koic, K., Sakata, Y., and Kobakura, M.,** Congenital deficiency of alpha-2-plasmin inhibitor associated with severe haemorrhagic tendency, *J. Clin. Invest.,* 63, 877, 1979.

49. **Müllertz, S. and Clemmensen, I.,** The primary inhibitor of plasmin in human plasma, *Biochem. J.,* 159, 545, 1976.

50. **Aoki, N., Moroi, M., Matsuda, M., and Tachiya, K.,** The behaviour of alpha-2-plasmin inhibitor in fibrinolytic states, *J. Clin. Invest.,* 60, 361, 1977.

51. **Jones, J. M., Creeth, J. M., and Kekwick, R. A.,** Thiol reduction of human alpha-2-macroglobulin, The subunit structure, *Biochem. J.,* 127, 187, 1972.

52. **Harpel, P. C. and Cooper, N. R.,** Studies on human plasma C1 inactivator-enzyme interactions. I. Mechanisms of interaction with C1, plasmin and trypsin, *J. Clin. Invest.,* 55, 593, 1975.

53. **Stenbjerg, S.,** Inherited alpha-2-macroglobulin deficiency, *Thromb. Res.,* 22, 491, 1981.

54. **Mega, T., Lujan, E., and Yoshida, A.,** Studies on the oligosaccharide chains of human alpha-1-protease inhibitor, *J. Biol. Chem.,* 255, 4053, 1980.

55. **Boswell, D. R., Jeppsson, J.-O., Brennan, S. O., and Carrell, R. W.,** The reactive site of alpha-1-antitrypsin is C-terminal, not N-terminal, *Biochim. Biophys. Acta,* 744, 212, 1983.

56. **Crawford, G. P. M. and Ogston, D.,** The influence of alpha-1-antitrypsin on plasmin, urokinase and Hageman cofactor, *Biochim. Biophys. Acta,* 354, 107, 1974.

57. **Carrell, R. W., Jeppsson, J.-O., Laurell, C.-B., Brennan, S. O., Owen, M. C., Vaughan, L., and Boswell, D. R.,** Structure and variation of human alpha-1-antitrypsin, *Nature,* 298, 329, 1982.

58. **Heide, K., Heimburger, N., and Haupt, H.,** An inter-alpha trypsin inhibitor of human serum, *Clin. Chim. Acta,* 11, 82, 1965.

59. **Hochstrasser, K., Feuth, H., and Steiner, O.,** Zur Characterisierung der sáuerstabilen Proteaseninhibitoren aus Humanplasma, *Hoppe-Seyler's Z. Physiol. Chem.,* 354, 927, 1973.

60. **Hochstrasser, K., Bretzel, G., Feuth, H., Hilla, W., and Lempart, K.,** The inter-alpha-trypsin inhibitor as precursor of the acid-stable proteinase inhibitors in human serum and urine, *Hoppe-Seyler's Z. Physiol. Chem.,* 357, 153, 1976.

61. **Dietl, T., Dobrinski, W., and Hochstrasser, K.,** Human inter-alpha-trypsin inhibitor. Limited proteolysis by trypsin, plasmin, kallikrein and granulocyte elastase and inhibitory properties of cleavage products, *Hoppe-Seyler's Z. Physiol. Chem.,* 360, 1313, 1979.

62. **Lambin, P., Fine, J. M., and Steinbuch, M.,** Inhibition of plasmin by a small molecular weight inhibitor derived from human inter-alpha-trypsin inhibitor, *Thromb. Res.,* 13, 563, 1978.

63. **Schreiber, A. D., Kaplan, A. P., and Austen, K. F.,** Inhibition by C1 INA of Hageman factor fragment activation of coagulation, fibrinolysis and kinin generation, *J. Clin. Invest.,* 52, 1402, 1973.

64. **Haupt, H., Heimburger, N., Kranz, T., and Schwick, H. G.,** Ein Beitrag zur Isolierung und Characterisierung des C1-Inaktivators aus Humanplasma, *Eur. J. Biochem.,* 17, 254, 1970.

65. **Trumpi-Kalshoven, M. M.,** The relevance of C1 inhibitor and the inhibition of the fibrinolytic activity of plasmin, in *Progress in Chemical Fibrinolysis and Thrombolysis,* Vol. 3, Davidson, J. F., Rowan, R. M., Samama, M. M., and Desnoyers, P. C., Eds., Raven Press, New York, 1978, 257.

66. **Koide, T.,** Isolation and characterization of antithrombin III from human, porcine and rabbit plasma, and rat serum, *J. Biochem.,* 86, 1841, 1979.

67. **Highsmith, R. F. and Rosenberg, R. D.,** The inhibition of human plasmin by human antithrombin-heparin cofactor, *J. Biol. Chem.,* 249, 4335, 1974.

68. **Collen, D.,** Identification and some properties of new fast reacting plasmin inhibitor in human plasma, *Eur. J. Biochem.,* 69, 209, 1976.

Chapter 8

THE PLASMINOGEN ACTIVATOR-PLASMIN SYSTEM OF THE TEARS

Part B
Plasminogen Activators, Plasmin, and the Cornea

The first observations on the fibrinolytic activity of human corneas were made by Paldolfi and colleagues, who used the histochemical fibrin slide technique. Normal corneas of man and several mammals were found to be fibrinolytically inactive, while following corneal injury fibrinolytic activity of corneal sections was detected in relation to the epithelial and endothelial linings, in and around detached or damaged epithelial and endothelial cells. Severely injured corneal cells showed higher fibrinolytic activity.[1,2]

In further experiments Pandolfi et al.[3] showed that fetal corneal explants released fibrinolytic agents when fibrin clots were present in the culture medium. These agents were considered to be plasminogen activators (PAs), since tranexamic acid, an inhibitor of plasminogen activation into plasmin, completely blocked the dissolution of the clot in the culture. It was suggested that the function of these activators is presumably to keep the cornea free from fibrin deposits that may occur after injury or inflammation.

Lantz and co-workers[4] demonstrated that rabbit corneas released PA continuously into the culture medium. The addition of prednisone did not modify the fibrinolytic activity, while the death of explants was followed by cessation of fibrinolysis. It was suggested that the main source of PA was the epithelium of the cornea and that continuous synthesis of this activator was going on in the corneal epithelial cells, at least in tissue cultures containing fibrin clots.

The next important step was made by Berman, who demonstrated that PA was released in relatively large quantities from ulcerating corneas partly from the destroyed and affected epithelial cells and also from the ulcerating stroma. He suggested that plasmin-dependent processes could be initiated in the corneal stroma subsequent to the activation of plasminogen by PA. These processes could include the generation of factor C3a from factor C3 that is chemotactic for polymorphonuclear leukocytes (PMNs), the activation of latent collagenase synthetized by and released from the fibroblasts of the ulcerating cornea, and the generation of vasoactive kinins known to cause permeability changes in perilimbal vessels and inflammatory leak into the corneal stroma.[5]

PA released by *in vitro* cultures of the cornea was isolated from the culture media by Pandolfi and Lanz.[6] It was shown to produce the same pattern of limited proteolysis on plasminogen as the urinary activator, urokinase (u-PA) did. The molecular mass of the corneal PA was found to be around 55,000, the same range as the molecular mass of u-PA. Like u-PA but unlike t-PA corneal

PA showed a biphasic activity response to increasing concentrations of epsilon aminocaproic acid (EACA) and was scarcely bound to fibrin during clotting. On the basis of these findings they suggested that the term keratokinase should be used, with the intention to express that it was an activator very much like urokinase that had been released from the cornea.

PA activity of corneal endothelial cells was demonstrated by Fehrenbacher et al.[7] in tissue culture by a solid-phase assay based on the hydrolysis of ^{125}I-labeled fibrin. Only a small fraction of PA was found to be associated with the membrane; 90% was shown to be located in the cytoplasm.

Further evidence proving that the PA-plasmin system plays an important role in the inflammatory processes of the cornea, especially corneal ulceration, was provided by Berman and colleagues, who found that plasmin activated latent collagenase from organ cultures of ulcerating rabbit corneas and from fibroblast cultures derived from such corneas. The activation process of latent collagenase caused by plasmin was shown to be very similar to that caused by trypsin, resulting in the conversion of the 40,000 molecular mass latent collagenase to the active form with a molecular mass of 23,000 Da.[8,9]

The sites of PA release from ulcerating rabbit corneas could be determined by the help of sections of such corneas put on fibrin films. These studies demonstrated that the lysis begins in the superficial stroma near the perifery of the cornea. Multiple freeze-thawed ulcerating corneas showed initial lysis at the ulcer region containing PMNs. That the lysis started periferally in corneas with centrally positioned ulcers suggested that PA was normally contained in the cells and were not made only in response to tissue injury. There was no relation between the location of the blood vessels and the plasminogen-dependent lysis. PA from the ulcerating cornea was shown by sodium dodecyl sulfate-gel electrophoresis to cleave plasminogen into heavy- and light-chain fragments similar to those produced from plasminogen by u-PA. The addition of plasminogen daily to cultures of ulcerating corneas resulted in earlier rises of PA, collagenase, and collagen degradation fragments in the culture media. On the basis of these findings it has been suggested that the release of PA from the corneal epithelial cells and fibroblasts due to injury is followed by the activation of the preexisting plasminogen into plasmin that in turn, if not stopped by inhibitors, will start a number of processes leading to the ulceration of the cornea. These processes include the release of inactive collagenase from corneal fibroblasts, the activation of latent collagenase, the activation of the complement and the kinin systems resulting in vascular permeability changes and the accumulation of PMNs. PMNSs on the other hand could provide further enzymes that can contribute to the enzymatic degradation of collagen fibrils and extracellular matrix in ulcerating corneas.[10,11]

The peripheral zone of normal rabbit cornea was shown to contain, presumably intracellular, PAs. It was suggested that the activator, once released, might regulate the permeability of limbal vessels and angiogenesis, by plasmin-dependent pathways.[11] Following the injection of u-PA into the stroma, sprouts arose from the circumlimbal vessels, beginning on the third day and grew into the cornea over the next several days. PMNs were observed in association with growing vessels. Contralateral corneas injected with u-PA previously by active site inhibitor, Phe-Ala-Arg-chloromethyl ketone, showed no vascularization. The observations were correlated with the possible release of intracellular PAs that in turn elicit neovascularization and vascular permeability changes in the perilimbel zone. These changes are supposed to result in the eventual arrest of corneal ulceration.[12]

In further experiments Lantz and Andersson studied the release of PAs from *in vitro* cultures of human and animal corneas. The corneal epithelium was found to be the main source. The addition of a synthetic tripeptide, salycilate, chloroquine, and phorbolester to the culture medium increased the fibrinolytic activity of the cultures. Corticosteroids on the other hand decreased the fibrinolytic activity. Explants of human conjunctiva were also shown to release fibrinolytic activators.[13]

Although a correlation between ulceration of the corneal stroma after alkali burns and the release of PAs has been suggested earlier,[11] the first direct evidence to prove that the plasminogen-plasmin system plays a role in the healing of the epithelial and stromal defects was provided by Berman et al. in 1983. They demonstrated that fibrin and fibronectin appear on the corneal surface after alkali burn and disappear in correlation with the appearance of PA followed by resurfacing of the stromal or the epithelial defect. That plasmin generated by PAs, besides fibrin cleavage, can degrade fibronectin and laminin components of the subepithelial basal membrane suggested that long-standing release of PAs in persistent epithelial defects could initiate a row of plasmin-dependent processes leading to corneal ulceration.[14]

Wang et al.[15] using a plasminogen-dependent fluorescent assay demonstrated that PA was present mostly in a latent (trypsin- or plasmin-activable) form (proactivator) in cultures of rabbit corneal epithelial cells or normal corneas. Cultures of ulcerating corneas demonstrated only active PA in the early phases, whereas latent PA levels increased later in culture. The authors suggested that ulceration was correlated with the conversion of latent PA to active PA. The profiles of proactivator, of latent collagenase, and of active collagenase were shown to be similar *in vitro,* suggesting that the production and activation of PA and collagenase are under coordinate control. Cultures of normal epithelial cells and nonulcerating corneas were shown to contain PA molecular weight species of 72,000 and 46,000 Da, and ulcerating corneas species of 72,000, 46,000, and 35,000 Da. A double-diffusion analysis indicated that rabbit epithelial cells, fibroblasts, and ulcerating corneas produced u-PA-like PA. Human corneal extracts and tears were also found to contain PA, immunoreactive with anti-u-PA antibodies.

Further characteristics of corneal and conjunctival PAs from human and rabbit eyes were determined in tissue culture by Lantz and Pandolfi. The fibrinolytic activity of culture fluid from human corneas was found to be low, roughly one eighth that of rabbits. The main activity was shown to be of u-PA type, as demonstrated by crossed immunoelectrophoresis against α-2-

antiplasmin after incubation with mixed plasma. The fibrinolytic activity of culture fluid from human conjunctival fluid was higher and a mixed secretion of t- and u-PA was demonstrated. The activators were separated by affinity chromatography against Sepharose-immobilized antibodies against t-PA.[16]

Chan[17] studied the PA production and release by cultured corneal epithelial cells during differentiation and epithelial wound closure. Rabbit corneal epithelial cells were cultured and the PA activity was measured in the harvested culture media using a spectrophotometric assay originally described by Coleman and Green.[18] The PA activity measured in the serum-free culture medium was plasminogen dependent and acid stable. When serum was added to the medium samples were acid treated before PA assay in order to inactivate plasma-derived inhibitors. The cells released a low level of PA during the first week of growth and confluency. PA activity gradually increased during the next weeks of differentiation, reaching a peak (18-fold increase) in the third week when the first multilayered foci appeared in the epithelial sheet. This was followed by a decrese in PA activity associated with the globular aggregation of the cells. The effect of wound closure on the PA levels of the culture medium was studied in an experimental model. The wound was produced by the mechanical removal of cells from the center of epithelial cell layer. A complete wound closure was observed within 2 to 4 d when the epithelial cells covered the defect zone. PA activity during the first day of wound closure was four times that of the activity measured in the unwounded culture. Based on the results the author suggested that PA secretion by the cultured corneal epithelium was associated with processes that occur during the differentation of the corneal epithelium and epithelial wound closure.[17]

In further experiments Chan investigated the effect of chemical injury on the release of PA of cultured corneal epithelial cells. Cultures of corneal epithelial cells from albino rabbits were briefly exposed to various concentrations of NaOH or formaldehyde. A differential release of PA was observed during the 6-d period of recovery depending on the concentration of the chemical and duration of exposure. The cells released increasing levels of PA in a concentration-dependent manner up to a critical concentration of the chemical. Higher concentrations resulted in the detachement of cells from the bottom of the culture dish and the inhibition of PA production, though most of the cells still remained viable. These experiments provided evidence supporting the assumption that the production of PA by these cells is an active secretion rather than a passive release from injured or destroyed epithelium.[19]

Geanon and colleagues[20] analyzed the t-PA content of the corneal epithelium, edothelium, and stroma in dog, calf, and monkey eyes. Epithelium and endothelium were scraped of the freshly removed corneas; these tissues and the remaining stroma were homogenized and lysed in a medium containing 1.0 mM $NaHCO_3$, 0.5 mM $CaCl_2$, and Tween-20. The activity of t-PA was measured in the tissue extracts by the (^{125}I)fibrin-coated well assay and by an enzyme-linked immunosorbent assay (ELISA) method. t-PA activities in the epithelium, stroma, and endothelium were aproximately the same with the first method, though slight differences between the three species were observed. The values, expressed in urokinase (UK) units of activity per milligram of protein, ranged from 0.8 ± 0.22 to 1.03 ± 0.23 for the epithelium, 0.47 ± 0.13 to 0.98 ± 0.2 for the stroma, and 0.48 ± 0.11 to 0.93 ± 0.22 for the endothelium. Using the ELISA technique the PA activity measured in the corneal epithelium was 1.43 ± 0.29 ng t-PA per milligram of protein. The authors pointed out that the presence of significant PA activity in the extracts of the cornea and other avascular tissues in the eye supports the hypothesis that t-PA plays a role in nonfibrinolytic processes such as mitosis and cell migration, as well as the destructive events involved in the remodeling of the corneal stroma and the turnover of collagen and glycoproteins of the extracellular matrix.[20]

PA activity was studied by Hayashi and co-workers[21] during epithelial wound healing in vitamin A-deficient rat corneas. Corneas were removed 1, 4, 8, 16, 24, 36, and 48 h after removal of the epithelium. The distribution of PA activity was demonstrated in the cryostat sections of the wounded corneas using fibrin layers according to the method of Todd.[22] Antibodies to t-PA (anti-t-PA) or to u-PA (anti-u-PA) were also incorporated in the overlaid fibrin film to determine the type of the detected activator. Corneas of nondeficient control animals showed t-PA-dependent lysis in association with the regenerating epithelium as well as in the defect region during epithelial wound healing. Corneas from vitamin A-deficient rats also demonstrated t-PA activity in association with corneal epithelium postscrape but showed no detectable t-PA activity in the defect region. Histological examination of the vitamin A-deficient corneas showed the presence of a pseudomembrane in the wounded area composed of PMNs, cell debris, and fibrinous exudate. It was suggested by the authors that the formation, or rather the persistance, of the pseudomembrane, which delayed reepithelization, might have resulted from the absence of PA activity in the defect region. They also pointed out that both excessive and inadequate levels of PA activity may result in impaired epithelial wound healing.[21]

In another study investigating the origin of tear PAs Hayashi and Sueishi performed Todd's fibrinolysis autography[22a] and immunohistological studies using an avidin biotin complex on normal bovine cornea and conjunctiva. PA activities were present in the corneal epithelium in and around the small blood vessels of the conjunctiva. The activity was particularly high in association with the corneal epithelium. PA of the bovine corneal epithelium was neutralized by adding anti-human t-PA immunoglobulin to the fibrin film, but the addition of anti-human u-PA antibody had no inhibitory effect. In the immunohistological studies the bovine corneal epithelial cells stained with anti-human t-PA reagents. Based on their findings the authors suggested that, besides other possible sources, corneal epithelial cells may also contribute to the PA content of human tears. Parallel biochemical analysis performed by the same authors demonstrated the presence of both t-PA and u-PA in normal human tears. As considerable differences in the reactivity of PAs and of their antibodies were observed in different species[11,14,23] similar investigations using human corneas are needed before the possible corneal origin of PAs will be fully understood.

Recent immunohistological studies provided further evidence concerning the presence and activities of PAs in the cornea under normal and pathological conditions. Tripathi et al.[24] reported the distribution of t-PA in human and in monkey eyes. They found a clear staining for t-PA of the corneal epithelium in both species. Tervo and co-workers[25] detected PAs (t-PA and u-PA) and a PA inhibitor (PAI-1) in experimental corneal wound. Rabbit corneas were investigated after anterior keratectomy by immunohistological and histochemical methods. A strong immunoreaction for u-PA, a weaker reaction for PAI-1, and a very weak t-PA-like immunoreaction appeared in the anterior stroma of the wounded area. None of the used antisera showed any immunostaining in the rabbit cornea outside the wound. These results suggest that corneal wounding leads to an expression of u-PA gene in the epithelium especially at the wound margin where epithelial regeneration starts.[25]

Recently Berman reviewed the role of PAs and plasmin in the development and healing of corneal epithelial defects.[26–28] Based on experimental data showing the production of PAs by regenerating corneal epithelial cells, Berman proposed a mechanism for epithelial resurfacing. After mechanical scraping the corneal epithelium is thought to require a temporary matrix composed of fibronectin and fibrin, on which the regenerating epithelium migrates covering the defect area. In the fibronectin/fibrin layer fibronectin molecules must be intact so that the migrating epithelial cells can temporarily attach to the cell-binding domains of the fibronectin molecules that are embedded in the surrounding fibrin matrix. Berman and co-workers demonstrated that the migrating epithelium secretes PAs (PA activity was demonstrated by fibrin layer technique around the leading edge of the regenerating epithelium and along the surrounding seemingly intact epithelium in the histological sections of alkali-burned ulcerated rabbit corneas). Berman suggested that PA released from the epithelial cells at the focal contact where fibronectin is attached to the cell membrane, through the action of generated plasmin, may cleave the fibronectin molecule in a limited and regulated manner. The fibronectin receptor of the cell membrane thus freed may become able to attach to the adjacent fibronectin molecule and the whole process may start again. This cyclic process together with intracellular contractile mechanisms may lead to the migration of the epithelial cell.[28]

In further experiments Berman demonstrated that PA activity at the leading edge of the regenerating epithelium can be blocked by amyloride, a selective inhibitor of u-PA, while the inhibitor did not affect the activity of the surrounding epithelium. Anti-t-PA antibodies in contrast prevented lysis along the epithelium except at the leading edge. Based on these observations Berman ephasized the role of u-PA, plasmin generated by u-PA, in the limited proteolysis at the focal contacts of fibronectin and fibronectin receptors of the epithelial cell membrane.[27] It has been pointed out that antibodies to u- or t-PA and inhibitors of u- or t-PA may not be effective on single-chain forms of these activators, and that such molecules may not have an access to membrane-bound or cell surface activators. Such differences may account for the variability of the effects of antibodies and inhibitors on epithelial PA activities found under different experimental conditions. Prolonged and unregulated secretion of PA, basically u-PA, may lead to uncontrolled and unlimited proteolysis that through an excessive plasmin action, may lead to the digestion of the temporary matrix and also of the subepithelial basal membrane and eventually can result in stromal ulceration.[26,27]

REFERENCES

1. **Pandolfi, M. and Astrup, T.,** A histochemical study of the fibrinolytic activity. Cornea, conjunctiva, and lacrimal gland, *Arch. Ophthalmol.,* 77, 258, 1967.
2. **Pandolfi, M. and Kwaan, H. C.,** Fibrynolysis in the anterior segment of the eye, *Arch. Ophthalmol.,* 77, 99, 1967.
3. **Pandolfi, M., Astedt, B., and Dyster-Aas, K.,** Release of fibrinolytic enzymes from human cornea, *Acta Ophthalmol.,* 50, 199, 1972.
4. **Lanz, E., Dyster-Aas, K., and Pandolfi, M.,** *In vitro* release of fibrinolytic activators from cornea, *Graefe's Arch. Clin. Exp. Ophthalmol.,* 206, 157, 1978.
5. **Berman, M.,** Regulation of collagenase. Therapeutic considerations, *Trans. Opthamol. Soc. U.K.,* 98, 397, 1978.
6. **Pandolfi, M. and Lantz, E.,** Partial purification and charactarization of keratokinase, the fibrinolytic activator of the cornea, *Exp. Eye Res.,* 29, 563, 1979.
7. **Fehrenbacher, L., Gospodarowicz, D., and Shuman, M. A.,** Synthesis of plasminogen activator by bovine corneal endothelial cells, *Exp. Eye Res.,* 29, 219, 1979.
8. **Berman, M., Leary, R., and Gage, J.,** Latent collagenase in the ulcerating rabbit cornea, *Exp. Eye Res.,* 25, 435, 1977.
9. **Berman, M., Leary, R., and Gage, J.,** Collagenase from corneal cell cultures and its regulation by phagocytosis, *Invest. Ophthalmol. Vis. Sci.,* 18, 588, 1979.
10. **Berman, M. B.,** Collagenase and corneal ulceration, in *Collagenase in Normal and Pathological Connective Tissues,* Woolley, D. E. and Evanson, J. M., Eds., John Wiley & Sons, New York, 1980, 141.
11. **Berman, M., Leary, R., and Gage, J.,** Evidence for a role of the plasminogen activator-plasmin system in corneal ulceration, *Invest. Ophthalmol. Vis. Sci.,* 19, 1204, 1980.
12. **Berman, M., Winthorp, S., Ausprunk, D., Rose, J., Langer, R., and Gage, J.,** Plasminogen activator (urokinase) causes vascularization of the cornea, *Invest. Ophthalmol. Vis. Sci.,* 22, 191, 1982.
13. **Lantz, E. and Andersson, A.,** Release of fibrinolytic activators from the cornea and conjunctiva, *Graefe's Arch. Clin. Exp. Ophthalmol.,* 219, 263, 1982.
14. **Berman, M., Manseau, E., Law, M., and Aiken, D.,** Ulceration is correlated with degradation of fibrin and fibronectin at the corneal surface, *Invest. Ophthalmol. Vis. Sci.,* 24, 1358, 1983.
15. **Wang, H.-M., Berman, M., and Law, M.,** Latent and active plasminogen activator in corneal ulceration, *Invest. Ophthalmol. Vis. Sci.,* 26, 511, 1985.
16. **Lantz, E. and Pandolfi, M.,** Fibrinolysis in cornea and conjunctiva: evidence of two types of activators, *Graefe's Arch. Clin. Exp. Ophthalmol.,* 224, 393, 1986.
17. **Chan, K. Y.,** Release of plasminogen activator by cultured corneal epithelial cells during differentiation and wound closure, *Exp. Eye Res.,* 42, 417, 1986.

18. **Coleman, P. L. and Green, G. D. J.,** A sensitive, coupled assay for plasminogen activator using a thiol ester substrate for plasmin, *Ann. N.Y. Acad. Sci.,* 370, 617, 1981.

19. **Chan, K. Y.,** Chemical injury to an *in vitro* ocular system: differential release of plasminogen activator, *Curr. Eye Res.,* 5, 357, 1986.

20. **Geanon, J. D., Tripathi, B. J., Tripathi, R. C., and Barlow, G. H.,** Tissue plasminogen activator in avascular tissues of the eye: a quantitative study of its activity in the cornea, lens, and aqueous and vitreous humors of dog, calf, and monkey, *Exp. Eye Res.,* 44, 55, 1987.

21. **Hayashi, K., Frangieh, G., Kenyon, K. R., Berman, M., and Wolf, G.,** Plasminogen activator activity in vitamin A-deficient rat corneas, *Invest. Ophthalmol. Vis. Sci.,* 29, 1810, 1988.

22. **Todd, A. S.,** The histological localisation of fibrinolysin activator, *J. Pathol. Bacteriol.,* 78, 281, 1959.

22a. **Hayashi, K. and Sueishi, K.,** Fibrinolytic activity and species of plasminogen activator in human tears, *Exp. Eye Res.,* 46, 131, 1988.

23. **Tözsér, J. and Berta, A.,** Urokinase-type plasminogen activator in rabbit tears. Comparison with human tears, *Exp. Eye. Res.,* 51, 33, 1990.

24. **Tripathi, B. J., Geanon, J. D., and Tripathi, R. C.,** Distribution of tissue plasminogen activator in human and monkey eyes, *Ophthalmology,* 94, 1434, 1987.

25. **Tervo, T., Tervo, K., van Setten, G. B., Virtanen, I., and Tarkkanen, A.,** Plasminogen activator and its inhibitor in the experimental corneal wound, *Exp. Eye Res.,* 48, 445, 1989.

26. **Berman, M.,** The pathogenesis of corneal epithelial defects, *Acta Ophthalmol.,* 192(suppl.), 55, 1989.

27. **Berman, M.,** The pathogenesis of corneal epithelial defects, in *Healing Processes in the Cornea,* Beurman, R. W, Crossen, C. E., and Kaufman H. E., Eds., Portfolio, Woodlands, TX, 1989, 14.

28. **Berman, M., Kenyon, K., Hayashi, K., and L'Hernault N.,** The pathogenesis of epithelial defects and stromal ulceration, in *The Cornea: Transactions of the World Congress on the Cornea III,* Cavanagh, H. D., Ed., Raven Press, New York., 1988, 35.

Chapter 8

THE PLASMINOGEN ACTIVATOR-PLASMIN SYSTEM OF THE TEARS

Part C
Plasminogen Activators, Plasmin, and the Tears

The very first publication on the fibrinolytic activity of human tears contains a surprisingly large amount of information concerning the components and function of the fibrinolytic system of the tear fluid.[1] Storm in his very simple but perfectly designed experiments studied the fibrinolytic activity of normal (emotional) tears by the original fibrin plate method[2] and the heated fibrin plate method,[3] with and without added streptokinase and bovine plasminogen. In the original form of the fibrin plate method the substrate (bovine fibrin) contained plasminogen, while in the heated plate method the plasminogen in the substrate was destroyed by heating the plates at 85°C for 45 min. Fresh untreated human tears produced fibrinolysis on the untreated fibrin substrate. When tears were placed on heated fibrin plates there was no fibrinolysis at all. If tears and equal amounts of bovine plasminogen solution (0.2 mg/ml) were put together on the heated plates, a digestion of the substrate was obtained. This led to the conclusion that normal tears contain plasminogen activators (PAs) but no active fibrinolytic enzyme (plasmin). Based on the observation of Müllertz and Lassen[4] that streptokinase transforms inactive proactivator in blood and milk to an active PA Storm tested the effect of a 0.002 mg/ml streptokinase solution on the fibrinolytic activity of normal tears on unheated fibrin plates. Following the addition of equal amounts of streptokinase solution to the tear samples, a pronounced increase in the fibrinolytic activity was obtained. Streptokinase alone on the control plates in corresponding concentration was only slightly active. According to Storm these results suggest that the normal tear fluid contains a small amount of active PA and large amount of inactive proactivator, which can be transformed by streptokinase to active PA. In a series of experiments attempting to study the stability of active activator in tears at different pH, temperature values, and under the effect of proteolytic enzymes, Storm detected an inhibitory substance that inhibited the enzymatic digestion by trypsin of heated fibrin. The inhibitor appeared to be destroyed at pH 1 to 2 by heating at 50°C for half an hour. This observation seems to be the first data concerning the trypsin inhibitor activity of human tears. In spite of the fact that in his experiments tear PA seemed to be more labile than the compound found in urine, and appeared more similar to compounds detected in blood, he came to the conclusion that the fibrinolytic systems in blood, milk, urine, and tears are very similar if not identical. He suggested that the physiological role of the fibrinolytic system of the tear fluid may be the "rapid resolution of clotted fibrin in the tear canals (excretory ducts and lacrimal ducts)."[1]

Rijken et al.[5] used a modification of the original fibrin plate method, and specific anti-urokinase and anti-tissue-type PA (anti-t-PA) antibodies to characterize by immunological methods the PAs in normal human tears and in other body fluids. Human emotional tears were tested. Tear samples were stored at –20°C, and centrifuged at 40,000 × g for 15 min before use. Fibrinolytic activities were assayed using plasminogen-containing fibrin plates.[6] Drops of 5-, 10-, and 35-µl tears were applied to the plates, after which the plates were incubated for 17 h at 37°C. Diameters of the lysed zones were measured; PA activities were expressed in Ploug units and international units (IVs).[7] Antibodies were incorporated in the fibrin plates by mixing the IgG fraction of the relevant antiserum with the fibrinogen solution before clotting. The extent of quenching was assessed by comparison of the activities measured on antibody-containing fibrin plates with the activities obtained on the original fibrin plates. The fibrinolytic activities of normal (emotional) tear samples and saliva were found to be low as compared to those of urine and seminal plasma. The activities were unaffected by anti-urokinase (anti-u-PA) antibodies, but they were completely quenched by antibodies against uterine t-PA. Because of the low activities longer incubation times (up to 64 h) were also tried. This resulted in the enlargement of lysis zones on antibody-free fibrin plates and on anti-u-PA containing plates, whereas no lysis was observed on anti-t-PA containing fibrin plates. All (normal) tear samples were found to be without plasminogen-independent fibrinolytic activity as tested on plasminogen-free bovine fibrin plates. The authors concluded that the PA activity of normal human tears is low compared to activities measured in other body fluids, and can be attributed solely to the presence of t-PA and not to the presence of u-PA.[5]

The results of Rijken et al.[5] were confirmed by Thörig and colleagues, who characterized the PAs in unstimulated and stimulated normal human tears. Tear samples of 5 to 20 µl were collected with glass capillaries from the lower conjunctival fornix without stimulation and following stimulation with air flow at the site of the cornea, tear gas (2-chloracetophenon), or local application of 2% pilocarpine. The samples were centrifuged (10,000 g) and stored at –20°C before assaying for enzyme activities. PA activities were determined in the same system (fibrin plates prepared according to Haverkate and Brakman[6]) and quenching of fibrinolytic activity was performed by anti-u-PA and anti-t-PA antibodies according to Rijken et al.[5] All PA activity was found to be t-PA related both in stimulated and unstimulated normal human tears, the activity values ranging from 0.7 to 5.0 IU/ml. The collected normal tear samples did not exhibit plasminogen-independent fibrinolytic activity as tested on plasminogen-free bovine fibrin plates. Important data concerning the origin of PAs in the tear fluid were provided by the same authors. They compared PA activities with the activities of three other enzymes (lysozyme, β-hexosaminidase, and lactate dehydrogenase) measured in the same tear samples and in tissue extracts of three possible sources: the lacrimal gland, conjunctiva, and cornea. Correlations between enzyme activities indicated that the major source for t-PA in normal tears is the lacrimal gland that secretes t-PA at constant levels at different flow rates. At low flow rates, in unstimulated tears, the variable contribution of the cornea and conjunctiva was suggested to be significant.[8–11]

In 1982 Lanz and Anderson,[12] in a paper dealing with the release of PAs from corneal and conjunctival explants in tissue culture, also published data concerning PA activities measured in normal human tears. Their tear samples were collected with filter paper strips. PA activities were determined on fibrin plates using a u-PA standard in 0.1 M phosphate buffer at pH 7.4. The activity values ranged from 14 to 60 PU/ml (mean = 25 PU/ml). They found lower activities in normal rabbit tears with a mean = 12 PU/ml.[12] In further experiments Lanz and Pandolfi demonstrated that the PA activity measured in the culture fluid of human and rabbit corneas was due to the release of u-PA and conjunctival explants released both u- and t-PA. It has been shown that most of the PA activity of normal human tears (measured on fibrin plates) can be quenched by anti-t-PA antibodies.[13]

Following the detection of plasmin activity in the tear fluid of a patient with therapy-resistant chronic corneal ulcer, Salonen et al.[14] analyzed the tear samples of 48 patients with different corneal lesions. Samples were collected with glass capillaries; 32 of them were found to be positive for proteolytic activity with the radial caseinolytic method of Saksela[15] using low-fat bovine powdered milk as a substrate. Four normal tears were used as controls and they did not show proteolytic activity in the same system. Antibodies to human plasminogen inhibited the proteolytic activity at a dilution of 1:25. The molecular size of the protease found in these tear samples was demonstrated by zymography to be in the range of 80,000 Da. Based on these data the authors concluded that the proteolytic activity found in pathological tears can be attributed to the presence of plasmin in the samples. In order to find out if the plasmin activity was the result of elevated PA activities, they assayed the tears of the patients for PA activity with the method of Saksela[15] and found no PA activity in the tear samples. The tears of eight normal subjects were shown to have a PA activity ranging from 0.6 to 9.8 PU/ml with the same method. The authors tested three different protease inhibitors *in vitro* for their ability to inhibit the proteolytic activity of the plasmin-positive tear samples, and found that aprotinin inhibited the proteolytic activity in all cases while L-cysteine and heparin did not have significant blocking effect; 18 of the patients were treated topically with aprotinin (20 to 40 IU/ml aprotinin eye drops). The treatment was followed by the disappearance of proteolytic activity from the tears of the patients and the healing of the corneal lesions in most of the cases.[14] The first publication was soon followed by reports of authors from the same group demonstrating the presence of plasmin in the tears under various pathological conditions. They demonstrated the presence of plasmin activity in the tear fluid following ocular allergen exposure.[16] Plasmin activity was also detected in the tears of patients with different corneal diseases,[17] in contact lens wearers,[18] and in normal subjects following the performance of Schirmer's test.[19]

We investigated proteases and their activators in normal and pathological human tears using casein-containing agarose gel and casein containing sodium dodecyl sulfate (SDS)-polyacrylamide gel electrophoresis in the presence and in the absence of plasminogen as a copolymerized substrate.[20] The tear samples were collected by capillary tubes without stimulation and with alcoholic stimulation of the nasal mucosa.[21] We detected two lytic zones using casein-agarose electrophoresis, which proved to be resistant, to a large extent, to protease inhibitors (EDTA, SBTI, PMSF, and DTT). By casein-SDS-polyacrylamide gel electrophoresis the appearance of plasminogen-independent proteinases of high molecular weight was also detected in some of the pathological tear samples. With the same method, using copolymerized plasminogen, closely spaced doublet bands of PAs (Mr 55 to 58 and 48 to 53 kDa) were demonstrated, suggesting the presence of urokinase-type PA (u-PA) both in normal and in pathological tears.[20]

In 1988 Hayashi and Sueishi provided further data proving that u- and t-PA coexist in normal human tears and characterized the molecular weights of both types of tear PAs. Samples were collected with Schirmer paper strips, and eluted into 0.01 M PBS containing 0.5% Triton X-100. The lytic bands of the two activators in fibrin enzymography appeared at 64 and 52 kDa, respectively. The first lysis at 64 kDa was shown to be blocked by adding anti-human t-PA IgG, and the second at 52 kDa by anti-human u-PA IgG. For the determination of t- and u-PA activities in tears they used an amidolytic assay, with anti-human u-PA rabbit IgG (46 μg) in the first, and anti-human t-PA rabbit IgG (7 μg) in the second case. Soluble fibrin (2 ml, 7 mg/ml) was always added to the reaction mixture "to accelerate the reaction". The activity of t-PA in normal tears was 5.2 ± 1.3 IU/ml (n = 6) and that of uPA was 2.8 ± 0.5 IU/ml (n = 6). Normal human tears did not show inhibitory activity to PAs or plasmin on reverse fibrinolysis enzymography. The authors demonstrated t-PA activity by Todd's fibrinolysis autography and the presence of t-PA antigen by immunohistochemistry in the corneal epithelium and in the vascular endothelial cells of the conjunctiva in the frozen sections of normal bovine eyes. They suggested that the major PA in tears is probably t-PA. Its main function, through converting plasminogen into plasmin in the presence of fibrin, is "to avoid the formation of even minimal fibrin clots and keep transparency of the outer surface of the cornea." Based on their histological findings gained on bovine corneas Hayashi and Sueishi suggested, that "corneal epithelium, possibly, produces tPA and takes part in the secretion of PA into tears."[22]

In another study we measured PA activities and plasminogen-independent amidolytic activities in human tears using human plasminogen and a chromogenic peptide substrate S-2251. Our assay, performed in the absence of fibrin, is sensitive predominantly to u-PA. The PA activity of normal tears collected with capillary tubes was found to be low, 0.06 ± 0.04 IU/ml. Elevated levels were measured in the tears of patients with various conjunctival and corneal disorders. Tear PA activities showed correlation with the severity of the epithelial diseases and an inverse relationship with tear flow rates. We sug-

TABLE 1

Tear Plasminogen Activator Activities in Different Corneal and Conjunctival Diseases Determined by Casein-Plate Method[a]

Diagnosis	Number of patients	Plasminogen activator activity	
		Elevated	Normal
Corneal ulcers	18	18 (100%)	—
Superficial keratitis	12	10 (83%)	2 (17%)
Persistent epithelial defect	2	2 (100%)	—
Recurrent erosion (active phase)	5	5 (100%)	—
Recurrent erosion (inactive phase)	7	4 (57%)	3 (43%)
Corneal erosion (fresh, traumatic)	13	4 (31%)	9 (69%)
Corneal foreign body	12	4 (30%)	8 (70%)
Chemical burn with corneal involvement	17	16 (94%)	1 (6%)
Marginal ulcer	3	2 (66%)	1 (33%)
Contact lens-associated erosion	11	8 (73%)	3 (27%)
Bullous keratopathy	2	2 (66%)	1 (33%)

[a] Data taken from Berta, A., Tözsér, J., and Holly, F. J., *Acta Ophthalmol.*, 68, 508, 1990.

gested that most of the u-PA in tears originates from the affected epithelial cells of the cornea and the conjunctiva. Plasminogen-independent amidolytic activity (measured with the same substrate in the absence of plasminogen) was found to be very low in the tear samples, except in cases with increased permeability of the conjunctival blood vessels.[23]

In further experiments we investigated the components of the fibrinolytic system of rabbit tears[24] and of human tears[24,25] by electrophoretic, immunoblot, and zymographic methods. The PA activity determined by a microtiter plate method using a chromogenic substrate in stimulated rabbit tears was 4.0 ± 2.5 IU ml^{-1}. This activity was about 100 times higher than that observed in normal human tears with the same method (0.03 ± 0.04 IU ml^{-1}). The plasminogen-independent amidolytic activity was negligible both in normal human and rabbit tears. Antibody against u-PA almost completely inhibited the PA in both species. However, anti-t-PA had no significant effect. Using fibrin-agar zymography the PA activity was found at the 40- to 50-kDa range, the same molecular weight range where the high molecular weight standard u-PA digested the detector gel. Small-intensity lytic bands in the range between 70 and 200 kDa were also detected. With immunoblotting a multiband pattern for u-PA antigen was detected ranging from 60 to 70 kDa to 200 kDa. While the 60- to 70-kDa protein was resistant to reduction, the higher molecular weight bands disappeared after incubation under reducing conditions. The antigen concentration found was more than a hundredfold higher than the concentration expected from the activator measurements, indicating that most of the urokinase in the tear fluid is in inactive forms, such as in inhibitor or receptor complexes.[24,26,27]

TABLE 2
Detection of Plasminogen Activators t-PA and u-PA in Human Tears by Different Authors

Authors	Year	Collection	With/without fibrin	t-PA	u-PA
Storm[1]	1955	?	With	Undefined	Undefined
Rijken et al.[7]	1979	?	With	+	−
Berman et al.[31]	1983	?	Without	−	+
Thörig et al.[11]	1983	Capillaries	With	+	−
Wang et al.[32]	1985	?	Without	−	+
Punyiczki et al.[20]	1988	Capillaries	Without	−	+
Hayashi and Sueishi[22]	1988	Schirmer papers	With	+	+
Tözsér and Berta[24]	1990	Capillaries	Without	+	+

In a clinical study we demonstrated elevated PA (u-PA) activities in the tears of patients with corneal and conjunctival inflammations (Table 1). The increase in the activities correlated with the severity of the diseases. We provided clinical data concerning the role tear of u-PA in the pathogenesis of various inflammatory eye diseases and the possibility of monitoring epithelial cell damage by measuring u-PA activities in tears of traumatic and inflammatory cases.[28] In rabbit experiments we studied the effect of alkali burn and *n*-heptanol and mechanical debridment of the corneal epithelium on the PA activity of the tear fluid. The activities increased within a few hours after the injury and gradually normalized by the time the epithelial defects healed.[29]

Barlati et al.[30] investigated fibronectin, PAs (u- and t-PA), plasminogen, and plasmin in human tears by means of SDS-polyacrylamide gel electrophoresis and immunoblotting. They collected tears with Schirmer paper strips from patients with ocular pathologies (thermal or chemical burns, foreign body ulcers, keratitis) and from normal controls. The protein composition of the tear samples was analyzed by SDS-polyacrylamide gel electrophoresis. Components of the fibrinolytic system were detected after transfer of the separated proteins with monospecific rabbit antisera against human t-PA, u-PA, plasminogen/plasmin, and fibronectin. They demonstrated intact fibronectin and fibronectin fragments in normal tears. A staining with anti-t-PA was detected in the same samples at 68 and 83 kDa. Normal tears also showed the presence of u-PA only in the region of 80 to 90 kDa, while urokinase standard was present at 50 kDa. A positive reaction was demonstrated with anti-plasminogen/plasmin antibodies at 66 and 97 kDa. A significant modification was observed in the protein profiles of fibronectin, t-PA, u-PA, and plasminogen/plasmin in cases of corneal ulcers or thermal or chemical burns compared to the patterns observed in normal tears. In herpetic keratitis only the plasminogen/plasmin antigen pattern showed slight variations. The altered protein patterns gradually normalized during treatment and at remission returned to normal composition and protein levels.[30]

THE FIBRINOLYTIC SYSTEM OF HUMAN TEARS

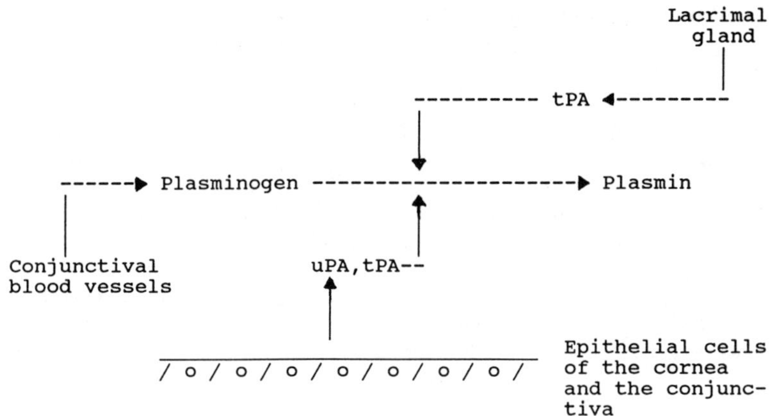

FIGURE 1. The schematic representation of the fibrinolytic system of human tears.

The exact mechanism of how PAs are secreted into in tears in not yet known. Several different sources seem to contribute to the PA content and PA activity of tears (Table 2, Figure 1). The possible sources include the lacrimal glands, the epithelial cells of the cornea and the conjunctiva, stromal fibroblasts, vascular endothelial cells of the conjunctiva, blood plasma, polymorphonuclear leukocytes (PMNs), and different microorganisms. Data suggesting the contribution of these sources either showing the release from or proving the ability to produce tear PAs has been published in the literature (Table 3). The amount of activators produced by these sources varies in normal and in various pathological states (Tables 4 and 5).

The PA activity in normal tears that we found to be 0.06 ± 0.04 IU is much lower than the values published by Lantz and Pandolfi (25 PU/ml, 1 PU = 0.7 IU), Hayashi and Suieshi (7.9 ± 2.0 IU/ml), and even lower than those found by Thörig et al. (1.92 ± 0.89 IU/ml in unstimulated and 1.26 ± 0.38 Iu/ml in stimulated tears). The method by which the samples were collected greatly influences the results of activity measurements. Tears collected with Shirmer papers showed higher activities probably due to the release of PAs from the epithelial cells due to the slight injury caused by the the collection itself. Other differences in tear PA activity may arise from the use of different methods. Most PA activities published in the literature were meausred in the presence of fibrin. The enhancement of t-PA activity by fibrin may account for the higher activities measured by the fibrin plate method. This however may be disadvantageous because the uniform activity due to the presence of t-PA in the lacrimal gland fluid, once enhanced, might mask the changes in the relatively low PA levels caused by epithelial cell destruction.[23]

Accumulating experimental and clinical data suggest that increased PA activity in the tears of patients is not only an accompanying phenomenon of epithelial cell destruction, but it may also play a pathogenetic role. Intrastromal

TABLE 3
Data Suggesting the Contribution of Different Sources to the
Plasminogen Activator Content and
Plasminogen Activator Activity of Tears

Author	Year	Source	Type of PA	Normal/ pathological
Pandolfi[33]	1967	Corneal epithelium	?	Injured
		Corneal endothelium	?	Injured
		Vascular endothelium	?	Normal
Pandolfi et al.[34]	1972	Whole cornea	?	Explant in culture
Lantz et al.[35]	1978	Corneal epithelium	?	Explant in culture
Pandolfi and Lantz[36]	1979	Whole cornea	u-PA	Explant in culture
Berman et al.[37]	1980	Corneal epithelium, stromal fibroblasts, PMNs	u-PA (?)	Ulcerating corneas, cultured fibroblasts
Lantz and Anderson[12]	1982	Corneal epithelium	u-PA (?)	Explant in culture
Berman et al.[31]	1983	Corneal epithelium	u-PA	Alkali-burned ulcerating corneas
Thörig et al.[9–11]	1983	Lacrimal gland	t-PA	Normal
	1984	Lacrimal gland, other sources	?	Normal
	1985	Lacrimal gland	t-PA	Normal
Wang et al.[32]	1985	Corneal epithelium, stromal fibroblasts	u-PA	Ulcerating corneas
Haeringen and Thörig[8]	1985	Lacrimal gland, conjunctiva	t-PA(?)	Normal
Lantz and Pandolfi[13]	1986	Cornea	u-PA	Explants
		Conjunctiva	t-PA,u-PA	in culture
Chan[38a,38b]	1986	Corneal epithelium	?	Alkali burn
	1986	Corneal epithelium	u-PA(?)	Wound healing
Geanon et al.[39]	1987	Corneal epithelium	u-PA,t-PA	Normal
		Corneal stroma	u-PA	
		Corneal endothelium	u-PA	
Hayashi and Sueishi[22]	1988	Corneal epithelium, vascular endothelium of the conjunctiva	t-PA	Normal
Berman[21]	1989	Corneal epithelium	u-PA,t-PA	Epithelial defect
Tervo et al.[40]	1989	Corneal stroma	u-PA,t-PA	Corneal wound
Hayashi et al.[41]	1991	Corneal epithelium	u-PA,t-PA	Epithelial defect and wound healing

injection of u-PA in the rabbit eye caused angiogenesis, initiated polymorphonuclear leukocyte (PMN) chemotaxis and had a regulatory effect on blood
vesel wall permeability. Since PMNs are able to produce u-PA themselves, u-
PA of leukocyte origin may contribute to the measured total PA activity in
pathological conditions when they are present in the anterior segment of the
eye. An increase in the u-PA activity in the tears and on the surface of the eye
by activating free plasmin, which besides its wide spectrum proteolytic activity

TABLE 4
Plasminogen Activator Activities in Normal and Pathological Human Tears

Authors	Year	Collection	Method	PA activity
Tsung and Holly[42]	1981	Capillaries	Fluorometry	36–56 pM/mg/h (normal tears)
Lantz and Anderson[12]	1982	Schirmer papers	Fibrin plate	14–60 PU/ml (normal tears)
Thörig et al.[11]	1983	Capillaries	Fibrin plate	0.7–5.0 IU/ml (normal tears)
Lantz and Pandolfi[13]	1986	Schirmer papers	Fibrin plate	Mean:25 PU/ml (normal tears)
Salonen et al.[14]	1987	Capillaries	Plg-casein plate	0.6–9.8 PU/ml (normal tears) 0 (pathological tears)
Tözsér et al.[23]	1989	Capillaries	Amidolytic	0.06(±0.04) IU/ml (normal tears) 0.02–0.82 IU/ml (pathological tears)
Berta et al.[28]	1990	Capillaries	Plg-casein plate	0–0.45 IU/ml (normal tears) 1.32–27.84 IU/ml (pathological tears)

Note: 1 International Unit (IU) = 0.7; Ploug Unit (PU).

is able to activate latent collagenase, may play a role in the destructive proceses of the eye.

To determine PAs and differentiate them from each other, from plasmin, and from activator-inhibitor complexes is not easy in a complex biological system like tears. Their activities can easily be masked by other enzymes. PA activity may appear as plasmin activity if the reagents used, or the samples themselves, are not completely plasminogen free, as activators are determined in systems where they are measured through plasmin activity they generate from added plasminogen. With immunological methods inactive activators (proactivators, activator-inhibitor complexes) can also be detected as PAs. The sensitivity of immunological methods (both in immunoblotting and immunohistology) is considerably less than that of the activity measurements. Therefore, it is possible that the enzyme is not detected by immunological methods but its activity still can be demonstrated.

PA, plasmin, plasmin inhibitor, and PA inhibitor determinations were suggested to be of diagnostic value in a number of diseases: different types of corneal ulcers, persitent epithelial defects, recurrent erosions, contact lens-associated lesions, herpetic keratitis, bullous keratopathy, keratoconjunctivitis sicca, and chemical burns of the cornea and the conjunctiva (Plates 1 to 13)*. Tear levels may reflect the degree of cellular damage, indicate the progress of

* Plates 1 to 13 appear after page 130.

TABLE 5
**The Presence of the Components of the Fibrinolytic System
in Normal and Pathological Human Tears**

	Plasminogen	t-PA	u-PA	Plasmin
Normal tears	–	+/–	+	–
Pathological tears	+	+	+++	++

the healing process, and be used as prognostic indicators. These studies, besides providing a better understanding of the pathological processes going on in the anterior segment of the eye, may also provide a possibility for selective therapeutical interventions.[28]

REFERENCES

1. **Storm, O.,** Fibrinolytic activity in human tears, *Scand. J. Clin. Lab. Invest.,* 7, 55, 1955.
2. **Astrup, T. and Müllertz, S.,** The fibrin plate method for estimating fibrinolytic activity, *Arch. Biochem. Biophys.,* 40, 346, 1952.
3. **Lassen, M.,** Heat denaturation of plasminogen in the fibrin plate method, *Acta Physiol. Scand.,* 27, 371, 1952.
4. **Müllertz, S. and Lassen, M.,** An activator system in blood indispensible for formation of plasmin by streptokinase, *Proc. Soc. Exp. Biol. Med.,* 82, 264, 1953.
5. **Rijken, D. C., Wijngaards, G., and Welberger, J.,** Immunological characterization of plasminogen activator activities in human tissues and body fluids, *J. Lab. Clin. Med.,* 97, 477, 1981.
6. **Haverkate, F. and Brakman, P.,** Fibrin plate assay, in *Progress in Chemical Fibrinolysis and Thrombolysis,* Vol. 1, Davidson, J. F., Samama, M. M., and Desnoyers, P. C., Eds., Raven Press, New York, 1975, 151.
7. **Rijken, D. C., Wijngaards, G., Zaal-de Jong, M., and Welbergen, J.,** Purification and partial characterization of plasminogen activator from human uterine tissue, *Biochim. Biophys. Acta,* 580, 140, 1979.
8. **van Haeringen, N. J. and Thörig, L.,** Enzymology of tear fluid, *Protides Biol. Fluids Proc. Colloq.,* 32, 399, 1985.
9. **Thörig, L., van Agtmaal, E. J., Glasius, E., Tan, K. L., and van Haeringen, N. J.,** Comparison of tears and lacrimal gland fluid in the rabbit and guinea pig, *Curr. Eye Res.,* 4, 913, 1985.
10. **Thörig, L., van Haeringen, N. J., and Wijngaards, G.,** Comparison of enzymes of tears, lacrimal gland fluid, and lacrimal gland tissue in the rat, *Exp. Eye Res.,* 38, 605, 1984.
11. **Thörig, L., Wijngards, G., and van Haeringen, N. J.,** Immunological characterization and possible origin of plasminogen activator in human tear fluid, *Ophthal. Res.,* 15, 268, 1983.
12. **Lantz, E. and Anderson, A.,** Release of fibrinolytic activators from the cornea and conjunctiva, *Graefe's Arch. Clin. Exp. Ophthalmol.,* 219, 263, 1982.
13. **Lantz, E. and Pandolfi, M.,** Fibrinolysis in cornea and conjunctiva: evidence of two types of activators, *Graefe's Arch. Clin. Exp. Ophthalmol.,* 224, 393, 1986.
14. **Salonen, E.-M., Tervo, T., Törmá, E., Tarkkanen, A., and Vaheri A.,** Plasmin in tear fluid of patients with corneal ulcers: basis for new therapy, *Acta Ophthalmol.,* 65, 3, 1987.

15. **Saksela, O.,** Radial caseinolysis in agarose: a simple method for the detection of plasminogen activator in the presence of inhibitory substances and serum, *Anal. Biochem.,* 111, 276, 1981.

16. **Salonen, E.-M., Lauharanta, J., Sim, P.-S., Stephens, R., and Vaheri, A.,** Rapid appearance of plasmin in tear fluid after ocular allergen exposure, *Clin. Exp. Immunol.,* 73, 146, 1988.

17. **Tervo, T., Salonen, E.-M., Vaheri, A., Immonen, I., van Setten, G.-B., Himberg, J J., and Tarkkanen, A.,** Elevation of tear fluid plasmin in corneal disease, *Acta Ophthalmol.,* 66, 393, 1988.

18. **Tervo, T., van Setten, G.-B., Andersson, R., Salonen, E.-M., Vaheri, A., Immonen, I., and Tarkkanen, A.,** Contact lens wear is associated with the appearance of plasmin in the tear fluid — preliminary results, *Graefe's Arch. Clin. Exp. Ophthalmol.,* 227, 42, 1989.

19. **van Setten, G.-B., Stephens, R., Tervo, T., Salonen, E.-M., Tarkkannen, A., and Vaheri, A.,** The Schirmer test leads to activation of the fibrinolytic system in the tear fluid, *Curr. Eye Res.,* 8, 1293, 1989.

20. **Punyiczki, M., Berta, A., and Tözsér, J.,** Study of neutral proteinases of human tear samples using gels containing substrates, *Clin. Chem. Enzym. Comns.,* 1, 115, 1988.

21. **Berman, M.,** The pathogenesis of corneal defects, *Acta Ophthamol.,* 192(suppl.), 55, 1989.

22. **Hayashi, K. and Sueishi, K.,** Fibrinolytic activity and species of plasminogen activator in human tears, *Exp. Eye Res.,* 46, 131, 1988.

23. **Tözsér, J., Berta, A., and Punyiczki, M.,** Plasminogen activator activity and plasminogen independent amidolytic activity in tear fluid from healthy persons and patients with anterior segment inflammation, *Clin. Chim. Acta,* 183, 323, 1989.

24. **Tözsér, J. and Berta, A.,** Urokinase-type plasminogen activator in rabbit tears. Comparison with human tears, *Exp. Eye Res.,* 51, 33, 1990.

25. **Tözsér, J. and Berta, A.,** Components of the fibrinolytic system of human tears, *Invest. Ophthalmol. Vis. Sci.,* 69, 426, 1991.

26. **Tözsér, J. and Berta, A.,** Plasminogen activator inhibitors in human tears, *Acta Ophthalmol.,* 69, 426, 1991.

27. **Tözsér, J., Punyiczki, M., and Berta, A.,** The fibrinolytic system of human tears (abstr.), *Fibrinolysis,* 1(suppl.), 110, 1988.

28. **Berta, A., Tözsér, J., and Holly, F.,** The determination of plasminogen activator activity in normal and pathological human tears, and the significance of tear plasminogen activators in the inflammations of the cornea and the conjunctiva, *Acta Ophthalmol.,* 68, 508, 1990.

29. **Berta, A., Holly, F. J., Tözsér, J., and Holly, T. F.,** Tear plasminogen activators — indicators of epithelial cell destruction. The effects of scraping, n-heptanol treatment, and alkali burn of the cornea on the plasminogen activator activity of rabbit tears, *Int. Ophthalmol.,* 15, 363, 1991.

30. **Barlati, S., Marchina, E., Quaranta, C. A., Vigasio, F., and Semerano, F.,** Analysis of fibronectin, plasminogen activators, and plasminogen in tear fluid as markers of corneal damage and repair, *Exp. Eye Res.,* 51, 1, 1990.

31. **Berman, M., Manseau, E., Law, M., and Aiken, D.,** Ulceration is correlated with degradation of fibrin and fibronectin at the corneal surface, *Invest. Ophthalmol. Vis. Sci.,* 24, 1358, 1983.

32. **Wang, H. M., Berman, M., and Law, M.,** Latent and active plasminogen activator in corneal ulceration, *Invest. Ophthalmol. Vis. Sci.,* 26, 511, 1985.

33. **Pandolfi, M. and Astrup, T.,** A histochemical study of the fibrinolytic activity. Cornea, conjunctiva, and lacrimal gland, *Arch. Ophthalmol.,* 77, 258, 1967.

34. **Pandolfi, M., Astedt, B., and Dyster-Aas, K.,** Release of fibrinolytic enzymes from human cornea, *Acta Ophthalmol.,* 50, 199, 1972.

35. **Lantz, E., Dyster-Aas, K., and Pandolfi, M.,** *In vitro* release of fibrinolytic activators from cornea, *Graefe's Arch. Clin. Exp. Ophthalmol.,* 206, 157, 1978.

36. **Pandolfi, M. and Lantz, E.,** Partial purification and charactarization of keratokinase, the fibrinolytic activator of the cornea, *Exp. Eye Res.,* 29, 563, 1979.

37. **Berman, M., Leary, R., and Gage, J.,** Evidence for a role of the plasminogen activator-plasmin system in corneal ulceration, *Invest. Ophthalmol. Vis. Sci.,* 19, 1204, 1980.

38a. **Chan, K. Y.,** Chemical injury to an *in vitro* ocular system: differential release of plasminogen activator, *Curr. Eye Res.,* 5, 357, 1986.

38b. **Chan, K. Y.,** Release of plasminogen activator by cultured corneal epithelial cells during differentation and wound closure, *Exp. Eye Res.,* 42, 417, 1986.

39. **Geanon, J. D., Tripathi, B. J., Tripathi, R. C., and Barlow, G. H.,** Tissue plasminogen-activator in avascular tissues of the eye — a quantitative study of its activity in the cornea lens and aqueous and vitreous of dog, calf and monkey, *Exp. Eye Res.,* 44, 55, 1987.

40. **Tervo, T., Tervo, K., van Setten, G.-B., Virtanen, I., and Tarkkanen A.,** Plasminogen activator and its inhibitor in the experimental corneal wound, *Exp. Eye Res.,* 48, 445, 1989.

41. **Hayashi, K., Berman, M., Smith, D., El-Ghatit, A., Pease, S., and Kenyon, K. R.,** Pathogenesis of corneal epithelial defects: role of plasminogen activator, *Curr. Eye Res.,* 10, 381, 1991.

42. **Tsung, P.-K. and Holly, F. J.,** Protease activities in human tears, *Curr. Eye Res.,* 1, 351, 1981.

43. **Johnson, A. J., Kline, D. L., and Alkjaersig, N.,** Assay methods and standard preparations for plasmin, plasminogen and urokinase in purified systems, *Thromb. Diath. Haemorrh.,* 21, 259, 1969.

THE PLASMINOGEN ACTIVATOR-PLASMIN SYSTEM OF THE TEARS

Part D
Plasminogen Activator Inhibitors and Plasmin Inhibitors in the Cornea and in the Tears

Plasminogen activation provides an important source of localized proteolytic activity in the blood during fibrinolysis and in tissues during various normal and pathological events that involve tissue degradation, tissue reorganization, and invasive processes. The secretion or release of plasminogen activators (PAs) is the starting point of the activation of the fibrinolytic cascade resulting in proteolytic activity in various compartments of the extracellular space. The regulation of PA activity was shown to play a key role in a number of biological processes, and abnormalities of this regulation are known to lead to the development various diseases. The control of PA activity is achieved at many levels. The synthesis of PAs by different types of cells is regulated by hormones, cytokines, and growth factors. Besides active secretion another possible mechanism is the release of these activators from the same types of cells due to mechanical, chemical, or other injuries. Once PAs are released from the cells their activity is further modulated through interactions with components of the extracellular matrix, with plasma cofactors, and with cell surface receptors. The most important physiological regulators of PA activity are specific PA inhibitors (PAIs). Plasmin, the proteolytic enzyme generated from plasminogen through the action of PAs is also regulated by physiological inhibitors, first α-2-antiplasmin and α-2-macroglobulin.[1,2]

The results of recent investigations indicate that in humans at least four immunologically distinct proteins exist that are capable of inhibiting PAs. These are the endothelial PAI (PAI-1), the placental PAI (PAI-2), a urinary inhibitor of urokinase-type activator (PAI-3), and protease nexin (PN).[3] PAI-3 and PN, however, are not specific to PAs.[3a] Most of the research effort so far has been directed on PAI-1 and -2. Various types of cells have been found to synthetize and secrete one or both of these two inhibitors.[4]

PAI-1 is the major physiological inhibitor of tissue-type PA (t-PA), at least in blood. Complexes of single-chain t-PA with PAI-1 but not with other inhibitors were detected in the blood. The second-order rate constant for the interaction of single-chain t-PA with PAI-1 is approximately 20,000 times higher than the constant for the interaction of t-PA with PAI-2 or PN. Elevated PAI-1 levels were detected in patients suffering from various thrombotic diseases, disseminated intravascular coagulation, and systemic Gram-negative bacterial infections. Elevated PAI activity in blood is often detected in association with deep vein thrombosis, myocardial infarction, acute pancreatitis, and pregnancy. High PAI activities and elevated levels of PAI-1 were also detected in diffuse liver diseases and after surgery or severe trauma, suggesting that PAI-1 can be considered to be an acute-phase reactant. The normal plasma concentration of PAI-1 is 5 to 20 ng/ml that can rise to 50 to 200 ng/ml in sepsis.[2]

PAI-2 was first detected in and purified from human placenta. Human leukocytes and phorbol ester-stimulated U-937 cells contain even more PAI-2 than human placenta. The molecular weight of the purified PAI-2 molecule is 47 kDa. PAI-2 is an efficient inhibitor of u-PA (second order rate constant: 10^6 $M^{-1}s^{-1}$) and the two-chain form of t-PA (second-order rate constant: 2×10^5 $M^{-1}s^{-1}$). It is therefore the second-best inhibitor of u-PA and of two-chain t-PA. PAI-2 is however 10 times less potent an inhibitor of u-PA and 50 times less effective on two-chain t-PA than PAI-1. In contrast to PAI-1, PAI-2 is a poor inhibitor of single-chain t-PA. PAI-2 is also present in pregnancy in plasma. In placenta PAI-2 is synthetized by the trophoblast and is considered to be a marker of placental function. Besides this PAI-2 has important regulatory activity on the differentiation of monocytes, macrophages, endothelial cells, and fibroblasts. The inhibition of fibrinolytic activity by PAI-2 on the surface of leukocytes and tumor cells affects the ability of these cells to pass through tissue boundaries.[1]

PAI-3, also called activated protein C (APC) inhibitor, is a 57-kDa glycoprotein that is present in plasma at a concentration of 2 to 5 μg/ml.[5] The physiological role of PAI-3 is not certain, but it is likely to be the major inhibitor of APC in plasma. The inhibitory activity of PAI-3 towards u-PA and APC is stimulated by heparin. PAI-3 is also found in considerable amounts in normal human urine, both in active form and in complexes with urokinase (u-PA). The second-order rate constant for PAI-3 toward u-PA is 8×10^3 $M^{-1}s^{-1}$ in the absence and 9×10^4 $M^{-1}s^{-1}$ in the presence of heparin.[6]

PN is a 50-kDa glycoprotein normally present in human plasma at very low concentrations. Different cell types produce PN in culture. Its physiological function is not clear, but it is capable of inhibiting trypsin, plasmin, and thrombin. Its inhibitory function is also stimulated by heparin.[7] PN inhibits u-PA with a second-order rate constant of 1.5×10^5 $M^{-1}s^{-1}$ and is a much less potent inhibitor of single- and two-chain forms of t-PA. Its low level in plasma suggests that it is not an important PAI in the circulation, though it probably has important regulatory functions on the cellular level.[8]

Hayashi and Sueishi[9] were the first to study PAI activity in human tears. They did not find PAI activity in normal human tears collected with Schirmer paper strips. In 1990 at the World Congress of Ophthalmology in Singapore we reported PAI activities and the presence of PAI-1 in the tears of patients suffering from keratoconjunctivitis sicca with or without Sjögren's syndrome.[10]

Tears of 21 patients with keratoconjunctivitis sicca (KCS) (17 of them with Sjögren's syndrome) and were studied. The Copenhagen Criteria for KCS and Sjögren's syndrome[11] were used for patients in this study. All of the patients were females with a mean age of 53.2 (SD: 12.8) years. An age-matched group

TABLE 1
PLI and PA Activities in the Tears of Patients with KCS and Sjögren's Syndrome[a]

Subjects (number of patients)	Tear PLI activity (CU/ml)	Tear PA activity (IU/ml)
Normal (13)	0	0.08 (±0.07)
KCS (21)	0.15 (±0.07)	0.48 (±0.12)
Sjögren's Syndrome (17)	0.14 (±0.06)	0.52 (±0.14)

[a] Data taken from Berta, A. and Tözsér, J. in *New Frontiers in Ophthalmology*, Khoo, C. Y., Ang, B. C., Cheah, W. M., Chew, P. T. K., and Lim, A. S. M., Eds., Excerpta Medica, Amsterdam, 1991, 777.

consisting of healthy females, cataract patients before surgery, and patients with macular degenerations was used as a control. Samples were collected with capillary tubes following the use of nasal stimulation. Tear samples collected at flow rates between 10 and 60 µl/min only were used in the study.[12,13] Samples were centrifuged (at 500 rpm for 5 min); the supernatants were frozen in plastic tubes and were thawed only when the assays were performed. PA and plasminogen independent (PLI) activities were determined in the tears by a microtiter plate method based on the spectrophotometric measurement of the light absorption of a colored product that is produced by plasmin from a chromogenic substrate. The details of the method were described by Tözsér et al. in 1989.[14] PAI activities of the samples were determined according to Kruithof et al.[15] The inhibitor activity values were expressed in IU per milliliter urokinase units. The serum albumin concentrations in tear samples were determined with NOR-Partigen radial immunodiffusion plates and the standard serum was purchased from Behringwerke AG, Germany.

Both PA and PLI activities were increased in the tears of the patients with KCS and Sjögren's syndrome (Table 1). PA activities showed a good correlation with the van Bijsterveld score and and inverse correlation with the tear film breakup time (BUT) values but did not correlate well with Schirmer values. PAs (u- and t-PA) and PAI-1 were detected in the tears of the same patients with immunoblot, fibrin zymography, and reverse fibrin zymography techniques. Surprisingly high levels of PAI were detected in most cases with Sjögren's syndrome, which were not proportional to the serum albumin levels measured in the same tear samples (Table 2).

In the tears of KCS patients high PA levels were found indicating epithelial cell damage. PAIs were also found in the tears of these patients. These findings, together with earlier observations that Sjögren patients have high α-1-antitrypsin levels in tears,[16] suggest that inhibiting mechanisms in eyes with KCS are highly active to prevent proteolysis by proteases also present at high levels in the tears of these patients. PAIs may be especially effective in the prevention of corneal ulceration since they are capable of limiting the cascade of tear

TABLE 2
PAI and Serum Albumin Levels in the Tears of Patients
with KCS and Sjögren's Syndrome[a]

Subjects (number of patients)	Tear PAI activity (IU/ml)	Tear albumin conc (mg/ml)
Normal (13)	0	0.45 (±0.57)
KCS (21)	8.92 (±7.30)	5.51 (±2.74)
Sjögren's Syndrome (17)	12.41 (±11.13)	6.32 (±1.83)

[a] Data taken from Berta, A. and Tözsér, J. in *New Frontiers in Ophthalmology*, Khoo, C. Y., Ang, B. C., Cheah, W. M., Chew, P. T. K., and Lim, A. S. M., Eds., Excerpta Medica, Amsterdam, 1991, 777.

proteases at an early stage. Enzymes, activators, and inhibitors of the plasminogen-plasmin system can be used as indicators of epithelial cell damage and of ulcerating tendency in KCS and Sjögren's syndrome.

In another study we investigated the plasminogen inhibitor content and activity in the tears of patients with various corneal and conjunctival diseases. Tears were collected with capillary tubes with and without stimulating the nasal mucosa. Secretion rate values were determined by dividing the volume of the tear sample by the time length of the sample collection.[13] Samples collected with 10- to 60-µl/min secretion rate were included in the study. The protein content was measured spectrophotometrically according to Bradford[17] and the albumin concentration by radial immunodiffusion according to Mancini[18] to determine the degree of transudation as it was described by Tözsér et al.[14] Neutral proteinase activity was determined by radial caseinolysis as described by Punyiczki et al.[19] and expressed in micrograms trypsin equivalent per milliliter values. PAI activities were measured by a photometric assay according to a modification of the original assay of Kruithof et al.[15] described in detail by Tözsér and Berta.[20] PA activity was determined in the same tear samples with a chromogenic assay.[14]

Total protein, serum albumin levels, as well as PA, PAI, and total proteolytic activities measured in the tear samples are listed in Table 3. No PAI was detected in the normal tear samples. No PAI activity or low activities (<1.0 IU/ml) were detected in the tears of patients with reflex tearing due to corneal foreign body. A lack of significant change in the permeability of the conjunctival blood vessels in these cases was indicated by the low albumin levels measured in these tear samples. The total tear protein concentration was also low compared to the normal level (3 to 7 mg/ml) obtained with this method.[14] No proteolytic activity was measured by radial caseinolytic method in these cases. Substantial PAI activities were measured in the tears of patients who had undergone eye surgery and most patients suffering from severe corneal and conjunctival diseaseas, characterized by extensive epithelial injuries including corneal abrasion, herpetic keratitis, bullous keratopathy, alkali burn, and lignotic conjunctivitis. These samples usually represented high protein content and

TABLE 3
PA, PAI, and Total Proteolytic Activities in the Tears of Patients with Different Corneal and Conjunctival Diseases[a]

No.	Diagnosis	Albumin (mg/ml)	Protein (mg/ml)	PA (IU/ml)	PAI (IU/ml)	Proteolytic (µg trypsin equivalent/ml)
1	Corneal foreign body	1.2	1.6	—	0.5	0
2	Corneal foreign body	0.7	1.2	—	0.3	0
3	Corneal foreign body	0.6	0.6	—	0	0
4	Corneal foreign body	—	0.6	0.06	0.9	—
5	Corneal abrasion	1.6	3.6	—	1.5	0.2
6	Corneal abrasion	5.0	10.0	—	4.0	0.2
7	Corneal abrasion	—	1.9	0.24	1.9	—
8	Herpetic keratitis	—	6.3	0.60	3.2	—
9	Bullous keratopathy	—	3.6	1.68	14.0	—
10	Alkali burn	—	18.4	1.89	16.0	—
11	Postsurgery 1 d	13.0	20.7	—	18.0	0.9
12	Postsurgery 1 d	1.6	12.3	—	0	0.2
13	Postsurgery 1 d	—	18.0	0.12	1.6	—
14	Postsurgery 5 d	—	8.3	0.32	0.3	—
15	Acute bacterial conjunctivitis	7.5	11.0	0.60	1.5	0.5
16	Acute bacterial conjunctivitis	3.3	7.4	0.50	1.0	0.1
17	Bacterial conjunctivitis	—	9.2	0.42	0.5	—
18	Chronic conjunctivitis	2.0	3.2	—	0	0.1
19	Chronic conjunctivitis	—	8.8	0.87	0.5	—
20	Lignotic conjunctivitis	4.2	9.3	—	3.0	0.9
21	Bacterial ulcer	—	8.1	0.81	0	—
22	Nonbacterial ulcer	—	7.2	1.20	0.3	—
23	Contact lens wearer (soft)	—	5.5	0.25	0.3	—
24	Contact lens wearer (soft)	—	7.7	0.34	0.1	—
25	Contact lens wearer (soft)	—	1.4	0.42	0.4	—

[a] Data taken from Tözsér, J. and Berta, A., *Acta Ophthalmol.*, 69, 426, 1991.

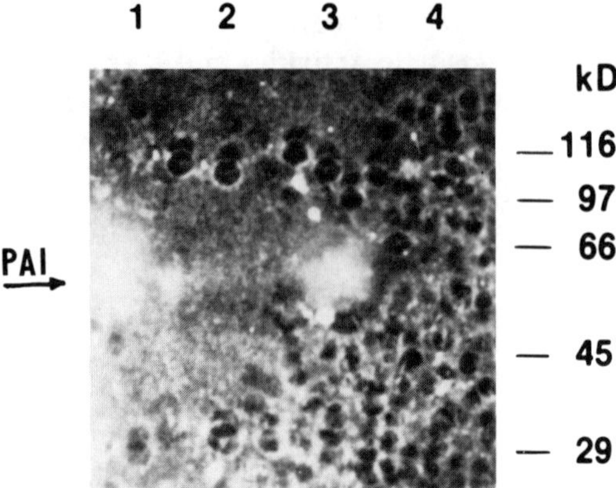

FIGURE 1. PAI activity in human tears detected by reverse fibrin zymography. Tear samples were obtained from patients with the following diagnoses: (1) postsurgery, 1 d; (2) marginal keratitis; (3) Sjögren's syndrome; (4) normal tears. (From Tözsér, J. and Berta, A., *Acta Ophthalmol.*, 69, 426, 1991. With permission.)

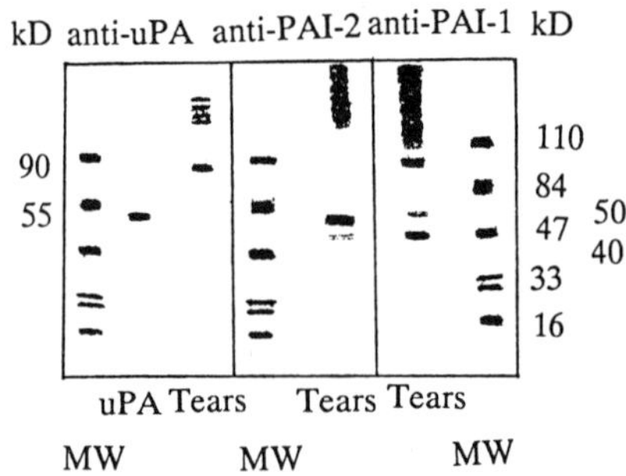

FIGURE 2. Immunoblot analysis of PAIs of the tears of a patient suffering from epithelial damage due to contact lens wear revealing the simultaneous presence of PAI-1 and PAI-2. Left, immunoblot with anti-u-PA antibody; right, immunoblot with anti-PAI-1 antibody; middle, immunoblot with anti-PAI-2 antibody. MW, molecular weight standards at 110, 84, 47, 33, and 16 kDa.

sometimes substantial PA or proteolytic activity. Serum albumin was present only in a few of these cases, and the highest PAI activities were measured in those samples that did not contain detectable albumin, suggesting that the measured PAI activity can only partly be the result of the transudation of circulating PAI-1 from the blood.

The PAI content and activity of the tear samples was also studied by reverse fibrin zymography. Normal tears did not show any PAI activity. In the tears of a patient suffering from marginal keratitis a faint 50-kDa band was detected; in the tears of a patient 1 d after cataract surgery and in another suffering from KCS associated with Sjögren's syndrome, PAI activity was clearly present in the form of a 50-kDa band, also indicating that active PAI is present in the tears of patients suffering from these pathological conditions (Figure 1).

Immunoblot analysis of PAIs of the tears of a patient suffering from epithelial damage due to contact lens wear revealed the simultaneous presence of both type 1 and 2 PAIs (Figure 2). Using anti-PAI-1 antibody a faint band was detected at 50 kDa. The presence of further bands at 35 and 60 kDa was also demonstrated. However, most of the PAI-1 antigen was found at 100 kDa and in the region above this band in form of a smear in the upper part of the gel. With anti-PAI-2 antibody, most of the antibody was detected in form of a 60-kDa band. A closely spaced doublet line at 50 kDa as well as a faint smear in the upper part of the gel was also demonstrated.[20]

PAI activity is not present in normal human tears. However, substantial PAI activity may be present in tears if the epithelial cells are injured and/or the permeability of the conjunctival vessels is increased. The presence of PAI activity was also demonstrated by reverse fibrin zymography (Figure 1). Immunoblotting revealed the presence of both PAI-1 and -2 in human tears (Figure 2). The 50- and 100-kDa PAI-1 antigens indicate the presence of free inhibitor and that of complexes of PAI-1 formed with PAs, as detected in plasma.[21] However, most of the PAI-1 antigen was detected in higher molecular weight regions (>120 kDa). Since these bands also reacted with anti-u-PA and anti-t-PA antibodies, they may be the result of dimerization or oligomerization of PAI (the electrophoresis was performed under nonreducing conditions).

PAI-I has been detected in tissue culture media and by histological methods in the sections of ulcerating (Berman, personal communication) and regenerating corneas.[22] The faint staining detected by immunohistological methods around the regenerating epithelium suggest that locally produced PAI-1 may play a role in epithelial wound healing.[22] The presence of PAI-1 in wounded corneas and in the culture fluid of ulcerating corneas suggests that PAI-1 in tears may originate from the injured or regenerating epihelial cells. Essentially two mechanisms have been proposed to explain the appearance of PAIs in the vicinity of corneal epithelial cells: (1) the passive release from damaged epithelial cells and (2) the active secretion during epithelial wound healing.[10] Another possible source of PAI-1 in tears is the blood, due to increased permeability of the conjunctival blood vessels, indicated by the elevated serum albumin and total protein content of the tear samples. The 60-kDa PAI-2 is probably the glycosylated form of the free activator, which may be secreted by an active process, while the 50-kDa bands may be nonglycosylated forms released from injured cells.[23] As in the case of PAI-1, the higher molecular weight bands may represent u-PA-PAI-2 complexes.[3a] PAI-2 is not a constant component of normal plasma; it is present in the blood of healthy pregnant women. Besides producing PAI-1, most epithelial cells are capable of producing PAI-2. A number of different cell types was found to be able to produce both types of PAIs.[4]

High levels of PAI activity in the tear fluid of KCS patients were detected together with increased tear u-PA activity.[10] Injured corneal and conjunctival epithelial cells were also suggested to be the source of increased u-PA levels in tears.[14,23] Leukocytes are also able to produce PAI-2;[3a] in inflammatory diseases leukocytes may also contribute to the PAI-2 content of tears, as was suggested for u-PA.[14] The simultaneous presence of u-PA and PAIs in tears may support the assumption that at least part of the u-PA may be present in tears in prourokinase form, which cannot be complexed by PAIs.[3a] The presence of prourokinase in tears was also suggested based on the detection of tear u-PA activity in the form of doublet lytic bands on casein zymograms.[19,24]

PAI and proteolytic activities were detected simultaneously in some of the tear samples. The proteolytic activity in the tear samples is predominantly the result of the appearence of active plasmin,[25] indicating that the activation by PAs may occur even in the presence of PAIs. Besides plasmin other proteolytic enzymes were detected in pathological tears using zymography after electrophoresis, probably due to the presence of proteinases of bacterial, viral, and leukocyte origin.[19] The proteolytic events involved in corneal wound healing and re-epithelization may be similar to those occurring in other re-epithelization and healing processes occurring in the human body involving a regulation based on the balance of u-PA and PAI-2. In corneal epithelial resurfacing, however, another balance between t-PA and PAI-1 also must be taken into consideration.[20]

REFERENCES

1. **Kruithof, E. K. O.,** Plasminogen activator inhibitor type 2: biochemical and biological aspects, in *Protease Inhibitors,* Takada, A., Samama, M. M., and Collen, D., Eds., Excerpta Medica, Amsterdam, 1990, 15.
2. **Loskutoff, D. J.,** The cell and molecular biology of type 1 plasminogen activator inhibitor, in *Protease Inhibitors,* Takada, A., Samama, M. M., and Collen, D., Eds., Excerpta Medica, Amsterdam, 1990, 3.
3. **Tor, N. Y., Lawrence, D., Mikus, P., Srandberg, L., and Astedt, B.,** Structure-functional analysis of plasminogen activator inhibitors type 1 and 2 using *in vitro* mutagenesis, in *Protease Inhibitors,* Takada, A., Samama, M. M., and Collen, D., Eds., Excerpta Medica, Amsterdam, 1990, 23.
3a. **Kruithof, E. K. O.,** Plasminogen activator inhibitors — a review, *Enzyme,* 40, 113, 1988.
4. **Schleef, R. R., Wagner, N. V., and Luskutoff, D. J.,** Detection of both type 1 and type 2 plasminogen activators in human cells, *J. Cell. Physiol.,* 134, 269, 1988.
5. **Suzuki, K., Deyashiki, Y., Nishioka, J., Kurachi, K., Akira, M., Yamamoto, S., and Hashimoto, S.,** *J. Biol. Chem.,* 262, 611, 1987.
6. **Strump, D. C., Thienpoint, M., and Collen, D.,** Urokinase-related proteins in human urine. Isolation and characterization of single-chain urokinase (pro-urokinase) and urokinase-inhibitor complex, *J. Biol. Chem.,* 261, 1267, 1986.
7. **Scott, R. W., Bergman, B. L., Bajpai, A., Hersh, R. T., Rodriguez, H., Jones, B. N., Barreda, C., Watts, S., and Baker, J. B.,** *J. Biol. Chem.,* 260, 7029, 1985.

8. **Hiramoto, S. A. and Cunningham, D. D.,** *J. Cell. Biochem.,* 36, 199, 1988.

9. **Hayashi, K. and Sueishi, K.,** Fibrinolytic activity and species of plasminogen activator in human tears, *Exp. Eye Res.,* 46, 131, 1988.

10. **Berta, A. and Tözsér, J.,** Plasminogen activator and plasminogen activator inhibitor determinations in the tears of patients with keratoconjunctivitis sicca and Sjögren's syndrome, in *New Frontiers in Ophthalmology,* Khoo, C. Y., Ang, B. C., Cheah, W. M., Chew, P. T. K., and Lim, A. S. M., Eds., Excerpta Medica, Amsterdam, 1991, 777.

11. **Manthorpe, R., Oxholm, P., Prause, J. U., and Schiodt, M.,** The Copenhagen Criteria for Sjögren's syndrome, *Scand. J. Rheumatol. Suppl.,* 61, 19, 1986.

12. **Berta, A.,** Collection of tear samples with or without stimulation, *Am. J. Ophthalmol.,* 96, 115, 1983.

13. **Berta, A.,** Standardization of tear protein determinations. The effects of sampling, flow rate, and vascular permeability, in *The Preocular Tear Film in Health, Disease, and Contact Lens Wear,* Holly, F. J., Ed., Dry Eye Institute, Lubbock, TX, 1986, 418.

14. **Tözsér, J., Berta, A., and Punyiczki, M.,** Plasminogen activator activity and plasminogen independent amidolytic activity in tear fluid from healthy persons and patients with anterior segment inflammation, *Clin. Chim. Acta,* 183, 323, 1989.

15. **Kruithof, E. K. O., Vassalli, J. D., Schleuning, W. D., Mattaliano, R. J., and Bachmann, F.,** Purification and characterization of plasminogen activator inhibitor from histolytic lymphoma cell line U-937, *J. Biol. Chem.,* 261, 11207, 1986.

16. **Berta, A., Vámosi, P., and Tözsér, J.,** Eine neue Methode zur semiquantitativen Bestimmung von Alpha-1-Antitrypsin Inhalt der menschlichen Tranenflüssigkeit, *Fortschr. Ophthalmol.,* 86, 773, 1989.

17. **Bradford, M. M.,** A rapid and sensitive method for the quantitation of microgram quantities of protein utilizing the principle of protein binding dye, *Anal. Biochem.,* 72, 248, 1976.

18. **Mancini, G., Garbonera, A., and Heremans, J.,** Immunochemical quantitation of single radial immunodiffusion, *Immunochemistry,* 2, 235, 1965.

19. **Punyiczki, M., Berta, A., and Tözsér J.,** Study of neutral proteinases of human tear samples using gels containing substrates, *Clin. Chem. Enzym. Comns.,* 1, 115, 1988.

20. **Tözsér, J. and Berta, A.,** Plasminogen activator inhibitors in human tears, *Acta Ophthalmol.,* 69, 426, 1991.

21. **Genton, C., Kruithof, E. K. O., and Schleuning, W.-D.,** Phorbol ester induces the biosynthesis of glycosylated and nonglycosylated plasminogen activator inhibitor 2 in high excess over urokinase-type plasminogen activator inhibitor in human U-937 lymphoma cells, *J. Cell. Biol.,* 104, 705, 1987.

22. **Tervo, T., Tervo, K., van Setten, G. B., Virtanen, I., and Tarkkanen, A.,** Plasminogen activator and its inhibitor in the experimental corneal wound, *Exp. Eye Res.,* 48, 445, 1989.

23. **Berta, A., Tözsér, J., and Holly, F. J.,** Determination of plasminogen activator activities in normal and pathological human tears. The significance of tear plasminogen activators in the inflammatory and traumatic lesions of the cornea and the conjunctiva, *Acta Ophthalmol.,* 68, 508, 1990.

24. **Tözsér, J. and Berta, A.,** Urokinase-type plasminogen activator in rabbit tears. Comparison with human tears, *Exp. Eye. Res.,* 51, 33, 1990.

25. **Salonen, E. M., Tervo, T., Torma, E., Tarkannen, A., and Vaheri, A.,** Plasmin in tear fluid of patients with corneal ulcers: basis of new therapy, *Acta Ophthalmol.,* 66, 3, 1987.

Chapter 9

METABOLIC ENZYMES IN THE TEARS

Part A
Lactate Dehydrogenase and Malate Dehydrogenase in the Tear Fluid

Lactate dehydrogenase (LDH), like many other enzymes, exists within a single species or a single cell in different molecular forms that are called isoenzymes. Although all five isoenzymes of lactate dehydrogenase catalyze the same enzymatic reaction, they have distinctly different KM values for their substrates. The isoenzymes all have the same molecular weight, 134,000 Da, and all contain four polypeptide chains, each with molecular weight of 33,500 Da. They consist of five different combinations of two different types of polypeptide chains, designated M and H chains. One of the isoenzymes, which predominates in muscle, has four identical M chains and is designated M4. Another that predominates in heart has four identical H chains and is designated H4. The other three isoenzymes consist of the other possible combinations of H and M chains. The LDH isoenzymes are also numbered and are often called LDH-1 (H_4), LDH-2 (H_3M), LDH-3 (H_2M_2), LDH-4 (HM_3), and LDH-5 (M_4). All five isoenzymes have been obtained and studied in pure forms. The single M and H chains, which have been isolated and found to be enzymatically inactive, differ significantly in amino acid content and sequence. The two peptide chains of LDH are coded by two different genes, the type of LDH isoenzyme produced by a given cell is under genetic control, and is characteristic rather for the type of cell or tissue than for the individual organism. Isoenzymes of LDH exist in different proportions in different tissues (Table 1). The LDH isoenzyme composition of human normal serum is rather constant in the case of a certain individual, and only slightly differs from one person to the other. In the case of extensive cell destruction in various organs of the human body, the LDH isoenzyme coposition of the serum changes in a manner characteristic for the type of tissues involved in the destruction. Changes in the serum levels of LDH isoenzymes can give information on the site and degree of cell destruction on the tissues and organs involved in the pathologic processes. Changes in the LDH isoenzyme levels in serum in different diseases are demonstrated in Table 2.

Based on the extensive knowledge concerning changes in the LDH content and composition of human serum, several attempts were made to study LDH and its isoenzymes in human tears in various pathological states and correlate their levels with those found in various ocular tissues. The first observation on LDH in tears was the detection of very high activities of LDH, in the range of 0.063 to 5.57×10^3 U/l, measured in tear samples collected with Schirmer strips from the conjunctival sac of healthy individuals.[1] Information on the LDH isoenzyme pattern of tears can be obtained by electrophoresis and specific

TABLE 1
Distribution of LDH Isoenzymes in Normal Tissues

Tissue	Percent of total activity				
	LDH-1	LDH-2	LDH-3	LDH-4	LDH-5
Heart muscle	60	30	5	3	2
Kidney	28	34	21	11	6
Brain	28	32	19	16	5
Liver	0.2	0.8	1	4	94
Skeletal muscle	3	4	8	9	76
Skin	0	0	4	17	79
Lung	10	18	28	23	21
Spleen	5	15	31	31	18

Data taken from Pfeidler, G. et al., *Advances in Clinical Enzymology,* Schmidt, E. et al., Eds., S. Karger, Hanover, Germany, 1979, 17. With permission.

TABLE 2
Clinical Conditions Associated with Elevations of Different LDH Isozymes in Serum

Disease	LDH-1	LDH-2	LDH-3	LDH-4	LDH-5
Hemolysis	+	+			
Myocardial infarction	+	+			
Pulmonary infarction		+	+		
Shock	+	+	+	+	+
Renal cortex infarction	+	+			
Hepatocellular damage				+	+
Sceletal muscle trauma					+

staining. Different LDH isoenzymes can be separated on the basis of different electrophoretic mobilities either on cellulose acetate membranes or on polyacrylamide gels and can be visualized by tetrazolium dyes that are turned into colored products as the result of electron transfer mechanisms involved in the enzyme action of LDH. van Haeringen and Glasius[2] have shown that normal human tears, unlike human serum, contain isoenzymes LDH-5, LDH-4, and LDH-3 in large quantities. Their results have been confirmed by Kahán and Ottoway[3] and Liotet et al.[4] These isoenzymes are built up mainly from M subunits and are called muscle-type isoenzymes responsible primarily for the anaerobic conversion of pyruvate to lactate. The same LDH isoenzyme pattern was demonstrated in the extracts of the lacrimal gland and corneal epithelial scrapings. The similarity in enzyme pattern of the lacrimal gland and the tear fluid just like the discordance of enzyme activities of serum and tear fluid, as in the case of lysozyme,[5] suggested a lacrimal source for these enzymes, too.[2]

LDH activities and isoenzyme patterns of human and rabbit tears, lacrimal gland, and corneal epithelium extracts were studied by spectrophotometric and electrophoretic methods by Kahán and Ottoway.[6] The LDH isoenzyme patterns of human tears, lacrimal gland, and corneal epithelium were found to be very similar consisting mainly of isoenzymes built up of subunits M, while the isoenzyme patterns of the same rabbit samples showed a predominance of isoenzymes consisting of subunits H. LDH activities were tenfold higher in the tears and corneal epithelium of rabbits than of humans, whereas in the lacrimal glands and sera the enzyme activities did not differ significantly in the two species.[6]

Besides the detection of LDH isoenzymes by electrophoretic methods, Kahán and Ottoway[3] used another principle to estimate the levels of various isoenzymes. On the basis of differences in the activities of LDH isoenzymes in the presence of substrates pyruvate or lactate the proportion of different isoenzymes in tissue extracts and body fluids could be estimated. These LDH isoenzymes were those involved mainly in anaerobic metabolism (in the conversion of pyruvate to lactate, the end product of anaerobic glycolysis) and those involved primarily in aerobic biochemical processes (in the conversion of lactate to pyruvate ready for further dehydrogenization in the tricarbonic acid cycle). LDH activities measured in body fluids and secretions in the presence of pyruvate and lactate are designated LDH-P and LDH-L activity, respectively. The ratio of these two activities, LDH-P/LDH-L, was suggested to be a parameter expressing the ratio of different isoenzymes in the specimens. The higher the value of this quotient the higher the ratio of LDH isoenzymes involved in the anaerobic conversion of pyruvate to lactate. The lower the value of this quotient the higher the ratio of LDH isoenzymes converting lactate to pyruvate making it ready to join in the processes of aerobic energy production.[3]

Kahán and Ottoway[6] studied the LDH activities in normal human and normal rabbit tears and corneal epithelial extracts. LDH activities were found to be similar in tears and in corneal epithelium within one species, both being tenfold higher in rabbits than in humans. Very similar correlation was found for LDH/α-HBD (α-hydroxybutyrate dehydrogenase) ratio in tears and corneal epithelium of the two species, the ratio being higher than 2.0 in human and lower than 1.0 in rabbit tears and corneal epithelium. These observations, together with the similarities between the LDH isoenzyme patterns of tears and corneal epithelium in both species, suggested that some of the LDH in tears derives from the corneal epithelium, and the isoenzymes are primarily of muscular type involved mainly in anaerobic energy-producing metabolic processes.[3]

Further data on the possible origin of LDH and other tear enzymes were produced by van Haeringen and Glasius,[7] who studied the activities of various enzymes in tear samples collected by standard Schirmer strips from healthy individuals. Eight successive specimens were collected, and lacrimation was slightly stimulated with vapor of formaldehyde in their experiments. The enzyme activities were determined in each of the successively collected samples. LDH, malate dehydrogenase (MDH), and pyruvate kinase (PK) activities ranged between 10 and 250% of the initial values, those measured in the first collected samples. The activity values fluctuated randomly in the succesively collected specimens. The activities of LDH, MDH, and PK changed parallel to each individual sample, suggesting a similar source for these three enzymes. The activities of amylase varied on a similarly wide scale but did not show a correlation with the enzymes of the energy-producing metabolism. Lysozyme activities measured in the same tear samples were found to be rather constant and independent from the degree of lacrimation in good accordance with earlier findings of Sapse et al.[8]

The enzyme patterns of the collected tear samples were shown to depend on the method of sampling. van Haeringen et al.[9] could demonstrate the presence of only lysozyme and amylase in samples collected by glass capillaries both in case of stimulation with tear gas and without stimulation. In tear samples collected by Schirmer paper strips, however, the enzymes of the energy-producing mechanisms were also detected in relatively large amounts. These enzymes were suggested to originate from conjunctival epithelial cells disrupted or desquamated by the insertion of Schirmer test paper. It has been suggested that the appearance of intracellular enzymes in the tear fluid may indicate the presence or degree of epithelial cell damage in patients with contact lens problems.[10]

A comparative study concerning the activities of the enzymes of the energy-producing mechanisms was performed by van Haeringen and co-workers in tear samples collected with glass capillaries and with Schirmer paper strips. The conclusion of the study was that LDH, MDH, PK, isocytrate dehydrogenase, aldolase, glucose-6-phosphate dehydrogenase, sorbitol dehydrogenase, glutamate dehydrogenase, and glutamate-oxalacetate transaminase, being present only in samples collected with Schirmer strips, are probably not secreted by the lacrimal gland, but can be liberated into the tear fluid after slight damage of the corneal or conjunctival epithelium. Other enzymes such as hexokinase, glutamate-pyruvate transaminase, and some of the lysosomal enzymes, α-galactosidase and β-hexosaminidase, were also detected in tears collected with capillary tubes after stimulation, suggesting that these latter enzymes might be secreted by the lacrimal gland, though their activities seem to increase in the case of epithelial damage, too.[9]

An increase in the LDH activity, a decrease in LDH-P/LDH-L ratio, and a slight increase in the level of LDH isoenzymes consisting of H subunits (a shift towards plasma LDH isoenzyme pattern) were observed in tears in various external inflammatory diseases of the eye.[4,11] These changes can be attributed to an increase in the permeability of the conjunctival blood vessels and the appearance in tears of LDH isoenzymes deriving from the blood. In the tears of patients suffering from superficial keratitis and in most contact lens wearers a moderate increase in LDH activity in tears was accompanied by an increase in LDH-P/LDH-L ratio and an increase of the level in tears of LDH isoenzymes consisting of M subunits.[4,12] These latter changes are probably brought about by the release of LDH isoenzymes from the damaged superficial epithelial cells of the ocular surface.

Fullard and Carney[13] used a fluorometric technique to measure the activities of LDH and MDH in tears to assess corneal response to contact lens wear. The absolute values of LDH and MDH proved to be very variable depending on the rate of stimulated lacrimation, while the LDH/MDH ratio proved to indicate changes in the levels of the enzymes in tears of epithelial cell origin. In this way the effects of environmental stresses and the severity of damage caused by them to the corneal epithelial cells could be estimated. Following short-term wear of contact lenses, the LDH and MDH activities were shown to alter so that the LDH/MDH ratio was elevated. The magnitude and time course of this elevation were found to be influenced by contact lens type, fit, and duration of wear.[13]

REFERENCES

1. **van Haeringen, N. J. and Glasius, E.,** Lactate dehydrogenase in tear fluid, *Exp. Eye Res.,* 18, 345, 1974.
2. **van Haeringen, N. J. and Glasius, E.,** Enzymes of the energy producing metabolism in human tear fluid, *Exp. Eye Res.,* 18, 407, 1974.
3. **Kahán, I. L. and Ottoway, É.,** The significance of tears lactate dehydrogenase in health, and external eye diseases, *Graefe's Arch. Clin. Exp. Ophthalmol.,* 194, 267, 1975.
4. **Liotet, S., Jacq, C., and Warnet, V. N.,** Untersuchung der Isoenzyme der Laktatdehydrogenase in der menschlichen Tranenflüssigkeit, *Contactologia,* 2, 245, 1980.
5. **Covey, W., Perillie, P., and Finch, S. C.,** The origin of tear lysozyme, *Proc. Soc. Exp. Biol. Med.,* 137, 1362, 1971.
6. **Kahán, I. L. and Ottoway, É.,** Lactate dehydrogenase of tears and corneal epithelium, *Exp. Eye Res.,* 20, 129, 1975.
7. **van Haeringen, N. J. and Glasius, E.,** Enzymatic studies in lacrimal secretion, *Exp. Eye Res.,* 19, 135, 1974.
8. **Sapse, A. T., Bonavida, B., Stone, W., Jr., et al.,** Human tear lysozyme. III. Preliminary study lysozyme levels in subjects with smog irritation, *Am. J. Ophthalmol.,* 66, 76, 1968.
9. **van Haeringen, N. J., Ensink, F., and Glasius, E.,** The origin of the enzymes in tear fluid, *Exp. Eye Res.,* 22, 297, 1976.
10. **van Haeringen, N. J. and Glasius, E.,** The origin of some enzymes in tear fluid, determined by comparative investigation with two collection methods, *Exp. Eye Res.,* 22, 267, 1976.
11. **Kahán, I. L.,** Prognostic value of tear lactate dehydrogenase isoenzyme determinations, in *The Preocular Tear Film in Health, Disease, and Contact Lens Wear,* Holly, F. J., Ed., Dry Eye Institute, Lubbock, TX, 1986, 182.
12. **Kahán, I. L., Pálfalvi, M., and Opauszki, A.,** Biochemical characteristics of the tear film in the dry eye syndrome, *IRCS Med. Sci. Clin. Biochem.,* 8, 827, 1980.
13. **Fullard, R. J. and Carney, L. G.,** Use of tear enzyme activities to assess the corneal response to contact lens wear, *Acta Ophthalmol.,* 64, 216, 1986.

Chapter 9

METABOLIC ENZYMES IN THE TEARS

Part B
Lysosomal Enzymes in the Tears

Most eukaryotic cells contain digestive organelles called lysosomes containing an array of hydrolytic enzymes capable of digesting various types of macromolecules, constituents of living cells. Lysosomes digest bacteria or other particles ingested by cells by phagocytosis. The membrane of the lysosome fuses with the membrane surrounding the captured particle to form a secondary lysosome that is responsible for the initial degradation of the particle. Once a harmful captive has been hydrolyzed, the monomers can be catabolized by other metabolic pathways outside the lysosome. When a cell dies, its lysosomes rupture and the enzymes that are released hydrolyze the dead cell, and the lysosomal enzymes appear in the extracellular space. Rupture of the lysosomes occurs when the cells are badly injured or totally damaged. Mild trauma causes a much earlier release of cytoplasmic enzymes from the cells without affecting the integrity of lysosomes.[1,2]

Lysosomal enzymes are hydrolytic enzymes used by the cells to digest proteins, nucleic acids, glycogen, mucopolysaccharides, lipids, proteoglycans, etc. Lysosomal enzymes can be detected in body fluids in cases with extensive tissue damage. Though the hydrolytic enzymes are basically the same in most cells, the levels of the various enzymes in the plasma in certain pathologic conditions may reflect their tissue or organ origin and the metabolic state of the ruptured cells as well.[3] A list of some common lysosomal enzymes also detected in tears is shown in Table 1.

Inborn errors of metabolism that occur in storage diseases can be characterized by the lack of production or disturbed function of certain lysosomal enzymes. Missing or low levels of these enzymes in blood or secretions can help the diagnosis of these diseases together with the clinical and laboratory signs of the accumulation of the substrates of the missing enzymes. Not only enzyme deficiences but also hyperactivities are of great diagnostic value as in mucolipidosis II (I-cell disease).[4] More details on tear enzymes in metabolic and storage diseases are given in Chapter 16.

The number of hydrolytic enzymes detected in tears is above 30. Most of them can be considered to be of lysosomal origin. The first of the lysosomal enzymes to be detected in tears were acid phosphatase,[5] β-hexosaminidase,[6] and α-galactosidase.[7] The majority of hydrolytic enzymes has been found to be active at low pH; therefore, they are often referred to as acid hydrolases. The source of these enzymes is thought to be the lacrimal gland, as demonstrated for α-galactosidase and β-hexosaminidase by van Haeringen and Glasius.[8] Conjunctival and corneal epithelium may act as a secondary source especially in pathological cases and when tears are collected with filter papers causing

TABLE 1
Lysosomal Enzymes Detected in Tears

Name	E.C. no.	Authors (year)
Acid phosphatase	3.1.3.2	Liotet (1967)
Alkaline phosphatase	3.1.3.1	Liotet (1967)
Aldolase	4.1.2.13	van Haeringen, Glasius (1974)
α-Amylase	3.2.1.1	van Haeringen, Glasius (1974)
α-D-Galactosidase	3.2.1.22	Del Monte et al. (1974)
Glycogen phosphorylase	2.4.1.1	Thoft, Friend (1975)
β-D-Galactosidase	3.2.1.23	van Haeringen, Glasius (1976)
β-D-Glucuronidase	3.2.1.31	van Haeringen, Glasius (1976)
α-L-Fucosidase	3.2.1.51	van Hoof et al. (1977)
α-D-Glucosidase	3.2.1.20	van Hoof et al. (1977)
β-D-Glucosidase	3.2.1.21	van Hoof et al. (1977)
α-L-Iduronidase	3.2.1.76	van Hoof et al. (1977)
α-D-Mannosidase	3.2.1.24	van Hoof et al. (1977)
β-*N*-Acetyl-D-glucosaminidase	3.2.1.30	van Hoof et al. (1977)
α-*N*-Acetyl-D-glucosaminidase	3.2.1.50	van Hoof et al. (1977)
Sulfatase A	3.1.6.1	van Hoof et al. (1977)
Sulfatase B	3.1.6.1	van Hoof et al. (1977)
Cathepsin D	3.4.23.5	van Hoof et al. (1977)
Cathepsin B	3.4.22.1	Tsung, Holly (1981)
Cathepsin C	3.4.14.1	Tsung, Holly (1981)
Leukocyte elastase	3.4.21.37	Tsung, Holly (1981)

mild trauma to the superficial epithelial cells.[9,10] The lacrimal origin of lysosomal hydrolases was partly based histological studies of the secretory cells of the lacrimal gland[11,12] and partly on the observation that stimulated tears (tears collected at high flow rates) still contained large amounts of these enzymes.[8,13]

In further experiments van Haeringen and Glasius[10] studied lysosomal enzymes in tissue extracts of human lacrimal glands. In the lacrimal gland extracts they differentiated free activities and activities gained after the breakdown of lysosomes and compared these values with those measured in normal tears. They found a correlation between the concentrations of ten acid hydrolyses in tears and in lacrimal gland extracts, providing further evidence for the lacrimal origin of these enzymes. The apparent differences between the levels measured in the "free" and in the "lysosomal" fraction of the tissue extracts support the suggestion that these enzymes originate from intracellular vesicles (presumably lysosomes) of the secretory cells of the lacrimal gland. The same authors demonstrated that acetylsalicylic acid, an antiinflammatory agent known to have a stabilizing effect on lysosomes, when given systemically, significantly lowers the secretion of β-hexosaminidase in tears.[10]

The activities of lysosomal enzymes were shown to be higher than those in serum,[9,13] which suggests a local production. There is accumulating evidence that these enzymes get into the tear fluid through active secretion and not by passive transport mechanisms. The high molecular weight of these enzymes makes it unlikely that they could pass the blood tear barrier under normal circumstances either at the lacrimal gland or through the conjunctival vessel walls. What is even more convincing is that higher concentrations in tears can be maintained only by active transport mechanisms. Repeated determinations of α-galactosidase, β-hexosaminidase, and acid phosphatase did not show parallel changes as one would have expected in case of simple release of enzymes from intracellullar lysosomes.[13] Moreover, the secretion of large amounts of enzymes in tears occurs without any signs of cellular damage and without the appearance of cytoplasmic enzymes in tears. On the basis of these considerations, the large number of vesicules detected in the acinar cells of the lacrimal gland[12] should be cosidered to be secretory vesicles and the process of the production of "lysosomal enzymes" should be looked upon as a secretory mechanism.

The physiological role of tear hydrolases is rather obscure. Most of these enzymes are active at low pH, which rarely if ever occurs in tears. These enzymes fulfill their physiological role, the cleavage of phagocytosed particles, inside digestive intracellular vesicles at low pH and are known to have no significant activity in the cytoplasm outside the lysosomes.

The same enzymes have been detected in other parts of the bulb and ocular adnexa,[14–16] including endothelial[17] and epithelial cells[18] of the cornea. Levels of lysosomal enzymes in tears were shown to increase in experimentally induced inflammation,[19] inflammatory and degenerative diseases of the cornea and the conjunctiva,[20–22] and in mucolipidosis type II,[4] and gradually decrease following the treatment by bone marrow transplantation of the same disease.[23] These observations indicate that, besides the unquestionable secretion by the lacrimal gland of intracellular hydrolyses, other sources such as corneal and conjunctival epithelial cells probably also contribute to the high activities measured in the tears of patients suffering from various systemic and eye diseases.

REFERENCES

1. **de Duve, Ch.,** The lysosome, *Sci. Am.,* 208, 64, 1963.
2. **Weissmann, J.,** Lysosomes, *Blood,* 24, 594, 1964.
3. **Barrett, A. J. and Heath, M. F.,** Lysosomal enzymes, in *Lysosomes,* Dinge, J. T., Ed., North-Holland, Amsterdam, 1977, 19.
4. **van Hoof, F., Libert, J., Aubert-Tulkens, G., and Sera, M. V.,** The assay of lacrymal enzymes and the ultrastructural analysis of conjunctival biopsies: new techniques for the study of inborn lysosomal diseases, in *Metabolic Ophthalmology,* Vol. 1, Pergamon Press, Elmsford, NY, 1977, 165.
5. **Liotet, S.,** Etude l'activite transaminasique et de l'activite phosphatasique des larmes humanies, *Ann. Oculist.,* 200, 1258, 1967.
6. **Singer, J.-D., Cotlier, E., and Krimmer, R.,** Hexosaminidase A in tears and saliva for rapid identification of Tay-Sachs disease and its carriers, *Lancet,* 2, 116, 1973.
7. **Del Monte, M. A., Johnson, D. L., Cotlier, E., Krivit, W., and Desnick, R. J.,** Diagnosis of Fabry's disease by tear alpha-galactosidase A, *N. Engl. J. Med.,* 290, 57, 1974.

8. **van Haeringen, N. J. and Glasius, E.,** The origin of some enzymes in tear fluid, determined by comparative investigation with two collection methods, *Exp. Eye Res.,* 22, 267, 1976.

9. **Kitaoka, M., Nakazawa, M., and Hayasaka, S.,** Lysosomal enzymes in human tear fluid collected by filter paper strips, *Exp. Eye Res.,* 41, 259, 1985.

10. **van Haeringen, N. J. and Glasius, E.,** Lysosomal hydrolases in tears and the lacrimal gland: effect of acetylsalicylic acid on the release from the lacrimal gland, *Invest. Ophthalmol. Vis. Sci.,* 19, 826, 1980.

11. **Kuhnel, W.,** Vergleichende histologische, histochemishe und elektronenmikroskopische Untersuchungen an Tranendrusen. VI. Menschliche Tranedruse, *Z. Zellforsch. Mikrosk. Anat.,* 89, 550, 1968.

12. **Ruskell, G. L.,** Nerve terminals and epithelial cell variety in the human lacrimal gland, *Cell. Tissue Res.,* 158, 121, 1975.

13. **van Haeringen, N. J. and Glasius, E.,** Characteristics of acid hydrolases in human tear fluid, *Ophthal. Res.,* 8, 367, 1976.

14. **Hayashaka, S.,** Distribution of lysosomal enzymes in the bovine eye, *Jpn. J. Ophthalmol.,* 18, 233, 1974.

15. **Hayashaka, S., Shiono, T., Hara, S., and Mizumo, K.,** Regional distribution of lysosomal enzymes in the retina and chorioid of human eyes, *Graefe's Arch. Clin. Exp. Ophthalmol.,* 216, 269, 1981.

16. **Yamaguchi, K., Hayasaka, S., Hara, S., Shiono, T., and Mizuno, K.,** Lysosomal enzyme activities in ocular tissues and adnexa of rabbits, *Ophthalmic Res.,* 21, 187, 1989.

17. **Hara, S., Hayasaka, S., and Mizumo, K.,** Lysosomal enzyme activities of the bovine corneal endothelium, *Graefe's Arch. Clin. Exp. Ophthalmol.,* 224, 384, 1986.

18. **Shiono, T., Hayasaka, S., and Mizumo, K.,** Acid hydrolases in the bovine corneal epithelium, *Graefe's Arch. Clin. Exp. Ophthalmol.,* 224, 467, 1986.

19. **Tsung, P.-K., Price, J., and Holly, F. J.,** Changes in lysosomal enzyme activities of eye tissues in endotoxin-induced inflammation, *Inflammation,* 7, 257, 1983.

20. **Hayashaka, S.,** Lysosomal enzymes in ocular tissues and disease, *Surv. Ophthalmol.,* 27, 245, 1983.

21. **Hayasaka, S., Tsuchiya, M., Sekimoto, M., Noda, S., Shibuya, Y., and Setogawa, T.,** Lysosomal enzymes in tear fluids from patients with Terrien's marginal corneal degeneration, *Graefe's Arch. Clin. Exp. Ophthalmol.,* 225, 335, 1987.

22. **Kitaoka, M., Yamaguchi, K., and Hayasaka, S.,** Lysosomal enzymes in tear fluids from patients with ocular diseases, *Jpn. J. Ophthalmol.,* 30, 154, 1986.

23. **Yamaguchi, K., Hayasaka, S., Hara, S., Kurobane, I., and Tada, K.,** Improvement of tear lysosomal enzyme levels after treatment with bone marrow transplantation in a patient with I-cell disease, *Ophthalmic Res.,* 21, 226, 1989.

Chapter 9

METABOLIC ENZYMES IN THE TEARS

Part C
Proteolytic Enzymes in Tears

"Proteolytic enzymes", "proteases", and "proteinases" are synonyms. Proteolytic enzymes and proteases are older terms widely used in the literature. The term proteinase has gained ground in recent years, after it had been adopted by the International Union of Biochemistry for the Enzyme Nomenclature in 1972, partly revised but in principle confirmed in 1978 and in 1984.[1-3]

Proteinases are divided into two main groups on the basis of the sites of their action in protein molecules: exopeptidases and endopeptidases. Endopeptidases cleave bonds in the inner parts protein molecule, in a certain distance from (often also near to) the ends of the polypeptide chain. Exopeptidases break peptide bonds only near (within two residues from) the end of the polypeptide chain.[4]

Recently the term "peptidase" has been proposed by Barrett and McDonald to be used in a general sense as peptide bond hydrolase. Peptidase in earlier literature had been used for enzymes cleaving small peptides, and by others as a synonyme for exopeptidase. The new proposal, still to be accepted by the Nomenclature Commission, recommends the term peptidase (peptide bond hydrolase) to be used for all enzymes capable of cleaving peptide bonds, and the terms endopeptidase (endo-acting peptide bond hydrolase) and exopeptidase (exo-acting peptide bond hydrolase) for enzymes breaking peptide bonds at differents sites of the protein or peptide molecule.[5]

The terms "tissue proteinase" and "intracellular proteinase" are also used, usually associated with the release of enzymes from injured tissues in connection with extensive cell damage. Neither of these terms is well defined and they do not allow differentiation between proteinases of lysosomes and other organelles that have intracellular functions from those that originate from secretory granules of different cells of various organs (not only secretory glands). There are a number of physiological processes, besides the functioning of secretory glands, in which intracellular enzymes are produced by or released from various cells. These intracellularly produced but pericellularly acting proteinases are responsible for a number of physiological and pathological processes in which cell-to-cell or cell-to-extracellular matrix adhesions are resolved and pathological disorganization or physiological reorganization of tissues takes place.

The division of proteinases into exopeptidases and endopeptidases is very useful from a practical point of view. Theoretically, however, the distinction is not totally justifiable. It has been known for a long time that endopeptidases often have some exopeptidase activity. It has been shown that some exopeptidases may also cleave peptide bonds inside the polypeptide chain under certain conditions; however, their major function under physiological conditions is the exopeptidase activity.[6,7]

Exopeptidases are generally divided into two groups: carboxypeptidases and aminopeptidases, depending on which end of the polypeptide chain they cleave the terminal peptide bonds. Further subdivision and individual names of exopeptidases are based on substrate specificity (e.g., glycylglycyne dipeptidase), intracellular localization of the enzyme (e.g., lysosomal carboxypeptidase), pH optimum of enzymatic action (e.g., acid carboxypeptidase), or metalloenzyme characteristics (e.g., metallocarboxypetidase). No satisfactory subdivision of endopeptidases could be worked out based on specific characteristics of their enzymatic action. Instead, the essential catalytic group of the enzyme was used to create four subclasses of endopeptidases — serine, thiol, carboxyl, and metallo proteinases — numbered as 3.4.21, 3.4.22, 3.4.23, and 3.4.24, respectively, by the Enzyme Nomenclature Committee.[7] A fifth category (3.4.99) has been reserved for endopeptidases with uncharacterized catalytic groups, as undefined catalytic class. The inclusion criteria to put an endopeptidase in one of these subclases is based on the inhibition by specific inhibitors of their catalytic groups rather than on substrate specificity. For this reason inhibition studies have gained special importance in the classification of endopeptidases.[8]

Basically two pricipal regulatory mechanisms exist in the human body and in other highly organized living organisms to control proteolytic enzyme action: (1) activation of inactive protease precursors (zymogens) by limited proteolysis and (2) inactivation of proteases by forming complexes with protein inhibitors. The physiological importance of both mechanisms has been demonstrated in different disease states that are related to deficiencies of zymogens or that of protease inhibitors. Different types of familial hemophilias are due to deficiences of one of the plasma proteases responsible for activating zymogens in the blood coagulation cascade.[9] Inherited forms of pulmonary emphysema are examples for the other type associated with the impaired secretion by the liver of an inhibitor α-1-antitrypsin. The abnormally low plasma concentrations of α-1-antitrypsin result in the destruction of connective tissue of the pulmonary alveoli by leukocyte elastase.[10] Proteolytic enzymes are not only a physiological necessity but also a potentional hazard, since, if uncontrolled, they can destroy the protein components of cells and tissues.

A comparison of the amino acid sequences and three-dimensional structures of proteases revealed striking similarities among proteases with different substrate specificities. On the basis of their structure and enzymatic action proteases can be grouped into families. The closer two proteases are on the family tree, the more similarities they show in molecular structure. It has been suggested that proteases have a common ancestry and are the products of an evolutionary process. Different proteases found in mammalian cells show similarities with the digestive enzymes of microorganisms. It is more and more accepted that the proteases of highly organized living organisms have originated some billion years ago, and in the course of evolutionary development the proteolytic enzymes that originally served purely digestive functions have adapted to the more specific and complex tasks they now fulfill in physiological regulation.[7,11,12]

Lysosomal hydrolases, including proteinases, have been detected in normal tears, and the primarily lacrimal gland origin of these enzymes seems to be well established.[13] Collagenases and enzymes of the plasminogen activator (PA)-plasmin system were also detected in the tear fluid under normal and pathological conditions. Collagenases originate from corneal epithelial and stromal cells,[14,15] PAs are known to be secreted both by the lacrimal gland and the epithelial cells of the cornea and the conjunctiva[16-18] Under pathological circumstances leukocytes invading the cornea and the conjunctiva also contribute to the proteinases of the tears by releasing some of their intracellular enzymes.[19,19a] Proteinases originating from various microorganisms (primarily bacteria but also fungi and some of the viruses) may also be detected in the tear fluid.[20] Proteinases of various origin, besides their diagnostic significance, also play a pathogenetic role in the development of corneal ulcers and other destructive diseases.[21-23]

In 1981 Tsung and Holly measured proteinase activities in normal human tears. Tears were collected by capillary tubes or with filter paper strips from healthy caucasian and healthy oriental people. The activities of cathepsin B, cathepsin C, cathepsin D, elastase, urokinase, and a "trypsin-like" proteolytic activity were determined using spectrophoretic methods and specific substrates. All these enzymes were active in pooled normal human tears. There were large individual variations, and differences were found in the levels of some of the enzymes between caucasians and orientals. No comparison was made, however, concerning tear enzyme activity differences that probably exist between tear samples that were collected by capillary tubes or by filter paper strips. The authors suggested that various proteinases may affect the surface activity of aqueous tears by cleaving various tear proteins to produce low molecular weight substances that may act as tear surfactants.[24]

Based on the detection of collagenolytic enzymes by biochemical methods in the culture fluid of ulcerating corneas and by immunohistological and enzymographical methods in the frozen sections of ulcerating and nonulcerating corneas, several authors investigated the collagenolytic activity of the tear fluid both in patients and in animal experiments. In an extensive study, Prause studied the levels of polymophonuclear leukocyte collagenase (PML-c-ase), serum albumin (SA), α-1-antitrypsin (AT), and α-2-macroglobulin (MG) in the tears of healthy persons (Prause 1983a) and in the tears of patients with corneal ulcers before treatment[25] and following the treatment with *n*-butylcyano acrylate glue.[25] Using an electroimmunoassay he detected PML-c-ase in the tears of patients with central melting corneal ulcers. The enzyme was not present in the tears of healthy persons. The PML-c-ase gradually disappeared from the tear fluid by the time the healing was complete. The levels of SA and serum antiproteases were high in the tears of corneal ulcer patients. The level of these proteins gradually normalized during the healing process. Prause demonstrated that the source of this collagenolytic enzyme are the PMNs invading the ulcerating cornea, and suggested that the measurements of the tear fluid concentrations of PML-c-ase and serum antiproteases (AT and MG) can be used as indicators of corneal healing.[19]

We studied the collagenolytic activity in normal and pathological human tears by a microtiter plate method. Acid-soluble type I collagen was gelled and dried in the wells of microtiter plates and was used as substrate for the determination of collagenolytic activity. After incubation with tear samples the digested, soluble protein fractions were washed away and the remaining substrate was stained with Coomassie brillant blue. Absorbance values at 550 nm were determined by an automatic microtiter plate reader photometer and proved to be linear with the increasing amount of collagen in the range of 3 to 24 µg. Tear samples assayed both with and without previous activation by trypsin. Trypsin is known to activate various proenzymes including procollagenase into the active forms of the enzymes. Collagenolytic activity measured without activation may be considered to be the manifest collagenolytic activity, while that detected after trypsin activation may be considered the sum of the manifest and the latent collagenolytic activity. Collagenolytic activity in the tears of healthy individuals was detected only in a few cases, and only after trypsin activation; however, tear samples of patients with corneal ulcers, chemical burns, and herpetic keratitis solubilized a sizable amount of collagen also without previous activation. We found this method suitable to measure latent and manifest collagenolytic activity in small volumes of tears both for diagnostic and scientific purposes.[26]

Freshly collected normal tears do not exhibit proteolytic activity when assayed by the radial caseinolytic method. Tears collected from patients suffering from various corneal diseases, however, produce measurable proteolysis digesting various proteins that are included in the detector gel. Salonen and co-workers[23] demonstrated such proteolytic activity in the tears of patients with nonhealing corneal ulcers. Using electrophoretic and enzymographic methods and antiplasmin antibodies demonstrated that this proteolytic activity is the result of the presence of plasmin in the tears of corneal ulcer patients. The proteolytic activity of tear samples, in all cases studied, could be inhibited *in vitro* by aprotinin. The authors reported several cases of otherwise uncurable corneal ulcers that healed when treated by aprotinin eyedrops.[23] Such plasmin activity was also demonstrated in the tears of patients suffering from various corneal diseases, of contact lens wearers, of a patient with allergic conjunctivitis, and in the tears of normal persons following the performance of Shirmer's test.

In 1989 and 1990 we demonstrated high PA activities in the tears of patients suffering from various corneal and conjunctival diseases including those that were found to show plasmin activity. Based on these results we suggested that the appearance of plasmin in the tears of patients with corneal pathology is probably the result of the liberation of urokinase-type PA from the affected corneal and conjunctival epithelial cells.[18,22] Another study using sodium dodecyl sulfate-polyacrylamide electrophoretic methods and casein as a copolymerized substrate demonstrated the presence of several proteinases with various molecular weights in the tears of patients suffering from corneal ulcers. Besides collagenases and plasmin several other proteinases may play a role in the development of corneal ulcers and their presence must be taken into consideration, both in diagnostic tests and in inhibitory therapy.[26]

REFERENCES

1. International Union of Biochemistry, *Recommendations of the Nomenclature Committee of the International Union of Biochemistry,* Academic Press, New York, 1978.
2. International Union of Biochemistry, *Enzyme Nomenclature 1984,* Academic Press, Orlando, FL, 1984.
3. International Union of Pure and Applied Chemistry and the International Union of Biochemistry, *Enzyme Nomenclature. Recommendations,* Elsevier, Amsterdam, 1972.
4. **Barrett, A. J.,** *Proteinases in Mammalian Cells and Tissues,* North-Holland, Amsterdam, 1977.
5. **Barrett, A. J. and McDonald, J. K.,** Nomenclature: protease, proteinase, peptidase, *Biochem. J.,* 237, 1986.
6. **Alfroz, H., Otto, K., Müller, R., and Fughe, P.,** On the specificity of bovine spleen cathepsin B2, *Biochim. Biophys. Acta,* 452, 503, 1976.
7. **James, M. N.,** An X-ray crystallographic approach to enzyme structure and function, *Can. J. Biochem.,* 58, 251, 1980.
8. **Hartley, B. S.,** Proteolytic enzymes, *Annu. Rev. Biochem.,* 29, 45, 1960.
9. **Ratnoff, O. D.,** The molecular basis of hereditary clotting disorders, *Progr. Hemostasis Thromb.,* 1, 39, 1972.
10. **Kidd, V. J., Wallace, R. B., Itakura, K., and Woo, S. L. C.,** Alpha-1-antitrypsin deficiency detection by direct analysis of the mutation in the gene, *Nature,* 304, 230, 1983.
11. **Neurath, H.,** Evolution of proteolytic enzymes, *Science,* 224, 350, 1984.
12. **Neurath, H. and Walsh, K. A.,** Role of proteolytic enzymes in biological regulation (a review), *Proc. Natl. Acad. Sci. U.S.A.,* 73, 3825, 1976.
13. **van Haeringen, N. J. and Glasius, E.,** Lysosomal hydrolases in tears and the lacrimal gland: effect of acetylsalicylic acid on the release from the lacrimal gland, *Invest. Ophthalmol. Vis. Sci.,* 19, 826, 1980.
14. **Berman, M., Leary, R., and Gage, J.,** Latent collagenase in the ulcerating rabbit cornea, *Exp. Eye Res.,* 25, 435, 1977.
15. **Berman, M., Leary, R., and Gage, J.,** Collagenase from corneal cell cultures and its regulation by phagocytosis, *Invest. Ophthalmol. Vis. Sci.,* 18, 588, 1979.
16. **Berman, M., Leary, R., and Gage, J.,** Evidence for a role of the plasminogen activator-plasmin system in corneal ulceration, *Invest. Ophthalmol. Vis. Sci.,* 19, 1204, 1980.
17. **Thörig, L., Wijngards, G., and van Haeringen, N. J.,** Immunological characterization and possible origin of plasminogen activator in human tear fluid, *Ophthal. Res.,* 15, 268, 1983.
18. **Tözsér, J., Berta, A., and Punyiczki, M.,** Plasminogen activator activity and plasminogen independent amidolytic activity in tear fluid from healthy persons and patients with anterior segment inflammation, *Clin. Chim. Acta,* 183, 323, 1989.
19. **Prause, J. U.,** Serum antiproteases and polymorphonuclear leucocyte neutral collagenolytic protease in the tear fluid of patients with corneal ulcers treated with N-butylcyano acrylate glue, *Acta Ophthalmol.,* 61, 283, 1983.

19a. **Prause, J. U.,** Cellular and biochemical mechanisms involved in the degradation and healing of the cornea, *Acta Ophthalmol.,* 168(suppl.), 7, 1984.

20. **Punyiczki, M., Berta, A., and Tözsér, J.,** Determination of collagenolytic activity in human tears by microtiter method, *Clin. Chem. Enzym. Comns.,* accepted.

21. **Berman, M. B.,** Collagenase and corneal ulceration, in *Collagenase in Normal and Pathological Connective Tissues,* Woolley, D. E. and Evanson, J. M., Eds., John Wiley & Sons, New York, 1980, 141.

22. **Berta, A., Tözsér, J., and Holly, F. J.,** Determination of plasminogen activator activities in normal and pathological human tears. The significance of tear plasminogen activators in the inflammatory and traumatic lesions of the cornea and the conjunctiva, *Acta Ophthalmol.,* 68, 508, 1990.

23. **Salonen, E.-M., Tervo, T., Törmá, E., Tarkkanen, A., and Vaheri, A.,** Plasmin in tear fluid of patients with corneal ulcers: basis for new therapy, *Acta Ophthalmol.,* 65, 3, 1987.

24. **Tsung, P.-K. and Holly, F. J.,** Protease activities in human tears, *Curr. Eye Res.,* 1, 351, 1981.

25. **Prause, J. U.,** Serum albumin, serum antiproteases and polymorphonuclear leucocyte neutral collagenolytic protease in the tear fluid of patients with corneal ulcers, *Acta Ophthalmol.,* 61, 272, 1983.

26. **Punyiczki, M., Berta, A., and Tözsér, J.,** Study of neutral proteinases of human tear samples using gels containing substrates, *Clin. Chem. Enzym. Comns.,* 1, 115, 1988.

Chapter 9

METABOLIC ENZYMES IN THE TEARS

Part D
Peroxidase, Catalase, Hydrogen Peroxide, and Free Oxygen Radicals in the Tears and the Cornea

Peroxidase is present in human tears. Peroxidase in normal tears is thought to originate from the lacrimal gland, not from the conjunctiva. Mechanical irritation with filter paper did not result in higher peroxidase levels compared to those measured in tears collected with capillary tubes. Peroxidase, however, was also detected in the conjunctiva. It has been suggested that the enzyme might migrate from the tears into the conjunctiva. Peroxidase activity was found to be different in the tears of various species. In rats the activity of this enzyme was very high, 3.7×10^5 U/l, while in humans only 10^3 U/l was measured, and in normal rabbit tears no activity was detected.[1]

The first observations concerning the peroxidase content in and peroxidase secretion by the lacrimal gland were made by Morrison and Allen, who using immunohistological methods detected the enzyme in Harderian and lacrimal glands.[2] In 1973, Herzog, using an electron microscopic and cytochemical methods, demonstrated the presence of peroxidase and catalase in the secretory cells of bovine lacrimal gland.[3] Gachon et al.,[4] in rabbit experiments, using histochemical and biochemical methods, found detectable peroxidase activity only in the cubic epithelial cells of the ducts of the lacrimal gland. Peroxidase was demonstrated by histochemical and immunological methods also in the salivary glands and in the saliva of different species.[5-7] The biological significance of peroxidase was associated with the thiocyanate-hydrogen peroxide antimicrobial system, which was suggested to play a role in the control of the bacterial flora in the oral cavity and in the eye. The suggested antibacterial effect of this system is independent from lysozyme, lactoferrin, and immunoglobulins.[8,9]

Iwata and co-workers studied the peroxidase activity in the conjunctiva of rats. They detected peroxidase activity in many conjunctival goblet cells using the diaminobenzidine method. The reaction was present in secretory granules as well as in the perinuclear cisternae, the rough endoplasmic reticulum, several Golgi saccules, and the adjacent small smooth-surfaced vesicles and condensing vacuoles. Based on the observation that the various secretory granules within a goblet cell showed different degrees of reaction, they concluded that some goblet cells probably produce both mucous substance and peroxidase in one cell. The mode of production of peroxidase in goblet cells is thought to be similar to that of secretory proteins in pancreatic exocrine cells. The authors suggested that peroxidase together with thiocyanate and endogenous hydrogen peroxide might serve as an antibacterial protective mechanism in the tear fluid.[9]

In 1979, van Haeringen and colleagues studied the peroxidase-thiocyanate-hydrogenperoxide system in the tear fluid and in the saliva of different species. Tears were collected with glass capillaries or with filter paper strips. Unstimulated whole saliva was also collected with glass capillaries. Peroxidase activity was determined with the colorimetric diaminobenzidine assay originally described by Herzog.[3] Thiocyanate was also determined in the samples with a spectrophotometric method.[10] Peroxidase activities found in the tears and saliva differed markedly between the species investigated. The level of this enzyme was very high in rat tears, and no peroxidase was detected in rabbit tears. Tear collection with filter paper in humans and in rats did not alter the peroxidase activities in the tear fluid. Thiocyanate levels determined in the same samples proved to be high in human and guinea pig saliva in contrast to low tear concentrations. There was no correlation between salivary or lacrimal peroxidase activities and the corresponding thiocyanate concentrations. Peroxidase and thiocyanate levels in the tears of all and in the saliva of some of the species investigated were too low to support the proposed widespread action of this antibacterial system. Only human saliva and guinea pig saliva fulfilled the requirements for an effective antibacterial action. The tears of all species investigated and the saliva of some of the species either showed no demonstrable peroxidase activity or too low thiocyanate levels for the proposed mechanism to function. Therefore the peroxidase-thiocyanate-hydrogen peroxide antibacterial system does not seem to be active in the tears of humans and of the studied animal species.[10]

Thörig et al.[11] determined the activity of a variety of enzymes, including peroxidase, in tears and lacrimal gland fluid of rats. They collected the lacrimal gland fluid with a tapered glass canula right from the opening of the excretory duct of the lacrimal gland. The same enzyme activities were also determined in tissue extracts of the lacrimal gland. Thus the secretion of the lacrimal gland was not contaminated by secretions of the Harderian gland and by desquamating conjunctival and corneal epithelial cells. Based on the comparison of enzyme activities in tears, lacrimal gland fluid, and lacrimal gland tissue, they concluded that the lacrimal gland is the primary tissue source for peroxidase, and amylase in rat tears, while plasminogen activator and lactate dehydrogenase may be present in tears primarily as secretion products from other ocular tissue sources. β-Hexosaminidase was found to be released from the lacrimal gland and from other ocular tissue sources, too. Similar observations were made concerning the secretion of peroxidase by the lacrimal gland in the rabbit and guinea pig.[12] In another study, Thörig et al.[13] tested tear fluid production and lacrimal peroxidase secretion in rats without and with drug consumption. Twelve frequently prescribed drugs (a chlorpheniramine-pseudoephedrine combination, promethazine, atropine, timolol, aspirin, diazepam, and furosemide) were given in recommended therapeutic and excessive doses to rats. Doses equivalent to those used for adult humans on a drug-to-body weight basis or excessive doses resulted in about 20 to 60% reduction of tearing. Parallel changes in lacrimal peroxidase secretion were found after administration of atropine, aspirin, furosemide, indomethacin, and pilocarpine. Generally, tear production and lacrimal peroxidase secretion returned to baseline levels after withdrawal of the drugs.[13] van Agtmaal and colleagues reported that in rat experiments acetazolamide (Diamox), given once a day orally in a dose comparable to the dose used for adult humans on a drug-to-body weight basis, decreased the lacrimal peroxidase secretion by 60% of the baseline level, while tear production remained unaffected. After withdrawal of acetazolamide the peroxidase secretion returned to the baseline level.[14]

Bromberg and Welch studied the peroxidase secretion of the lacrimal gland in rats. Tissue fragments of lacrimal glands from healthy 4- and 24-month-old rats were incubated in a perfusion chamber and stimulated to secrete by carbachol. Fractions of the perifusate were collected and assayed for protein concentration and peroxidase activity. The response to carbachol was a well-defined peak of secreted total protein and peroxidase. The response could be inhibited by atropine. No differences were found between the responses of lacrimal gland from male and female rats. The mean activity of the secreted peroxidase was similar in the two age groups, while the aged glands were significantly more variable in their response. The total protein concentration was significantly lower in the perfusate of lacrimal glads from aged rats.[15]

In further experiments Bromberg and et al.[16] studied the effect of sympathetic stimulation on the peroxidase secretion of the lacrimal gland. Perfusates of lacrimal gland tissue from young and aged rats were tested for peroxidase activity. Secretion was stimulated by phenylephrine or isoproterenol, alone or in combination with IBMX. Upon stimulation of tissue from young animals with phenylephrine secretion above baseline levels of protein and 45.2 delta A X min-1/g tissue peroxidase, secretion was observed. The response of the aged tissue to phenylephrine was not significantly different from that of the young tissue. β-Adrenergic stimulation by isoproterenol evoked only a modest secretion of protein and no consistently measurable peroxidase from young tissue. IBMX alone and in combination with isoproterenol evoked an increase in protein secretion, and a modest secretion of peroxidase, 9.5 delta A X min-1/ g tissue by young tissue. The aged tissue, upon stimulation with the combination of IBMX and isoproterenol, secreted significantly less protein and peroxidase than the young tissue. The authors suggested that the decrease in the secretion of tears and tear proteins in aged lacrimal gland may be related to different responses to physiologic neurohormonal stimuli, and changes in the peroxidase production of the lacrimal gland might be used as an indicator of age-related lacrimal gland atrophy.[16]

The cytotoxic effect of oxygen-derived free radicals and their possible role in tissue damage have been described and investigated extensively.[17,18] Catalase, an enzyme involved in the enzymatic conversion of hydrogen peroxide, was also detected by biochemical methods in the extracts of corneal epithelium and endothelium but not in the stroma. In corneal epithelium catalase was associated with a subcellular fraction isolated by centrifugation at 13,000 and 100,000 g.[19] The physiologic function of catalase is thought to be the protection of tissues from the oxidative effect of hydrogen peroxide by breaking the H_2O_2 molecule producing H_2O and O_2. Catalase is present in a variety of animal and plant cells. Most of the enzyme activity is located in intracellular vesicules called peroxisomes.[20] Besides catalase superoxide dismutase is also able to scavenge the oxygen-derived free radicals.[21]

Hydrogen peroxide has been shown to be an effective disinfectant and is widely used for hydrogel contact lenses.[22] There is much concern, however, that residual H_2O_2 dissolved from the contact lens may have an unfavorable effect on the cornea.[23,24] Direct exposure of the eye to H_2O_2 at a concentration (3%) used in contact lens disinfecting systems results in tearing, hyperemia, edema, and sometimes in permanent corneal damage.[25] Similar symptoms may result from the misuse of lens-cleaning solutions or from inadequate neutralization.[26] Whether H_2O_2 at much lower concentrations, without causing irritation, may have a long-term toxic effect on the epithelium is still debated. Recently Tripathi and Tripathi[27] reported that hydrogen peroxide at a concentration as low as 30 ppm caused cell retraction as well as cessation of cell movement and mitotic activity of cultured corneal epithelial cells. With a concentration of 50 ppm, normal cell activity ceased almost instantaneously. The effects on corneal epithelial cells of similar or somewhat higher concentrations of H_2O_2 may be different *in vivo*. Tears may further dilute the peroxide and may also have a neutralizing effect. The basal layer of the corneal epithelium is protected in part by the wing and superficial layers. The effectiveness of H_2O_2 as a contact lens disinfectant is unquestioned. How contact lens wearers can be best saved from possible long-term effects of residual hydrogen peroxide and whether there are such effects are problems that still remain to be investigated.[27]

REFERENCES

1. **van Haeringen, N. J.,** Clinical biochemistry of tears, *Surv. Ophthalmol.,* 26, 84, 1981.
2. **Morrison, M. and Allen, P. Z.,** Lactoperoxidase: identification and isolation from Harderian and lacrimal glands, *Science,* 152, 1626, 1966.
3. **Herzog, V.,** Elektronenmikroskopisch-cytochemische Untersuchungen von Katalase und Peroxydase in der Tranendruse der Rattle, *Acta Histochem.,* 13(suppl.), 203, 1973.
4. **Gachon, A. M., Kpamegan, G., Kantelip, B., and Dastugue, B.,** Relationship between lacrimal gland, isolated cells (lacrimocytes) and tears: biochemical and histological studies in the rabbit eye, *Curr. Eye Res.,* 5, 647, 1986.
5. **Garrett, J. R. and Kidd, A.,** Effects of secretory nerve stimulation on acid phosphatase and peroxidase in submandibular saliva and acini in cats, *Histochem. J.,* 9, 435, 1977.
6. **Nickerson, J. F., Kraus, F. W., and Perry, W. I.,** Peroxidase and catalase in saliva, *Proc. Soc. Exp. Biol. Med.,* 95, 405, 1957.
7. **Thomson, J. and Morell, D. B.,** The structure, location and distribution of salivary gland peroxidase, *J. Biochem.,* 62, 483, 1967.
8. **Essner, E.,** Localisation of endogeneous peroxidase in rat exorbital lacrimal gland, *J. Histochem. Cytochem.,* 19, 216, 1971.

9. Iwata, T., Ohkawa, K., and Uyama, M., The fine structural localization of peroxidase activity in goblet cells of the conjunctival epithelium of rats, *Invest. Ophthalmol.,* 15, 40, 1976.

10. van Haeringen, N. J., Ensink, F. T., and Glasius, E., The peroxidase-thiocyanate-hydrogenperoxide system in tear fluid and saliva of different species, *Exp. Eye Res.,* 28, 343, 1979.

11. Thörig, L., van Haeringen, N. J., and Wijngaards, G., Comparison of enzymes of tears, lacrimal gland fluid and lacrimal gland tissue in the rat, *Exp. Eye Res.,* 38, 605, 1984.

12. Thörig, L., van Agtmaal, E. J., Glasius, E., Tan, K. L., and van Haeringen, N. J., Comparison of tears and lacrimal gland fluid in the rabbit and guinea pig, *Curr. Eye Res.,* 4, 913, 1985.

13. Thörig, L., Halperin, M., and van Haeringen, N. J., Cotton-thread tear test: an experimental study for testing drugs suspected of side effects on lacrimation, *Doc. Ophthalmol.,* 58, 307, 1984.

14. van Agtmaal, E. J., Thörig, L., and van Haeringen, N. J., Effect of acetazolamide (Diamox) on tear secretion, *Doc. Ophthalmol.,* 59, 77, 1985.

15. Bromberg, B. B. and Welch, M. H., Lacrimal protein secretion: comparison of young and old rats, *Exp. Eye Res.,* 40, 313, 1985.

16. Bromberg, B. B., Cripps, M. M., and Welch, M. H., Sympathomimetic protein secretion by young and aged lacrimal gland, *Curr. Eye Res.,* 5, 217, 1986.

17. Bulkley, G. B., The role of oxygen free radicals in human disease processes, *Surgery,* 94, 407, 1983.

18. Fridovich, I., The biology of oxygen radicals, *Science,* 201, 875, 1978.

19. Bhuyan, K. C. and Bhuyan, D. K., Catalase in ocular tissue and its intracellular distribution in corneal epithelium, *Invest. Ophthalmol.,* 69, 147, 1970.

20. de Duve, C. and Baudhuin, P., Peroxisomes (microbodies and related particles), *Physiol. Rev.,* 46, 323, 1966.

21. McCord, J. E., The superoxide free radicals: its biochemistry and pathophysiology, *Surgery,* 94, 412, 1983.

22. Krezansoki, J. Z. and Houlsby, R. D., A comparison of new hydrogen peroxide disinfection systems, *J. Am. Optom. Assoc.,* 59, 193, 1988.

23. Gyulai, P., Dziabo, A., Kelly, W., Kiral, R., and Powell, C. H., Relative neutralization ability of six hydrogen peroxide disinfection systems, *Contact Lens Spectrum,* 2, 61, 1987.

24. Yuan, N. and Pitts, D. G., Residual H_2O_2 compromises deswelling function of *in vivo* rabbit cornea, *Acta Ophthalmol.,* 69, 241, 1991.

25. Knopf, H. L. S., Reaction to hydrogen peroxide in a contact lens wearer, *Am. J. Ophthalmol.,* 97, 796, 1984.

26. Chalmers, R. L., Tsao, M., Scott, G., and Roth, L., The rate of *in vivo* neutralization of residual H_2O_2 from hydrogel lenses, Contact Lens Spectrum, July, 21, 1989.

27. Tripathi, B. J. and Tripathi, R. C., Hydrogen peroxide damage to human corneal epithelial cells *in vitro,* Arch. Ophthalmol., 107, 1516, 1989.

Chapter 10

TEAR ENZYMES AND ENZYME INHIBITORS IN CORNEAL ULCERATION AND BACTERIAL AND FUNGAL KERATITIS

I. CLINICAL FEATURES OF AND PATHOLOGICAL FINDINGS IN BACTERIAL AND FUNGAL KERATITIS

The term keratitis is used to describe various corneal inflammations, characterized by the loss of luster and transparency and the cellular infiltration of the cornea. Inflammation of the cornea is different from that in other tissues because the cornea is normally avascular. Inflammatory cells originate from the blood vessels of the limbal parts of the conjunctiva and the sclera and reach the affected region of the cornea via the tear film, migrating through the corneal stroma, and less commonly from the anterior chamber. Physiological or pathological membrane-like structures — basement membranes, stromal lamellas, Descemet's membrane, retrocorneal membranes, or anterior synechiae — may lead the invasion of leukocytes toward the central parts of the cornea.[1]

Microbial keratitis is caused by specific microorganisms (bacteria, viruses, or fungi). The inflammatory reaction in these cases is a direct response to the invading microorganism. Specific symptoms are brought about by the presence of the causative organism, by specific and nonspecific host defense mechanisms, and result from tissue destruction due to the combined action of these two causative factors. Tissue damage is mediated by toxic effects of the microorganisms, by humoral and cellular immunological mechanisms, and by proteolytic enzymes deriving from necrotic cells. Those cases in which no causative microorganism can be detected are probably caused by immunological, mostly autoimmune, mechanisms.[2]

Several classifications of keratites have been developed. Most are based on morphological and pathological features of the diseases. Keratitis can be superficial or deep, ulcerative or nonulcerative, microbial or nonmicrobial. Most superficial keratites are caused by external agents, viruses, or physical influences such as UV light or disorders of the tear film. Deep keratites develop either due to the progression of superficial keratitis involving the stroma and are caused by bacteria or fungi or affect only the stroma and are caused by immunological reactions. Ulcerating keratitis is characterized by extensive tissue destruction involving the corneal epithelium, the Bowman's membrane, the superficial, and the deeper parts of the stroma and usually leads to vascularization and excessive scar formation. Nonulcerative, superficial keratitis often heals without sequelae or leaves small scars. Deep forms of nonulcerating keratitis, however, when they involve collagen destruction, persisting stromal edema, or endothelial decompensation often lead to scarring and vascularization even in the presence of an intact epithelium.[3]

Clinical symptoms of ulcerating keratitis are uniform regardless of the ethiology, including decreased visual acuity, pain, photophobia, redness and swelling of the lids and of the conjunctiva, profuse tearing, and conjunctival discharge. Additional signs are infiltration, edema and tissue destruction in the cornea, hypopyon, and myosis due to consecutive iritis. The severity depends on the infective organism, the reactions of the host, and the effectiveness of the therapy. Bacteria that most often cause ulcerating keratitis are staphylococci, *Diplococcus pneumoniae,* streptococci, and *Pseudomonas aeruginosa.* Less-frequent causative agents of bacterial keratitis are *Klebsiella, Proteus, Moraxella lacunata, Neisserie gonorrheae,* and other Gram negatives.[4]

Ulcerative keratitis may follow different courses, depending primarily on the strain causing the infection and to a less extent on the responses of the host. The ulceration may progress rapidly leading to perforation and the loss of the eye in a relatively short time. The other possibility is slow progression or even indolent ulceration. The first possibility is more frequent with Gram-negative bacteria and is characterized by the elaboration of destructive enzymes such as proteinases, lipase, collagenase, and elastase. The most rapidly evolving infections are caused by *Pseudomonas* strains which elaborate proteases (collagenase, elastase) and endotoxin A.[5]

Ulcers caused by Gram-positive bacteria tend to remain localized; the infiltration is gray-white, with only slight adjacent epithelial and stromal edema. Diffuse graying away from the main infiltration is typical for Gram-negative infections, firstly *Pseudomonas* keratitis. Staphylococcal and streptococcal infections often occur in patients with compromised corneas, a history of herpetic keratitis, dry eyes, aphakic or pseudophakic bullous keratitis, rosacea keratitis, or atopic disease. *Staphylococcus aureus* causes a more-severe infiltration with mild to moderate anterior chamber reaction, while *S. epidermidis* causes a more-indolent ulceration with mild anterior chamber reaction. Staphylococcal keratitis may be multifocal with a large and several small infiltrates. Long-standing staphylococcal ulcers may lead to the formation of corneal abscess with a bad prognosis.[6]

Pneumococcal ulcers often develop following corneal injuries. The infiltration starts at one point usually at the border of the injury and rapidly progresses toward the center. The infiltration is gray, the rest of the corneal stroma is usually clear, fibrin is deposited on the back side of the cornea, the anterior chamber reaction is severe, and there is a hypopyon. Pneumococcal keratitis often leads to perforation. Pneumococcal conjunctivitis does not typically cause a pneumococcal ulcer because it is usually caused by a different type of Pneumococcus which is less virulent than type 4, the causative agent of pneumococcal ulcer.[7]

Keratitis caused by the particularly aggressive *P. aeruginosa* are characterized by extensive tissue necrosis, severe irritative symptoms, and rapid progression (perforation may occur within a few hours). The histologically large part of the corneal stroma is edematous; the necrotic part is infiltrated with debris-laden macrophages containing inclusion bodies. There is a marked loss of keratocytes and a heavy fibrin deposition in the stroma. Ring abscess often develops in early stages of Pseudomonas infection.[8,9]

Corneal infiltrates in mycotic keratitis have typical appearance. They are either stromal infiltrates with feathery, hyphate edges or dry and gray infiltrates somewhat elevated above the level of the corneal surface.[10] Other features such as satellite infiltrates, immune rings, and endothelial plaques frequently occur but are not specific for fungal keratitis. Besides the typical forms seen in keratitis caused by filamentous fungi, other manifestations can be seen in stromal keratitis caused by *Candida albicans* such as "collar button" infiltrates or small circumscribed ulcers.[11]

In spite of these "typical" features, there are no absolute clinical signs that distinguish between keratitis caused by different microorganisms. True differentiation is based on the detection and identification of microorganisms in conjunctival smears and in cultures. The latter also makes possible the determination of the sensitivity of the infective agents toward various antibiotics or chemotherapeutic drugs. The actual clinical appearance of the keratitis, besides the virulence of the infective agent, is determined by the status of the cornea, the duration of the infection prior to therapy, and the response of the host to the insult.[12]

Most cases of ulcerative keratitis begin with epithelial defect which renders the cornea susceptible to infection. Microorganisms are probably not able to pass the intact epithelium, but readily invade the corneal stroma. Besides corneal injury and the presence of the infective agent, other factors (disturbed or missing defense mechanisms) also must be present as not all infected corneal injuries lead to the development of a corneal ulcer. Disturbances of blinking, disorders of the tear film, alterations of the ocular surface, as well as the decrease in local or systemic immune defense mechanisms increase the potential for the onset of a corneal infection.[13] The epithelial defect becomes larger, and tissue destruction reaches the basement membrane, the Bowman's layer, and the superficial parts of the corneal stroma. The surrounding epithelium and the underlying stroma becomes edematous. In the early stages polymorphonuclear leukocytes (PMNs) and lymphocytes invade the affected parts of the cornea. The destruction of the stromal lamellae (collagen and glycosaminoglycans) is effected by hydrolytic enzymes deriving from necrotic cells (PMNs, corneal cells, and bacteria). Besides the passive release from disrupted cells, the active secretion by epithelial cells and keratocytes of collagenase and plasminogen activators (PAs) has been demonstrated in ulcerating corneas.[14-18] In later stages of corneal ulceration macrophages and fibroblasts appear in and around the necrotic area. Macrophages play a role in the removal of the necrotic tissue, while fibroblasts are the key cellular elements of stromal wound healing and scar formation. In favorable cases the destroyed corneal stroma is partly replaced by newly synthesized collagen. The new fibers are usually deposited in an irregular fashion. The regenerating epithelium gradually covers the scar tissue as the ulcer heals. In unfavorable cases the process of tissue destruction progresses until corneal perforation occurs. Following corneal perforation the eye is usually lost due to infection (endophthalmitis or panophthalmitis).[3]

II. BIOCHEMICAL MECHANISMS INVOLVED IN CORNEAL ULCERATION

Corneal ulceration is characterized by the progressive destruction of the cornea. In the ulceration process the extracellular matrix of the stroma is removed (degraded) enzymatically. Since collagen is the major structural protein of the cornea, proteases that degrade collagen fibers in the cornea are most important in effecting the destructive process.

Collagen is abundant in the cornea and constitutes approximately 90% of its total protein content.[19] The majority of the collagen of the corneal stroma is type I and type III,[20] but type V and type VI collagens are also present in the human cornea.[21,21a] Under physiological circumstances collagen fibrils of the cornea are very resistant to common proteases and can be digested only by specific collagenases. Collagenases cleave collagen molecules within their helical region (specific collagenolysis), while some proteases are capable of breaking peptide bonds in the nonhelical telopeptide region (nonspecific collagenolysis), and can further degrade cleavage products of specific collagenolysis. (For details see Chapter 7, Part A.)

The first observation on corneal collagenase was the demonstration that epithelial explants derived from ulcerating corneas lyse reconstituted collagen gel.[4,22,23] Corneal (tissue) collagenase was shown to be produced also by activated keratocytes.[24-27] It is secreted, at least partly, in the form of latent (pro)collagenase.[14,26] Latent collagenase in human cornea was shown to be activated by proteolytic enzymes, first by plasmin generated from plasminogen by PAs.[14,15,26] Corneal collagenase was characterized by Berman and co-workers to be a metalloenzyme, requiring Ca^{++}, and Zn^{++} to perform its enzyme function.[28,29] Collagenases were found in different tissues also in the form of enzyme-inhibitor complexes.[30] Aggregates of corneal collagenase formed with serum antiproteases were also detected in ulcerating corneas, indicating that physiological inhibitors play a role in the regulation of collagenase activity in ulcerating corneas.[15,26] Corneal collagenase was also demonstrated and characterized by immunohistological methods. Ulcerating human corneas were found to react with antibodies against skin collagenase, whereas normal corneal tissue proved to be nonreactive.[31]

Leukocytes play a key role in effecting both the histological and the biochemical changes seen in corneal ulceration resulting from bacterial infection.[32] PMNs heavily infiltrate the cornea around the bacterial ulcer. The necrotic tissue is also full of disrupted and intact leukocytes.[2] The main sources of various hydrolytic enzymes including the proteinases are the destroyed leukocytes in bacterial ulceration.[33] From these enzymes, those with collagenolytic action were studied most intensively. Biochemical analysis of human PMNs has shown the presence in and the production by these leukocytes of various collagenolytic enzymes.[34-37] Prause investigated the role of one of the leukocyte collagenolytic enzymes, the polymorphonuclear leukocyte neutral collagenolytic protease (PMN-c-ase).[38] Crude PMN-c-ase injected into normal rabbit corneas resulted in a collagen destruction similar to the destruction produced by clostridio-peptidase A.[37] The role of PMNs in the degradation of stromal collagen was demonstrated in mouse corneas following the injection of heat-inactivated *P. aeruginosa*[39] and in human corneas following the infection with *Proteus mirabilis*.[40] PMNs were found to mediate histological and biochemical changes responsible for corneal ulceration also in diseases with nonbacterial origin such as chemical and thermal burns,[41,42] keratomalatia due to vitamin A deficiency,[43-46] and in rabbit corneas following the injection of herpes simplex virus.[47] Leukocytes also play a key role in the development of corneal ulcers in patients with corneal immune diseases such as Mooren's ulcer,[17,48,49] ulcers associated with rheumatoid diseases,[50,51] and in patients suffering from Sjögren's syndrome.[52] A possible mechanism for the production of proteolytic enzymes by PMNs in immunological diseases was suggested by Ranadive and Morvat,[53] who demonstrated that immune complexes activate PMNs to release proteinases. Besides PMNs in the cornea, leukocytes invading the conjunctiva were also found to produce proteolytic enzymes in corneal ulceration associated with rheumatoid disease[54] and in Mooren's ulcer.[17]

Besides leukocytes and corneal cells, the infecting organisms may also be direct sources of hydrolytic enzymes. In this respect, *Pseudomonas aeruginosa* and different fungi play the most prominent role. Pseudomonas infections are characterized by rapid and extensive tissue destruction. This bacterium itself is known to collagenolytic enzymes and at least one hydrolytic enzyme that degrades corneal proteoglycans.[55] There exists some controversy whether it also produces a true collagenase.[56] The studies of Kessler and co-workers[55] have shown that the proteoglycan-degrading enzyme produced by *P. aeruginosa* also cleaves through the nonhelical ends of collagen, but does not cleave intrahelically nor solubilize cross-linked collagen. Others emphasized that the activation of host-defense mechanisms also contributes to the enzymatic action and tissue destruction in bacterial ulceration. The endotoxin produced by the bacterium activates the alternate (properidin) pathway of complement-generating chemotactic factors for PMNs.[49] In rabbit experiments topical corticosteroid medication prevented ulceration probably by suppressing the infiltrative response of leukocytes to heat-killed Pseudomonas.[55] Considerable data suggest that both the Pseudomonas itself and the host inflammatory cells are the sources of hydrolytic enzymes that produce tissue destruction in ulceration. The therapeutic use of corticosteroids is controversial as they can arrest the immune response and by doing so prevent host-mediated tissue destruction, but on the other hand they also may promote the spread of the infection within and outside the cornea. High tissue levels of active collagenase were shown to be present in the infected corneas during the active stage of the ulceration, but the topical use of collagenase inhibitors did not prove to be very effective in arresting the ulceration process.[56] The situation concerning the production of hydrolytic enzymes and the activation of the host response mechanism is very similar in case of fungi, though the clinical appearance of the keratitis and the degree of inflammatory signs are different. Fungi, such as *C. albicans* are known to produce proteolytic enzymes in large quantities. This is one reason mycotic keratitis often leads to ulceration and perforation.[8,11]

III. ENZYMES IN TEARS IN CORNEAL ULCERATION

In spite of the early recognition of the enzymatic mechanism of corneal destruction in bacterial ulcers[57] and the large number of publications concerning the detection of proteinases in the histological sections, in tissue extracts, and in culture fluids of ulcerating corneas, there is a dearth of information concerning the presence and activity of these enzymes in the tear fluid.

The first attempts to detect collagenase activity in the tears in the case of bacterial ulcers were unsuccessful, partly because the methods used were not sensitive enough, partly because a large part of the total quantity of tear collagenases is present in inactive form, usually in the form of complexes with serum antiproteases. A third explanation may be that besides true collagenases a number of other proteolytic enzymes is present in the ulcerating corneas and in the tears of patients with corneal ulcers. The differentiation between true collagenases and other collagenolytic proteases is not easy, especially if the collagen substrate used to determine collagenase activity is not intact and nondenatured. When that is the case, the substrate can be cleaved by a number of proteolytic enzymes. These other proteases on the other hand are just as important as the true collagenases because in corneal ulceration they also play an important pathogenetic role by digesting the denatured collagen fibers and collagen fragments.

Prause, in his academical thesis,[38] mentioned that he had attempted to measure collagenase activity in the tears of patients with melting corneal ulcers, using the method of Berman et al.[58] and his collagenolytic assay,[59] without success. He could demonstrate, however, the presence of a collagenolytic enzyme (PMNL neutral collagenolytic protease) in the tears of corneal ulcer patients using a micro-electroimmune assay,[60] probably because this method is more sensitive, and because it is based on the immunological detection on the enzyme protein rather than measuring its enzyme activity.[61] The enzyme was not present in normal tears[62] and in the tears of patients successfully treated with histoacryl glue membrane barrier to prevent the infiltration of the ulcer with PMNs via the tear film.[63] Using the original method of Berman et al.,[58] we detected collagenolytic activity in the tears of patients suffering from Mooren's ulcer in the active phase of the disease.[64,65]

There are few data available concerning the detection of collagenases in the tears of animals with bacterial keratitis and corneal ulcers. Kagonyera et al.[66] studied corneal histology and tear enzymes in healthy and immunomodulated calves following corneal inoculation of *Moraxalla bovis*. Some of the animals were treated with hydroxyurea, others with dexamethasone. Collagenase activity was not present in either of the tear samples in spite of the fact that ulceration was in progress until perforation, and large number of bacteria and leukocytes were detected histologically in the ulcerating corneas. Another enzyme, myeloperoxidase, was present in the tear samples during the active phase of the ulceration and was significantly lower in the hydroxyurea-treated calves.[66]

In a series of experiments, Berman and co-workers measured physiologic (serum) proteinase inhibitors in the tears of patients with corneal ulcers. Both α-1-antitrypsin and α-2-macroglobulin were detected in the tears of the patients with corneal ulcers, and these levels gradually decreased as the ulcers healed. They did not measure collagenase or other proteases in the tear samples, but demonstrated that these inhibitors can form complexes with corneal collagenase. They proposed that the inhibitors derive from the blood through the vessel walls of the conjunctiva and act as a physiologic defense mechanism, inactivating collagenase in the tears and cornea of these patients.[67,68]

We developed a microtiter-plate method[69] for the measurement of collagenolytic activity in tears. The method is based on the measurement of the absorption (optical density) of Commassie brilliant blue-stained collagen gels, reconstituted in the wells of microtiter plates. Collagenolytic activity was calculated from changes in the quantity of the collagen substrate and was given in microgram $\times$ hour $\times$ milliliter values. Tear samples were tested without and with a treatment with trypsin to activate latent collagenase. Trypsin and collagenase inhibitors were used to test the specificity of the assay and to inhibit other proteinases. Normal tears did not contain active collagenase; however, the enzyme was present in a latent form in most of the normal tear samples (the mean value of 16 normal tears was 34.97 μg $\times$ h^{-1} $\times$ ml^{-1}). Three cases with bacterial ulcer showed high active (537.60 to 845.34 μg $\times$ h^{-1} $\times$ ml^{-1}), somewhat lower but also very high inactive (388.18 to 473.00 μg $\times$ h^{-1} $\times$ ml^{-1}) collagenase activities in the tears. To our best knowledge these are the first values for collagenase activities measured in the tears of patients suffering from bacterial ulcers of the cornea. The details of the method and the results of collagenolytic activities measured in the tears of patients with other, nonbacterial ulcers and keratitis are shown and discussed in Chapter 7, Part D.

In 1987, Salonen and co-workers published their revolutionary paper about the appearance of plasmin in the tears of patients with corneal ulcers. Some of their patients were suffering from deep bacterial ulcers or corneal abscess. The plasmin level in the tears of 16 of their keratitis patients was higher than 0.1 μg/ml; in 11 of the cases tear plasmin levels were below 0.1 μg/ml or undetectable. Patients with bacterial keratitis usually exhibited very high tear plasmin levels, 6 to 7 μg/ml. They could not detect PA activities in the tears of their patients; however, such an activity was present in the tears of their healthy controls.[70]

Using a sodium dodecyl sulfate-electrophoresis and copolymerized casein substrate, we demonstrated the presence of various proteolytic enzymes in the tears of a patient suffering from bacterial ulcer of the cornea, and emphasized that besides plasmin other proteolytic enzymes are important both from a pathogenetic and from a diagnostic point of view.[71] In other experiments we detected high PA activities in the tears of patients suffering from various corneal inflammations including corneal ulcers and bacterial keratitis, and demonstrated that the starting point of the pathological process is the release of urokinase type PA from the affected cells.[16,18] This activator then activates plasminogen into plasmin, which, besides its wide proteolytic activity to digest various cell membrane and matrix proteins, is also able to activate procollagenase into collagenase.[14,15]

IV. ENZYME INHIBITOR THERAPY IN CORNEAL ULCERATION

Based on the detection of collagenase in the ulcerating corneas and on experimental data suggesting that they play a key role in the ulcerating process, it has been suggested that the topical application of collagenase inhibitors might be helpful in arresting corneal ulceration. Clinical trials have shown that collagenase inhibitors are not as effective as it had been supposed based on the results of *in vitro* experiments.

Kenyon[72] suggested that the treatment of external eye diseases of bacterial or nonbacterial origin should be approached on three levels: (1) the determination of the specific etiology and initiation of primary therapy, (2) limitation or prevention of ulceration, and (3) promotion of healing and repair.

In bacterial keratitis and corneal ulceration the specific etiology is the infective organisms and the specific therapy is the administration of appropriate antibiotics. Antibiotics can be given topically, subconjunctivally, and systemically. Specific antibiotic therapy should be based on the identification of the infecting bacterium using smears and cultures, and the determination of the susceptibility of the infecting strain to the available antibiotics. To obtain the exact microbiological diagnosis takes time; the initial antibiotic therapy is usually started with broad-spectrum antibiotics effective against the supposed infective agent or against the microorganisms that most often cause the particular type of bacterial corneal disease. The specific therapy can be started as soon as the exact diagnosis and the antibiogram is obtained. The use of conjunctival or corneal smears and specific stainings may be very helpful as they can provide useful information in a very short time.

There is considerable controversy concerning the use of proteinase inhibitors in the treatment of bacterial ulcers. According to Francois, "although collagenase inhibitors are inefficacious in case of noncontrolled infection, they are of great value in trophic, noninfectious ulcerations of the cornea." No well-controlled clinical trial is available concerning the effectiveness of collagenase inhibitors in bacterial ulcers as an adjunct to antibiotic therapy. Even those few clinicians who nowadays use inhibitory eyedrops think of this form of therapy in desperate cases and tend to evaluate the effectiveness based on the general outcome of the disease, whether the ulcer heals or the eye is lost due to perforation. We are of the opinion that proteinase inhibitors (Ca EDTA, cysteine, *N*-acetyl-cysteine, penicillamine, aprotinin) may be helpful in the active stage of bacterial ulceration as a form of adjunct therapy. It is advisable, however, to use them only in cases in which high proteolytic enzyme activity is present in the tears of the patients.

Proteinase inhibitors have no direct beneficial effect on epithelial healing or on the stromal repair processes. On the contrary, due to their toxic effects and owing to the inhibition of enzymes involved in corneal wound healing, they may prevent corneal wound healing. Their administration therefore should be stopped as soon as the excessive protease production is over and the cornea is no longer in danger of progressive melting and perforation. This can be judged by the careful evaluation of the clinical symptoms, by the lack of progression in the ulceration, by the appearance of the first signs of healing, and by the disappearance of proteolytic enzymes from the patients' tears.

Francois et al.[73] provided important data concerning the effects of collagenase inhibitors (cysteine and penicillamine) on the ulceration and the healing of the cornea. The observation that the topical use of collagenase inhibitors resulted in excessive scarring and a decrease in the number of leukocytes invading the cornea indicates the multiple physiological role of proteases in corneal ulceration and informs about the mechanism by which inhibitors may affect the healing process. Inhibitors may cause excessive scarring by preventing the removal of the necrotic tissue. This can be beneficial in preventing ulceration but may have an unfavorable effect on the visual outcome when the ulcer heals. The prevention of leukocyte infiltration of the ulcerating cornea suggests other possible mechanisms for the action of some of these inhibitors. Further details concerning the therapeutic use of various protease inhibitors and their possible side effects are discussed in Chapter 19.

REFERENCES

1. **Yanoff, M. and Fine, B. S.,** *Ocular Pathology,* Harper & Row, Hagerstown, MD, 1975, 261.
2. **Thomas, C. I.,** *The Cornea,* Charles C Thomas, Springfield, IL, 1955, 394.
3. **Hinzpeter, E. N. and Naumann, G. O. H.,** Cornea and sclera, in *Pathology of the Eye,* Naumann, G. O. H. and Apple, D. J., Eds., Springer-Verlag, New York, 1986, 317.
4. **Itoi, M., Gnadinger, M. C., Slansky, H. H., Freeman, M., and Dohlman, C. H.,** Collagenase in the cornea, *Exp. Eye Res.,* 8, 369, 1969.
5. **Vastine, D.,** Infections of the ocular adnexa and cornea, in *Principles and Practice of Ophthalmology,* Peyman, G. A., Sanders, D. R., and Goldberg, M. F., Eds., W.B. Saunders, Philadelphia, 1980.
6. **Hyndiuk, R. A., Nassif, K. F., and Burd, E. M.,** Infectious diseases, in *The Cornea. Scientific Foundations and Clinical Practice,* Smolin, G. and Thoft, R. A., Eds., Little, Brown and Company, Boston, 1983, 147.
7. **Ostler, H. B., Thygeson, P., and Okumoto, M.,** Infectious diseases of the eye. III. Infections of the cornea, *J.C.E. Ophthalmol.,* September, 13, 1978.
8. **Jones, D. B.,** Pathogenesis of bacterial and fungal keratitis, *Trans. Ophthalmol. Soc. U.K.,* 98, 367, 1978.
9. **Raber, I. M., Laibson, P. R., Kurz, G. H., and Bernardino, V. B.,** Pseudomonas corneal scleral ulcers, *Am J. Ophthalmol.,* 92, 353, 1981.

10. **Kaufman, H. E. and Wood, R. M.,** Mycotic keratitis, *Am. J. Ophthalmol.,* 59, 993, 1965.
11. **Foster, R. K.,** Fungal diseases, in *The Cornea. Scientific Foundations and Clinical Practice,* Smolin, G. and Thoft, R. A., Eds., Little, Brown and Company, Boston, 1983, 168.
12. **Liesegang, T. J.,** Bacterial and fungal keratitis, in *The Cornea,* Kaufman, H. E., McDonald, M. B., Barron, B. A., and Waltman, S. R., Eds., Churchill Livingstone, New York, 1988, 217.
13. **Grayson, M. and Keates, R. H.,** *Manual of Diseases of the Cornea,* Little, Brown and Company, Boston, 1969.
14. **Berman, M., Leary, R., and Gage, J.,** Latent collagenase in the ulcerating rabbit cornea, *Exp. Eye Res.,* 25, 435, 1977.
15. **Berman, M. B., Leary, R., and Gage, J.,** Evidence for a role of plasminogen activator-plasmin system in corneal ulceration, *Invest. Ophthalmol. Vis. Sci.,* 19, 1204, 1980.
16. **Berta, A., Tözsér, J., and Holly, F. J.,** Determination of plasminogen activator activities in normal and pathological human tears. The significance of tear plasminogen activators in the inflammatory and traumatic lesions of the cornea and the conjunctiva, *Acta Ophthalmol.,* 68, 508, 1990.
17. **Brown, S. I.,** Mooren's ulcer. Histopathology and proteolytic enzymes of the adjacent conjunctiva, *Br. J. Ophthalmol.,* 59, 670, 1975.
18. **Tözsér, J., Berta, A., and Punyiczki, M.,** Plasminogen activator activity and plasminogen independent amidolytic activity in tear fluid from healthy persons and patients with anterior segment inflammation, *Clin. Chim. Acta,* 183, 323, 1989.
19. **Maurice, D. M. and Riley, M. V.,** The biochemistry of the cornea, in *The Biochemistry of the Eye,* Graymore, C. N., Ed., Academic Press, New York, 1968.
20. **Francois, J. and Victoria-Troncoso, V.,** Molecular biology of the cornea, *Roy. Soc. Med. Ser.,* 40, 29, 1981.
21. **Davidson, P. F. and Canon, D. J.,** Heterogenity of collagens from basement membranes of lens and cornea, *Exp. Eye Res.,* 25, 129, 1977.
21a. **Newsome, A. D., Gross, J., and Hassell, J. R.,** Human corneal stroma contains three distinct collagens, *Invest. Ophthalmol. Vis. Sci.,* 22, 376, 1982.
22. **Brown, S. I., Weller, C. A., and Akiva, S.,** Pathogenesis of ulcers of the alkali-burned cornea, *Arch. Ophthalmol.,* 83, 205, 1970.
23. **Pfister, R. R., McCulley, J. P., Friend, J., and Dohlman, C. H.,** Collagenase activity of intact corneal epithelium in peripheral alkali-burns, *Arch. Ophthalmol.,* 86, 308, 1971.
24. **BenEzra, D. and Tanishima, T.,** Possible regulatory mechanisms of the cornea, *Arch. Ophthalmol.,* 96, 1891, 1978.
25. **Berman, M., Dohlman, C. H., Gnadinger, M., and Davison, P.,** Characterization of collagenolytic activity in the ulcerating cornea, *Exp. Eye Res.,* 11, 255, 1971.
26. **Berman, M., Leary, R., and Gage, J.,** Collagenase from corneal cell cultures and its modulation by phagocytosis, *Invest. Ophthalmol.,* 18, 588, 1979.
27. **Hook, R. M., Hook, C. W., and Brown, S. I.,** Fibroblast collagenase partial purification and characterization, *Invest. Ophthalmol.,* 12, 771, 1973.
28. **Berman, M. B., Kerza-Kwiatecki, A. P., and Davison, P. F.,** Characterization of human corneal collagenase, *Exp. Eye Res.,* 15, 367, 1973.
29. **Berman, M. B. and Manabe, R.,** Corneal collagenases: evidence for zinc metalloenzymes, *Ann. Ophthalmol.,* 5, 1193, 1973.
30. **Sellers, A., Cartwright, E., Murphy, G., and Reynolds, J. J.,** Evidence that latent collagenases are enzyme inhibitor complexes, *Biochem. J.,* 163, 303, 1977.
31. **Gordon, J. M., Bauer, E. A., and Eisen, A. Z.,** Collagenase in human cornea. Immunologic localization, *Arch. Ophthalmol.,* 98, 341, 1980.
32. **Prause, J. U. and Jensen, O. A.,** PAS-positive polymorphonuclear leukocytes in corneal ulcers, *Acta Ophthalmol.,* 58, 556, 1980.
33. **Prause, J. U.,** Healing and repair in corneal tissues. The role of the polymorphonuclear leucocyte, in *The Cornea in Health and Disease,* Trevor-Roper, P. D., Ed., Academic Press, London, 1981, 187.

34. **Cawston, T. E.,** Connective tissue catabolism in relation to the cornea, *Trans. Ophthalmol. Soc. U.K.,* 98, 329, 1978.
35. **Lazarus, G., Daniels, J., Brown, R., Balden, H., and Fullmer, H.,** Degradation of collagen by a human granulocyte collagenolytic system, *J. Clin. Invest.,* 47, 2622, 1968.
36. **Robertson, P. B., Ryel, R. B., Tayler, R. E., Shyre, K., and Fullmer, H.,** Collagenase: localization in polymorphonuclear leukocyte granules in the rabbit, *Science,* 177, 64, 1972.
37. **Rowsey, J. J., Nisbet, R. M., Swedo, J. L., and Katona, L.,** Corneal collagenolytic activity in rabbit polymorphonuclear leukocytes, *J. Ultrastruct. Res.,* 57, 10, 1976.
38. **Prause, J. U.,** Cellular and biochemical mechanisms involved in the degradation and healing of the cornea, *Acta Ophthalmol.,* 168(suppl.), 7, 1984.
39. **Hazlett, L. D., Rosen, D. D., and Berk, R. S.,** Murine corneal response to heat-inactivated *Pseudomonas aeruginosa, Ophthalmic Res.,* 10, 73, 1978.
40. **Mondino, B. J., Brown, S. I., Rabin, B. S., and Lemp, M. A.,** Autoimmune phenomenons of the conjunctiva and the cornea, *Arch Ophthalmol.,* 95, 468, 1977.
41. **Kenyon, K. R., Berman, M. B., Rose, J., and Gage, J.,** Prevention of stromal ulceration in the alkali burned rabbit cornea by glued on contact lenses. Evidence for the role of polymorphonuclear leukocytes in collagen degradation, *Invest. Ophthalmol. Vis. Sci.,* 18, 570, 1979.
42. **Kenyon, K. R., Gipson, I. K., and Hanninen, L.,** Effects of colchicine on experimental corneal burns, *Invest. Ophthalmol. Vis. Sci.,* 19(suppl.), 227, 1980.
43. **Pirie, A.,** Effect of vitamin A-deficiency on the cornea, *Trans. Ophthalmol. Soc. U.K.,* 98, 357, 1978.
44. **Pirie, A., Werb, Z., and Burleigh, M. C.,** Collagenase and other proteinases in the cornea of the retinol-deficient rat, *Br. J. Nutr.,* 34, 297, 1975.
45. **Seng, L. W., Wolf, G., and Kenyon, K. R.,** Collagenase release from corneas of vitamin A-deficient rats, *Invest. Ophthalmol. Vis. Sci. Suppl.,* 19, 227, 1980.
46. **Seng, W. L., Glogowski, J. A., Wolf, G., Berman, M. B., Kenyon, K. R., and Kiorpes, T. C.,** The effect of thermal burns on the release of collagenases from corneas of vitamin A-deficient rats, *Invest. Ophthalmol. Vis. Sci.,* 19, 1461, 1980.
47. **Meyers, R. L. and Pettit, T. H.,** Chemotaxis of polymorphonuclear leukocytes in corneal inflammation: tissue injury in herpes simplex infection, *Invest. Ophthalmol. Vis. Sci.,* 13, 187, 1974.
48. **Foster, R. K., Kenyon, K. R., Greiner, J., Greineder, D. K., Friedland, B., and Allansmith, M. R.,** The immunpathology of Mooren's ulcer, *Am. J. Ophthalmol.,* 88, 149, 1979.
49. **Mondino, B., Rabin, B., Kessler, E., Gallo, J., and Brown, S.,** *Arch. Ophthalmol.,* 95, 2222, 1977.
50. **Brown, S. I. and Grayson, M.,** Marginal furrows: a characteristic corneal lesion of rheumatoid arthritis, *Arch. Ophthalmol.,* 79, 563, 1968.
51. **Jayson, M. I. V. and Easty, D. L.,** Ulceration of the cornea in rheumatoid arthritis, *Ann. Rheum. Dis.,* 36, 428, 1977.
52. **Pfister, R. R. and Murphy, G. E.,** Corneal ulceration and perforation associated with Sjögren's syndrome, *Arch. Ophthalmol.,* 98, 89, 1980.
53. **Ranadive, N. S. and Morvat, H. Z.,** Tissue injury and inflammation induced by immune complexes, in *Inflammation, Immunity and Hypersensitivity,* Morvat, H. Z., Ed., Harper & Row, New York, 1979, 409.
54. **Eiferman, R. A., Carothers, D. J., and Yankeelov, J. A., Jr.,** Peripheral rheumatoid ulceration and evidence for conjunctival collagenase production, *Am. J. Ophthalmol.,* 87, 703, 1979.
55. **Kessler, E., Mondino, B., and Brown, S.,** The corneal response to *Pseudomonas aeruginosa:* histopathologic and enzymatic characterization, *Invest. Ophthalmol. Vis. Sci.,* 16, 116, 1977.
56. **Berman, M. B.,** Collagenase and corneal ulceration, in *Collagenase in Normal and Pathological Connective Tissues,* Woolley, D. E. and Evanson, J. M., Eds., John Wiley & Sons, New York, 1980, 141.

57. **Fischer, E. and Allen, J. H.,** Corneal ulcers produced by cell-free extracts of *Pseudomonas aeruginosa, Am. J. Ophthalmol.,* 46, 21, 1958.

58. **Berman, M. B., Manabe, R., and Davison, P. F.,** Tissue collagenase: a simplified semiquantitative enzyme assay, *Anal. Biochem.,* 54, 522, 1973.

59. **Prause, J. U.,** Collagenolysis of rat tail tendons by crude corneal collagenase and clostridiopeptidase A, *Graefe's Arch. Clin. Exp. Ophthalmol.,* 199, 249, 1976.

60. **Prause, J. U.,** Immunoelectrophoretic determination of tear fluid proteins collected by the Schirmer I test, *Acta Ophthalmol.,* 57, 959, 1979.

61. **Prause, J. U.,** Serum albumin, serum antiproteases and polymorphonuclear leucocyte neutral collagenolytic protease in the tear fluid of patients with corneal ulcers, *Acta Ophthalmol.,* 61, 272, 1983.

62. **Prause, J. U.,** Serum albumin, serum antiproteases and polymorphonuclear leucocyte neutral collagenolytic protease in the tear fluid of normal healthy persons, *Acta Ophthalmol.,* 61, 261, 1983.

63. **Prause, J. U.,** Serum antiproteases and polymorphonuclear leucocyte neutral collagenolytic protease in the tear fluid of patients with corneal ulcers treated with *N*-butylcyano acrylate glue, *Acta Ophthalmol.,* 61, 283, 1983.

64. **Berta, A., Zajácz, M., and Jaross, N.,** A könnyvizsgálatok jelentösége az ulcus rodens corneae (Mooren fekély) diagnosztikájában. [Significance of tear protein tests in the diagnosis of ulcus rodens corneae (Mooren's ulcer)], *Szemészet,* 124, 220, 1987.

65. **Zajácz, M. and Berta, A.,** Proteolytische Enzymaktivitat und chirurgusche eingriffe bei ulcus Mooren, *Klin. Mbl. Augenheilk.,* 187, 401, 1985.

66. **Kagonyera, G. M., George, L. W., and Munn, R.,** Light and electron microscopic changes in corneas of healthy and immunomodulated calves infected with *Moraxella bovis, Am. J. Vet. Res.,* 49, 386, 1988.

67. **Berman, M. B., Barber, J. C., Talamo, R. C., and Langley, C. E.,** Corneal ulceration and the serum antiproteases. I. Alpha-1-antitrypsin, *Invest. Ophthalmol. Vis. Sci.,* 12, 759, 1973.

68. **Berman, M., Gordon, J., Gardia, L. A., and Gage, J.,** Corneal ulceration and the serum antiproteases. II. Complexes of corneal collagenases and alpha-macroglobulins, *Exp. Eye Res.,* 20, 231, 1975.

69. **Berta, A., Punyiczki, M., and Tözsér, J.,** Collagenase in normal and pathological human tears determined by microtiter plate method, in *Ophthalmology Today,* Ferraz de Oliveira, L. N., Ed., Elsevier Science, Amsterdam, 1988, 503.

70. **Salonen, E.-M., Tervo, T., Törma, E., Tarkkanen, A., and Vaheri, A.,** Plasmin in tear fluid of patients with corneal ulcers: basis for new therapy, *Acta Ophthalmol.,* 65, 3, 1987.

71. **Punyiczki, M., Berta, A., and Tözsér, J.,** Study of neutral proteinases of human tear samples using gels containing substrates, *Clin. Chem. Enzym. Comns.,* 1, 115, 1988.

72. **Kenyon, K. R.,** Decision-making in the therapy of external disease, *Ophthalmology,* 89, 44, 1982.

73. **Francois, J., Cambie, E., Fehér, J., and van den Eeckhout, E.,** Collagenase inhibitors (penicillamine), *Ann. Ophthalmol.,* 5, 391, 1973.

Chapter 11

TEAR ENZYMES AND THE WOUND HEALING OF THE CORNEA

Part A
Corneal Wound Healing

I. INTRODUCTION

Wound closure in skin and mucous membranes is affected by two mechanisms: (1) there is a temporary cover of the defect zone by insoluble protein exudate, which staunches the flow of body fluid and protects the wounded tissue from external infections and (2) there is a rapid activation of the surrounding epithelium to cover the defect zone.

The activity of the adjacent epithelial cells involves cellular movements as well as cellular division to restore the surface integrity of the wound. The migration of the epithelium covering the defect zone starts from the wound perifery in deep wounds. In superficial abrasions when some epithelium is retained within the wounded area the remaining parts of the basal layer are also points where cell division and migration begins.[1]

The migrating epithelium covering corneal epithelial or stromal defects consists not of dissociated cells but of a unified sheet preserving its continuity when it moves.[2] Epithelium covers the defect zone by a process of active horizontally directed movement, and not by growth pressure.[3]

Though the basic mechanisms of cell division and cellular movements are the same in all corneal repair mechanisms, important differences exist among simple resurfacing of superficial abrasions, the healing of deep stromal defects, and that of incisional (surgical) or perforating traumatic wounds. The latter involves the activation of keratocytes, corneal fibroblasts, and the synthesis of collagen and proteoglycans resulting in scar formation.

II. REEPITHELIZATION

Reepithelization or epithelial wound closure following the simple mechanical removal of the corneal epithelium is effected by the migration of the intact epithelial cells surrounding the defect area. The healing process of simple epithelial wounds consists of two phases: the latent phase and the actual healing phase.[4]

The initial, latent phase is characterized by the lack of change in the area of the wound, i.e., by the absence of measurable wound closure. The duration of this phase has been shown to be approximately 6 h in animal experiments. Important changes occur, however, during this "latent" phase on the cellular level. As a result of cellular and subcellular reorganization of the wing and columnar cells of the epithelium at the wound margin, the normally stratified epithelium changes into a single layer of motile cells at the leading edge. These cellular and subcellular changes include the desquamation of squamous cells, the loss of columnar appearance of basal cells, formation of cellular processes at the wound edge, and the gradual disappearance of hemidesmosomes between the basal cells and the basal membrane of the epithelium.[5]

The latent phase is followed by the healing phase. This phase is characterized by a gradual decrease in the wound area lasting from the end of the latent period until wound closure is complete. Wound closure occurs at a constant rate when measured as a decrease in the radius of the defect zone. The rate of decrease of the wound radius has been shown to be 64 μm/h in the rabbit, and 84 μm/h in the monkey following mechanical scraping of the corneal epithelium.[4,6] The leading edge of the epithelium is composed of a single layer of transformed epithelial cells. These transformed cells migrate across the defect zone to close the epithelial defect. Migrating cells were shown to originate from the basal cell or deep wing cell layers of the epithelium. Transformation is characterized by the appearance of ruffled membranes and filopodia at the advancing edge of the migrating cells demonstrated by scanning electron microscopy in rabbit experiments.[5]

III. HEALING OF STROMAL DEFECTS

Following a penetrating corneal injury the stromal lamellae adjacent to the wound become edematous, and the defect, if it is not too wide, soon fills with a fibrin clot. Keratocytes located near the defect zone are either killed or paralyzed, shown by the withdrawal of their cytoplasmic processes.[7] The surviving cells increase their RNA content and endoplasmic reticulum cisternae, and begin protein synthesis within 2 h.[8] Within 2 to 6 h polymorphonuclear leukocytes (PMNs) appear in the wound area. These cells are responsible for the enzymatic removal of the debris from the wound. The first PMNs come from the limbal blood vessels and reach the defect zone via the tear film. Another way for PMNs to approach the wound area is by migrating through the stroma. By 24 h the DNA synthesis and tritiated thymidine uptake of keratocytes reaches their maximum. The so-called reactive keratocytes line up in the stroma parallel with the wound margins and start secreting collagen (mostly type II collagen) and proteoglycan components of the extracellular matrix, beginning with keratan sulfate.[9-11]

By the end of the 1st week, fibroblasts and PMNs have invaded the fibrin plug. With increasing deposition of collagen the tensile strength of the fibrotic tissue in the wound gradually increases. The secreted collagen is randomly organized in the extracellular space. The fibers have a larger diameter than those in the intact corneal stroma. The continuing wound healing process involves additional collagen secretion. By the end of 8 weeks, there are no inflammatory cells in the scar tissue. The only cells present are keratocytes showing signs of fibroblastic conversion. The wound strength continues to increase for 3 to 6 months. Macromolecules and collagen cross-links in the developing scar tissue after penetrating stromal injury of the adult rabbit cornea were shown to be similar to those found in developing corneas.[12] The scar tissue is more compact and less transparent than the adjacent stroma, though the transition to the surrounding tissues is gradual and continuous. With time the pattern of macromolecules and cross-links becomes more similar to that in normal stromal tissue, and the transparency improves, but the uniformity of fibrillar organization is never completely restored.[13]

A simple mechanical removal of the corneal epithelium also affects the keratocytes in the superficial layers of the corneal stroma. It has been demonstrated that stromal keratocytes disappear following scraping of the corneal epithelium, most likely due to the mechanical disruption of these cells. The disruption of cells was suggested to be brought about by the increased tension of the anterior stromal lamellae occurring due to the edema of the anterior stroma.[5] The repopulation with keratocytes of the superficial layers of the corneal stroma soon occurs when the epithelial defect heals and normal stromal hydration is restored. This may be one of the mechanisms by which reepithelization enhances the repair of stromal defects in corneal ulceration. These data also support the concept that restoration of normal corneal hydration enhances stromal wound healing.

IV. NEUROPEPTIDES AND CORNEAL WOUND HEALING

The cornea is densely innervated. Nerve fibers enter the corneal epithelium. The innervation density in some portions of the cornea is 300 to 600 times that of the skin and 20 to 40 times that of the tooth pulp.[14] Various diseases and traumatic lesions of the trigeminal nerve or the ciliary ganglion, like keratectomy or excimer laser photoablation, often result in sectoral hypesthesia cornea. In hypesthetic corneas delayed wound healing often occurs and keratitis neuroparalytica may develop.

Neuropeptides besides being the initial mediators of neurogenic inflammation are involved in the regulation of the wound healing process. Neuropeptides (substance P, calcitonin gene-related peptide, and vasoactive intestinal peptide) are co-released by axon reflexes from the sensory neurons following noxious stimuli, inflammation, and trauma in various parts of the human body. These peptides directly, and indirectly, via the induction of other mediators, trigger a cascade of events, including increased blood flow, vascular permeability changes, plasma extravasation, polymorphonuclear and mononuclear phagocyte chemotaxis and activation, and lymphocyte stimulation, in addition to modulating various forms and stages of wound healing.[15]

Substance P, an undecapeptide of the tachykinin class, is the most extensively studied and characterized neuropeptide. It is present in the central and peripheral processes of small diameter, unmyelinated sensory neurons (C fibers) in most tissues, including neurons of the trigeminal ganglia supplying the eye. Although substance P is considered to be a neurotransmitter at the central terminals of the C fibers, more than 90% of substance P synthesized in the neural cell bodies is transported down the axon to the peripheral nerve terminals. Substance P plays an important role in wound healing by enhancing proliferation of fibroblasts and smooth muscle cells. These growth-promoting activities may be the result of similar amino acid sequences of substance P and growth factors. Substance P was also suggested to play a role in angiogenesis. It has been demonstrated to modulate growth factor gene regulation in human capillary endothelial cells. These studies suggest that substance P may play a role in the regulation of the latter stages of the wound healing process.[15-17]

Calcitonin gene-related peptide is an oligopeptide with 37 amino acid residues that is the product of a limited proteolysis of longer peptides, calcitonin gene transcripts.[18] Calcitonin gene-related peptide was demonstrated by immunochemical techniques in the neurons of most of the sensory ganglia including the trigeminal ganglia. It has been shown to coexist with and to potentiate the activity of substance P and other tachykinins. In addition to its activities relating to neurogenic inflammation, an enhancing effect on fibroblast and smooth muscle cell proliferation has been reported.[16]

Vasoactive intestinal peptide is a neuropeptide containing 28 amino acids. It was first detected in intestinal neurons, but it is also present in other nerve terminals. Vasoactive intestinal peptide seems to be involved in a number of different inflammatory reactions including the dilatation of small blood vessels, and immunoregulatory and immunosuppressive activities. Vasoactive intestinal peptide was found in PMNs and mast cells. These cells release the neuropeptide following stimulation. It is not clear, however, whether the peptide is synthesized within these cells or simply accumulated in store granules for later release.[19]

Corneal nerve fibers were shown to contain neuropeptides (substance P, and calcitonin gene-related peptide). Sasoake et al.[20] using a whole-mount procedure detected substance P-containing nerve fibers in rat cornea. These fibers enter the cornea from the middle layers of the sclera and from the episclera. The nerve fibers form bundles in the uppermost part of the corneal stroma with fibers entering the superficial layers of the epithelium where they terminate. The distribution of sensory nerve fibers containing calcitonin gene-related peptide in the rat cornea is similar to that of substance P-containing fibers. A higher percentage of nerve fibers in the cornea and in the trigeminal ganglion contain calcitonin gene-related peptide than substance P. Most substance P-positive fibers also contain calcitonin gene-related peptide, but only some of the calcitonin gene-related peptide containing neurons contain also substance P.[21] Vasoactive intestinal peptide was not detected in the corneas of different species. Nerve fibers containing this neuropeptide, however, were found in the limbal region most often in association with blood vessels.[22]

The physiological function of neuropeptides in the cornea is unclear. Stimulation of the trigeminal nerve causes the release of substance P and other neuropeptides resulting in anterior segment inflammation. Retrobulbar capsaicin injections were shown to have similar effects, inducing the local release of neuropeptides from the sensory nerves evoking a cascade of inflammatory changes in the cornea.[23] Surgical denervation of the cornea on the other hand caused a decrease in substance P-containing fibers of the cornea and a delay in corneal wound healing by preventing the reestablishment of corneal homeostasis in the process of wound healing.[20]

The role of neuropeptides in corneal wound healing is multifaceted. The release of neuropeptides from the sensory nerves of the cornea is triggered by the stimulus causing the injury of the cornea. A nociceptive reflex induces antidromic impulses leading to the release of neuropeptides (substance P and calcitonin gene-related peptide) at the nerve endings of the corneal sensory fibers. Substance P stimulates chemotaxis into the cornea, activates polymorphonuclear and mononuclear phagocytes, and promotes the release of inflammatory mediators (arachidonic acid metabolites, interleukins, etc.). Substance P activates T and B lymphocytes, promotes proliferation of keratocytes, and helps to initiate the remodeling processes. Calcitonin gene-related peptide potentiates the activity of substance P and inhibits its breakdown. Vasoactive intestinal peptide may act by modifying tear secretion and via effects on epithelial cells.[15]

V. PHYSIOLOGICAL GROWTH FACTORS AND CORNEAL WOUND HEALING

Epidermal growth factor (EGF) was first isolated from the submaxillary gland of mice.[24] It is a polypeptide of 53 amino acids with a molecular weight of 6045 Da. The structure of the molecule is stabilized by three disulfide bonds. Mouse EGF is a potent stimulator of proliferation and maturation of various cells of mesodermal and ectodermal origin.[25,26]

Human EGF differs from the mouse EGF in amino acid sequence and physicochemical properties.[27] Human EGF contains 53 amino acids and 3 disulfide bonds, but differs from its mouse counterpart in 16 amino acid positions.[28] Human EGF (urogastone) was found in various tissues and body fluids including plasma, urine, saliva, breast milk, and tears.[29] The biological activities of the two molecules (human and mouse EGF) are similar and both bind to the same membrane receptors.[30]

The positive effect of EGF on corneal proliferation was first demonstrated by Savage and Cohen.[31] Covelli et al.,[32] using isotope-labeled mouse EGF, demonstrated that corneal epithelial cells take up the growth factor in greater amounts than any other cell studied. Both human and mouse EGF were found to stimulate epithelial growth in culture experiments. Frati et al.[33] demonstrated that mouse EGF accelerated the healing of simple corneal scrape wounds in rabbit experiments. They also suggested that a different healing mechanism may occur in the presence of EGF. Topically applied mouse EGF (0.05 to 0.5 mg/ml) was found to accelerate the healing of mechanical wounds and alkali burns in rabbit experiments.[34,35] In humans the healing rate of epithelial lesions was doubled in traumatic abrasions, bacterial ulcers, and caustic injuries using topical mouse EGF (2 mg/ml).[36]

Most of the experiments carried out utilized mouse EGF. Recently the synthesis of human EGF has been solved using a recombinant DNA technique,[36a] providing sufficient quantities of human EGF for experimental and clinical evaluation. The few experiments concerning the effectiveness of human EGF showed that the topical administration of this polypeptide (0.1 mg/ml) in combination with an antibiotic and a steroid accelerated the epithelial healing rate in nonpenetrating scrape wounds and increased wound strength following full-thickness stromal incisions. The effect of EGF on the healing process does not seem to be dose dependent and the response to EGF is reduced with increasing stromal damage.[37]

Human fibroblast growth factor (FGF) is also an agent that has been found to promote corneal wound healing in rabbit experiments.[38] FGF is a polypeptide that has a high affinity for basal membrane, extracellular matrix components, and heparin. It has been detected in various tissues and organs, including the cornea, retina, pituitary gland, and human placenta. Two forms of FGF are known: acidic and basic, showing 55% absolute homology. FGF was shown to effect the morphology, proliferation, differentiation, and senescence of various proliferating cells in culture experiments.[39]

The two forms of FGF share an affinity for heparin, and show basically similar activities toward a variety of cells of mesodermal and ectodermal origin. Basic FGF has a higher affinity towards heparin and heparin-like structures than acidic FGF. Both forms are mitogenic and modulate the synthesis of extracellular matrix in connective tissues.[40,41]

FGF has been shown to bind to basement membranes in frozen sections of the embryonic eye. This binding could be prevented or reversed by the addition of heparin. Heparinase treatment of the basement membranes had a similar effect. A proteoglycan, heparin sulfate, was shown to be the basement membrane component that has a high affinity for both forms of FGF.[42]

Topically applied basic FGF extracted from human placenta was found to enhance the regeneration of corneal epithelium in rabbit experiments.[38] The healing rate of the corneal epithelium in the treated eyes was 120% of that in the untreated control animals. Intraocular injections of the same agent caused functional recovery and regeneration of the wounded endothelium of rabbit corneas. These reactions did not seem to be dose dependent; minute amounts of FGF were sufficient to obtain both of these effects. FGF had a beneficial effect on the healing of viral epithelial defects in experimental keratitis, similar to the positive effect of EGF demonstrated in human herpetic keratitis.[36] Growth factors, besides their direct effects enhancing reepithelization, may also counteract the adverse effects of antiviral drugs on cell division.

REFERENCES

1. **Kuwabara, T., Perkins, D. G., and Cogan, D. G.,** Sliding of the epithelium in experimental corneal wounds, *Invest. Ophthalmol.,* 15, 4, 1976.
2. **Gipson, I. K. and Kioprpes, T. C.,** Epithelial sheet movement: protein and glycoprotein synthesis, *Dev. Biol.,* 92, 259, 1982.
3. **Gipson, I. K., Westcott, M. J., and Brooksby, N. G.,** Effects of cytochalasins B and D and colchicine on migration of corneal epithelium, *Invest. Ophthalmol. Vis. Sci.,* 22, 633, 1982.
4. **Crosson, C. E., Klyce, S. D., and Beuerman, R. W.,** Epithelial wound closure in the rabbit cornea. A biphasic process, *Invest. Ophthalmol. Vis. Sci.,* 27, 464, 1986.
5. **Crosson, C. E.,** Cellular changes following epithelial abrasion, in *Healing Processes in the Cornea,* (Advances in Applied Biotechnology Series, Vol. 1), Beuerman, R. W., Crosson, C. E., and Kaufman, H. E., Eds., Gulf, Houston, 1989, 3.
6. **Pendroza, L., Crosson, C. E., and Beuerman, R. W.,** *Invest. Ophthalmol. Vis. Sci. Suppl.,* 29, 195, 1988.
7. **Wolter, J. R.,** Reactions of the cellular elements of the corneal stroma, *Arch. Ophthalmol.,* 59, 873, 1958.
8. **Smelzer, G. K. and Ozanics, V.,** Reaction of the cornea to injury and wound healing, *Trans. New Orleans Acad. Ophthalmol.,* 239, 1972.
9. **Baum, J. L.,** Source of the fibroblasts in central wound healing, *Arch. Ophthalmol.,* 85, 473, 1971.
10. **Dunnington, J. H. and Smelser, G. K.,** Incorporation of S^{35} in healing wounds in normal and devitalized corneas, *Arch. Ophthalmol.,* 60, 116, 1958.
11. **Robb, T. M. and Kuwabara, T.,** Corneal wound healing. II. An autoradiographic study of the cellular components, *Arch. Ophthalmol.,* 72, 401, 1964.
12. **Cintron, C., Hassinger, L. C., Kublin, C. L., and Cannon D. J.,** Biochemical and ultrastructural changes in collagen during corneal wound healing, *J. Ultrastruct. Res.,* 65, 13, 1978.
13. **Kenyon, K. R.,** Morphology and pathological responses of the cornea to disease, in *The Cornea. Scientific Foundations and Clinical Practice,* Smolin, G. and Thoft, R. A., Eds., Little, Brown and Company, Boston, 1983, 43.
14. **Rózsa, A. J. and Beuerman, R.,** *Pain,* 14, 105, 1982.
15. **Waldrep, J. C.,** Neurogenic inflammation of the cornea, in *Healing Processes in the Cornea,* (Advances in Applied Biotechnology Series, Vol. 1), Beuerman, R. W., Crosson, C. E., and Kaufman, H. E., Eds., Gulf, Houston, 1989, 27.
16. **Nilsson, J., von Euler, A. M., and Dalsgaard, C. J.,** Stimulation of connective tissue cell growth by substance P and substance K, *Nature,* 315, 61, 1985.
17. **Payan, D. G.,** Receptor-mediated mitogenic effects of substance P on cultured smooth muscle cells, *Biochem. Biophys. Res. Commun.,* 130, 104, 1985.
18. **Dockrey, G. J.,** *ISI Atlas Sci.: Pharmacology,* 2, 40, 1988.
19. **O'Dorsio, M. S., Wood, C. L., and O'Dorsio, T. M.,** Vasoactive intestinal peptide and neuropeptide modulation of the immune response, *J. Immunol.,* 135, 792, 1985.
20. **Sasoake, A. et al.,** *Invest. Ophthalmol. Vis. Sci.,* 25, 351, 1984.
21. **Stone, R. A., Kuwayama, Y., and Laties, A. M.,** Regulatory peptides in the eye, *Experientia,* 43, 791, 1987.
22. **Stone, R. A.,** Vasoactive intestinal polypeptide and the ocular innervation, *Invest. Ophthalmol. Vis. Sci.,* 27, 951, 1986.
23. **Burks, T. F., Buck, S. H., and Miller, M. S.,** *Fed. Proc.,* 44, 2531, 1985.
24. **Cohen, S.,** Isolation of a mouse submaxillary gland protein accelerating incisor eruption and eyelid opening in the newborn animal, *J. Biol. Chem.,* 237, 1555, 1962.

25. **Cohen, S., Carpenter, G., and Lembach, K. J.,** Interaction of epidermal growth factor (EGF) with cultured fibroblasts, *Adv. Metab. Disord.,* 8, 265, 1975.

26. **Cohen, S. and Elliott, G. A.,** The stimulation of epidermal keratinization by a protein isolated from the submaxillary gland of the mouse, *J. Invest. Dermatol.,* 40, 1, 1963.

27. **Cohen, S. and Carpenter, G.,** Human epidermal growth factor: isolation and chemical and biological properties, *Proc. Natl. Acad. Sci. U.S.A.,* 72, 1317, 1975.

28. **Savage, C. R., Jr., Inagami, T., and Cohen, S.,** The primary structure of epidermal growth factor, *J. Biol. Chem.,* 247, 7612, 1972.

29. **van Setten, G.-B., Viinika, L., Tervo, T., Pesonen, K., Tarkkanen, A., and Perheentupa, J.,** Epidermal growth factor is a constant component of normal tear fluid, *Graefe's Arch. Clin. Exp. Ophthalmol.,* 227, 184, 1989.

30. **Hollenberg, M. D. and Armstrong, G. D.,** *Polypeptide Hormone Receptors,* Marcel Dekker, New York, 1985.

31. **Savage, C. R., Jr. and Cohen, S.,** Proliferation of corneal epithelium induced by epithelial growth factor, *Exp. Eye Res.,* 15, 361, 1973.

32. **Covelli, I., Rossi, R., Mozzi, R., and Frati, L.,** Synthesis of bioactive ^{131}I-labeled epidermal growth factor and its distribution in rat tissues, *Eur. J. Biochem.,* 27, 225, 1972.

33. **Frati, L., Daniele, S., Delogu, A., and Covelli, L.,** Selective binding of epidermal growth factor and its specific effects on the epithelial cells of the cornea, *Exp. Eye Res.,* 14, 135, 1972.

34. **Culbertson, W. W. et al.,** *Invest. Ophthalmol. Vis. Sci.,* 19, 74, 1980.

35. **Singh, G. and Foster, C. S.,** Epidermal growth factor in alkali-burned corneal epithelial wound healing, *Am. J. Ophthalmol.,* 103, 802, 1987.

36. **Daniele, S., Frati, L., Fiore, C., and Santoni, G.,** The effect of the epidermal growth factor (EGF) on the corneal epithelium in humans, *Graefe's Arch. Clin. Exp. Ophthalmol.,* 210, 159, 1979.

36a. **Urdea, M. S., Merryweather, J. P., Mullenbach, G. T., Coit, D., Heberlein, U., Vanzuella, P., and Barr, P. J.,** Chemical synthesis of a gene for human epidermal growth factor urogastone and its expression in yeast, *Proc. Natl. Acad. Sci. U.S.A.,* 80, 7461, 1983.

37. **Stern, M. E., Brazzell, R. K., Beuerman, R. W., Aquavella, J. V., and Kirschner, S. E.,** The effects of human recombinant epidermal growth factor on epithelial wound healing, in *Healing Processes in the Cornea,* (Advances in Applied Biotechnology Series, Vol. 1), Beuerman, R. W., Crosson, C. E., and Kaufman, H. E., Eds., Gulf, Houston, 1989, 69.

38. **Assouline, M., Montefiore, G., and Pouliquen, Y.,** Fibroblast growth factor effects on corneal wound healing in the rabbit, in *Healing Processes in the Cornea,* (Advances in Applied Biotechnology Series, Vol. 1), Beuerman, R. W., Crosson, C. E., and Kaufman, H. E., Eds., Gulf, Houston, 1989, 79.

39. **Gospodarowicz, D., Neufeld, G., and Schweigerer, L.,** Molecular and biological characterization of fibroblast growth factor, *Cell Differ.,* 19, 1, 1986.

40. **Kem, P. et al.,** *Exp. Cell. Res.,* 149, 85, 1983.

41. **Moczar, E., Tassin, J., and Courtois, Y.,** Interaction of bovine epithelial lens (BEL) cells with extracellular matrix (ECM) and eye-derived growth factor (EDGF). III. Control of glycoprotein and proteoglycan synthesis, *Exp. Cell Res.,* 149, 95, 1983.

42. **Vigny, M., Ollier-Hartman, M. P., Lavigne, M., Fayein, N., Jeanny, J. C., Laurent, M., and Courtois, Y.,** Specific binding of basic fibroblast growth factor to basement membrane-like structures and to purified heparan sulfate proteoglycan of the EHS tumor, *Cell Physiol.,* 128, 475, 1988.

Chapter 11

TEAR ENZYMES AND THE WOUND HEALING
OF THE CORNEA

Part B
Fibronectin and the Wound Healing of the Cornea

Fibronectins are cell-adhesive and matrix-organizing glycoproteins. They occur in insoluble form in basal membranes and in connective tissue matrices and in soluble form in the blood plasma.[17] They also aid wound healing by stimulating the migration and the attachment of cells by forming (together with other molecules) the initial connective tissue matrix on the surface of wounds. Many types of cells were shown to produce fibronectins in tissue and cell cultures. *In vivo* fibronectins are secreted by and assembled into an insoluble matrix in the extracellular space around fibroblasts, and macrophages, and under epithelial cells. At wound healing sites, fibronectin is thought to originate from cells involved in wound healing, as well as from the deposition of plasma.[2]

Fibronectins are complex dimeric glycoproteins composed of two similar but nonidentical subunits of about 250,000 Da joined near their carboxyl terminal by disulfide bonds.[3] Amino acid sequence analysis revealed that fibronectin molecules contain repeated amino acid sequences that can be defined as three different types of polypeptide domains (types I, II, and III). These domains are partially (25 to 50%) homologous. Type I homology region occurs 12 times in the molecule, 9 times in the amino-terminal fibrin- and collagen-binding sequence, and 3 times in the carboxyl-terminal fibrin binding region.[3] Type II repeated regions are confined to the collagen binding domain in the amino terminal. The remaining central sequence contains either 15 (plasma fibronectin) or 16 (cellular fibronectin) type III repeated homologous polypeptide regions.[2]

Fibronectins are secreted by various types of cells in culture, and *in vivo,* and are also normally found in serum. At least 3 species of fibronectin are produced from a single gene by a pattern of alternative splicing: the "normal" fibronectin and 2 other variant lacking 95 and 120 amino acids, respectively. However, no functional differences between these species have yet been elucidated.[4] Fibronectins, besides being constant components of extra- and pericellular matrices, are present in various normal tissues in association with basement membranes. They are considered to be the major structural proteins that are responsible for the attachment of different cells to the surrounding structures and for the adhesion of epithelial cells to basal membranes. Fibronectins form fibrillar arrays on the surface of different types of cells and on the surface of basal membranes; they interconnect cells with each other and with basal membranes.[5]

Fibronectins were found to interact with a number of macromolecules such as collagen,[6] glycosaminoglycans,[7] plasminogen, plasminogen activators (PAs),[8] and C-reactive protein.[9] Fibronectins possess specific high-affinity binding sites for collagens, proteoglycans, cell membrane proteins, and matrix proteins and polysaccharides. The cell-binding domain of the fibronectin molecule contains a tetrapeptide sequence — Arg-Gly-Asp-Ser — which is the minimal structure required for attachment to the cell membrane. The cell membrane contains a 140-kDa glycoprotein that binds tightly and specifically to the cell-binding tetrapeptide region of fibronectin.[10] This receptor molecule also connects extracellular fibronectin to intracellular microfilaments. Vinculin and actin are also found in the cytoplasm near the place where fibronectin is attached to the plasma membrane. The interaction of fibronectin and intracellular contractile structures is believed to play a role in cellular movements in embryonal development and the healing processes.[11]

Fibronectin molecules immunologically and structurally very similar to insoluble fibronectin found on cell surfaces and extracellular matrices are present in plasma in soluble form. Plasma fibronectin is also designated as "cold insoluble protein" because of its tendency to coprecipitate with cryofibrinogen.[12] Both cellular and plasma fibronectins are able to bind to a number of other macromolecules including fibrin, collagen, heparin, and proteoglycans. Recent studies demonstrated that most extracellular matrices contain a tripeptide (Arg-Gly-Asp) (RGD) sequence and this sequence is probably responsible for binding fibronectin. Proteins carrying this tripeptide are also thought to play an important role in cell-matrix interactions. Besides extracellular matrix macromolecules fibronectin also binds to cell surface receptors. The cell-surface receptor for fibronectin recognizes a special tetrapeptide (Arg-Gly-Asp-Ser) (RGDS) or a pentapeptide (Gly-Arg-Gly-Asp-Ser) (GRGDS) sequence of the cell-binding domain of the fibronectin molecule.[13,14]

Fibronectins are supposed to be involved also in the spreading processes in cellular reorganization and regenerative processes following tissue injury. The presence of fibronectins in areas with intense cellular activity suggests their role in the attachment and reattachment of cells during embryogenesis, immunological reactions, wound healing, and neoplastic growth. There is some evidence that fibronectins may also induce cell migration and tissue reorganization. Fibronectin fragments are chemotactic for fibroblasts, and also promote the organization of endothelial cells *in vitro*.[15]

Fibronectins appear in large amounts on the surface of injured tissues.[16] They are thought to play an important role in the process of wound healing both in skin[17] and in corneal injuries.[18] In vascular lesions, when thrombin is generated and fibrin clots are formed, large quantities of plasma fibronectin are incorporated in the clot. Formation of covalent cross-links between fibronectin molecules and coagulation factor XIII-a is responsible for this process. A number of cells including fibroblasts, monocytes, vascular endothelial cells, epithelial cells in wounded areas, and other cells were shown to produce fibronectin.[1] When the epidermis is separated from superficial layers of the skin, both in case of epithelial defects and when blisters are formed, a rapid deposition of fibrin and fibronectin occurs in the defect zone.[19] In healing skin wounds, a large quantity of fibronectin mostly plasma derived appears on the surface of the defect zone and in the extracellular spaces of the surrouding tissue in the first 2 d after wounding; thereafter, fibronectin is actively synthesized by proliferating vascular endothelial cells, fibroblasts, and possibly regenerating epithelial cells in the wound area.[20]

Fibronectins are susceptible to proteases both in soluble and in insoluble form. Mild proteolytic treatment can clear the surfaces of cells or that of basal membranes of any fibronectin molecules or fragments attached to them. This task can be accomplished by a number of broad-spectrum proteases such as plasmin, trypsin, cathepsins, etc.[21]

Fibronectins are not easily detected in the extracellular matrices of human corneas and of other mature human tissues.[22] Berman et al.,[23] however, found that some of the studied nonulcerating (normal) human corneas, unlike the normal rabbit corneas, had detectable fibronectin under the epithelium in the basement membrane region. Fibronectin is involved in cell migration and tissue organization in embryonic development of the cornea.[24] Ben Zvi et al.[25] used the immunofluorescence technique with polyclonal antibodies directed against fibronectin and laminin in order to study their distribution in human embryonic corneas from 8 weeks of gestation and compare it with those in adult corneas. Immunostaining for fibronectin in Descemet's membrane was negative at 8 weeks of gestation, followed by patchy staining of corneal stroma and epithelial basal membrane up to the age of 37 weeks of gestation. Staining in the epithelial basal membrane was negative or variable in adults up to 70 years of age, but was found positive in a 77-year-old individual. Staining for laminin was positive in the epithelial basal membrane after 8 weeks of gestation. Staining was negative in the Descemet's membrane at that age, but became positive after 9 weeks of gestation.

Cintron et al.[26] provided data concerning the appearance of fibronectin in the developing rabbit cornea. Normal (adult) rabbit corneas were found to contain fibronectin by means of immunofluorescence microscopy at the Descemet's membrane but no staining for fibronectin was observed at the epithelium or the epithelial basal membrane zone.[23] However, fibronectin was seen beneath the migrating rabbit corneal epithelium both *in vivo*[23] and *in vitro*.[27] The attachment of cultured rabbit epithelial cells is mediated by fibronectin.[28]

Watanabe et al.[29] demonstrated that the number of attached epithelial cells increased in proportion to the concentration of fibronectin coated on the culture plate. The addition of an oligopeptide derived from sequence of the cell-binding domain of fibronectin (Gly-Arg-Asp-Ser) (GRGDS) inhibited the attachment of epithelial cells on the fibronectin-coated surface in a dose-dependent fashion.[30] Based on these data Nishida and co-workers suggested that the corneal epithelial cells have receptors for fibronectin which recognize the GRGDS sequence of the cell-binding domain of fibronectin.

Nishida and Nakagawa[31] investigated the expression of epithelial cell receptors for extracellular matrix components during epithelial wound closure. They studied the spreading activity of rabbit corneal epithelial cells on tissue culture plates coated with fibronectin, collagen types I and IV, and laminin matrices in different concentrations at different stages of healing after epithelial debridement. Freshly prepared cells obtained from normal intact corneas showed spreading activity only on the highest concentration examined; cells prepared 7 d after debridement exhibited spreading activity, even at lower fibronectin levels. Cells prepared from eyes 21 d after debridement did not show any response to fibronectin matrix. Epithelial cells responded with spreading on collagen types I and IV regardless of their stage of healing. Cells responded to laminin only at the highest concentration examined. Based on these experimental data the authors suggested that fibronectin receptors are expressed only in regenerating epithelial cells while receptors for collagen types I and IV are present on normal rabbit corneal epithelial cells, too.[31]

Nishida et al.[32] studied the effects of fibronectin and epithelial growth factor on the migration of epithelial cells on the cut stromal surface of rabbit corneal blocks in tissue culture. Histological and electron microscopic studies revealed the leading edge of the fibronectin-treated cornea was thin, single-layered epithelium, and the leading edge of the epithelial growth factor-treated cornea was hypertrophic, multilayered epithelium. By autoradiography with ^{3}H-thymidine, the authors demonstrated that epithelial growth factor stimulated corneal epithelial proliferation, whereas fibronectin had no effect on cell proliferation. Based on these experiments they proposed that fibronectin and epithelial growth factor promoted corneal stroma resurfacing in two different ways: fibronectin enhances the sliding of epithelium and epithelial growth factor facilitates epithelial cell proliferation.[29]

Phan et al. performed epithelial scrapes[33] and superficial keratectomies[34] on guinea pigs and prevented the formation of a matrix of fibronectin and fibrin by using specific antibodies, systemic administration of ancrod, or performing the wound *ex vivo*. Besides observing the healing rate *in vivo* they also excised and cultured corneas to study reepithelization *in vitro*. In contrast to the findings of Nishida et al.,[32] they found that the addition of exogeneous fibronectin did not accelerate *in vitro* the epithelial migration on bare stromal surfaces.[34] In this respect there seems to be no basic difference between superficial wounds where the basement membrane remains intact and deep wounds where stromal lamellae are exposed, as it had been suggested earlier.[33] The situation is probably different, however, concerning the need for adhesive proteins, when the epithelial cells grow on the surface of plastic dishes.[35] Other macromolecules, collagen types I and IV, laminin, vitronectin (serum-spreading factor), gelatin, and polylysine, can substitute fibronectin even under these experimental conditions.[36-40]

Further studies demonstrated that fibronectin plays an important role in corneal wound healing in rabbits *in vivo*. Fibronectin was observed at the wound site by immunofluorescent microscopy following simple mechanical scrape, caustic and thermal burns, and experimental bullous keratopathy. Fibronectin appeared immediately after the injury, and epithelial cells slid on the fibronectin-positive surface of the stroma. When the epithelial defect healed fibronectin disappeared from the wound site.[27,41-43] Fibronectin was found to affect the migration of rabbit corneal epithelial cells both *in vitro* and *in vivo*. Nishida and co-workers demonstrated that the administration of fibronectin eyedrops significantly stimulated the resurfacing of corneal epithelium in rabbit eyes after simple mechanical scrape and after debridement by iodine vapor.[44,45] Phan et al.[46] also made observations concerning the beneficial effects of fibronectin-containing eyedrops on the healing of epithelial defects and the decrease of the incidence of stromal ulceration in alkali-burned rabbit corneas. Others published results showing that the topical administration of fibronectin in rabbits does not necessarily promote corneal epithelial wound healing.[47] All experimental data must be handled and evaluated with caution, however, as the fibronectin used in topically applied preparations is in dimeric form and differs in some respects from the fibronectin found in the extracellular matrix.[48]

In human studies purified plasma fibronectin was successfully administered to patients with persistent epithelial defects after herpetic keratitis,[49] traumatic recurrent erosion, and chronic trophic ulcer.[45,50] The authors proposed that, when decreased attachment or adhesion of the epithelium is the main cause of the corneal disease, fibronectin would be the treatment of choice, whereas, if the decreased rate of epithelial proliferation is the cause of the disorder, the use of epithelial growth factor might be preferable.[29]

In an uncontrolled clinical trial Phan et al.[46] tested the effect of the topical application of fibronectin eyedrops on six patients with persistent corneal epithelial defects. Complete reepithelization was achieved in five patients: three healed within 17 d after the initiation of the treatment, two others healed 1 to 2 weeks after discontinuation of the topical fibronectin. Salonen et al.[51] suggested the use of fibronectin together with aprotinin in otherwise nonhealing corneal ulcers. Kono et al.[52] reported the beneficial effect of topically administered fibronectin on the healing of epithelial defects in keratoconjunctivitis sicca associated with Sjögren's syndrome. Spigelman et al.[53] used fibronectin eyedrops in the treatment of persistent epithelial defects.

In human studies autologous fibronectin is used to minimize immunologic reactions and to avoid the transmission of viral (HIV, hepatitis, etc.) infections. Fibronectin is purified from the patient's own plasma under sterile conditions. The original procedure of fibronectin preparation developed by Engvall et al.[54] was modified by Nishida et al. in 1982,[42] and by Phan et al. in 1987.[48] Individual CN-Br activated Sepharose 4B columns are used for each patient. The procedure in brief: 54 ml of citrated blood (0.38%) is centrifuged at 27,000 g. The supernatant is applied to a 5-ml lysine-coupled Sepharose 4B column to remove plasminogen, then passed through a 15-ml gelatin-coupled Sepharose 4B column to isolate fibronectin. After extensive washing with PBS (pH 7.2), and equilibration with isotonic saline solution (pH 7.2), fibronectin is eluted with 0.15 *M* NaCl (pH 2.2). The eluate is then dialyzed against PBS. Preparations are stored at 4°C until use containing 0.25 chloramphenicol for bacteristasis. Patients receive preparations containing 0.5 mg/ml fibronectin, in the form of eyedrops five times a day for 1 to 3 weeks. Fresh preparations must be made every 10 d.

Fibronectin was detected by immunohistological methods under normal and regenerating corneal epithelium and it has been suggested that fibronectin molecules play a role in the attachment epithelial cells to the basement membrane.[42,50,55] Fibronectin has been suggested to promote the healing of corneal epithelial defects, both in humans and in animal experiments.[44,45] Extensive studies have shown that a provisional fibronectin-fibrinogen matrix can be detected on the bare surface and under the migrating epithelium of corneal wounds. This matrix gradually disappears as the epithelial defect closes. Fibronectin and fibrin(ogen) are thought to migrate from the prilimbal conjunctival vessels through the tear film to the wound area. This mechanism was originally suggested by Fujikawa et al.[41] and was futher supported by Phan et al.,[33] who demonstrated that such a matrix does not form when the wound is made *ex vivo* after the experimental animal was sacrificed. Fibronectin is not a constant component of the tear fluid, but was shown to appear in tears following ocular injury.[56] Phan et al.[33] provided important data concerning the mechanism and time sequence of fibronectin and fibrin deposition in the corneal wound area. In their experiments they demonstrated that the formation of a fibronectin matrix is essential for subsequent deposition of fibrin, when fibronectin production was suppressed or fibronectin deposition preventing fibrin deposition did not occur either. In contrast, suppression of fibrin deposition on the scrape wound surface did not affect the formation of a fibronectin matrix.[33]

Exogeneous fibronectin was demonstrated to have a beneficial effect on the healing of epithelial defects of different origin by providing attachment for regenerating epithelial cells,[44,50] but substantial experimental data suggest that epithelial wounds can heal without fibronectin, too. Phan et al.[48] demonstrated that the deposition of fibronectin on the bottom of the wound is not essential for cell migration and wound closure. In their experiments resurfacing occurred following corneal abrasions and superficial keratectomies when there was no fibronectin in the wound area or when it was blocked by specific antibodies.[33,34] The possible explanations for their findings: (1) other extracellular matrix proteins can substitute fibronectin and serve as a matrix for epithelial cell sliding, (2) direct contacts between epithelial cells and the basal membrane or the stromal surface are enough for cell adhesion and there is no real need for adhesive proteins, (3) fibronectin promotes wound healing by some of its other effects, (4) the migrating epithelial cells synthesize their own fibronectin matrix.

1. Cameron et al.[57] studied the effects of matrix proteins on rabbit corneal epithelial cell adhesion and migration, and demonstrated that besides fibronectin type IV collagen effectively promoted cell adhesion. Laminin is a normal component of the corneal epithelial basement membrane. It

is firmly attached to the deeper layer (lamina densa) of the basement membrane. In superficial scrape wounds of the cornea the superficial layer (lamina lucida) may be removed with the scraped epithelial layer but the laminin molecules may remain in place and serve as attachment proteins for the regenerating epithelium. Laminin was also suggested to be a link between epithelial cells and type IV collagen of the basal membrane.[34,39,58]

2. Epithelial cells were shown to have receptor for various extracellular matrix proteins, glycoproteins, and glycosaminoglycans under experimental conditions.[31,57] It is very difficult to decide, however, which of these interactions actually play a role *in vivo*. The observations that epithelial cells are able to migrate on bare basement membrane, on stromal wound surface, and on artificial matrices covering the surface of culture dishes indicate that epithelial migration may also occur in the absence of attachment proteins.[34,46,48]

3. Fibronectin, besides promoting the attachment of different cells to basement membranes and collagen molecules,[59,60] has other functions by which it can enhance wound healing. These are chemotactic effects on PMNs and monocytes,[61] stimulation of macrophage growth factor activity,[62] and proliferation of endothelial cells of newly formed blood vessels.[20,63] Other possible effects include chemotaxis for corneal epithelial cells, and fibroblasts, alterations of cytoskeleton, or changes in intracellular processes causing increased protein synthesis and mitotic effect.[48] Besides, fibronectin also plays a role in the host-defense mechanism as a nonspecific, nonimmunological opsonic protein.[64] The fibronectin molecule has an array of domains that are capable of binding tissue debris, and fibrin aggregates and enhance the phagocytosis, and removal of these opsonized particles by the cellular elements of the reticuloendothelial system (RES).[65,66] It is therefore possible that in severe corneal wounds, involving extensive tissue injury and inflammatory cell infiltration, such as chemical or thermal burns and viral and bacterial infections, the role of fibronectin is multifaceted and more critical than in the healing of simple traumatic epithelial defects.[33]

4. It has been demonstrated that the resurfacing occurs on bare surfaces of the basement membrane or of corneal stroma without the previous formation of a continuous fibronectin-fibrin(ogen) layer in the defect zone.[33,34] These observations, however, do not rule out the possible role of fibronectin in the sliding of epithelial cells on these bare surfaces. It has been shown that fibronectin may appear under the migrating epithelium even in experiments when the deposition of fibronectin from tears and from plasma is prevented and no exogeneous fibronectin is administered. These experimental data suggest the migrating epithelial cells may have an intrinsic biochemical activity to synthesize and secrete their own fibronectin matrix providing proper attachment to the surface in cases when fibronectin from other physiological or extrinsic sources is missing.[34]

REFERENCES

1. **Hynes, R. O. and Yamada, K. M.,** Fibronectins: multifunctional modular glycoproteins, *J. Cell. Biol.,* 95, 369, 1982.

2. **McDonald, J. A.,** Fibronectin: a primitive matrix, in *The Molecular and Cellular Biology of Wound Repair,* Clark, R. A. F. and Henson, P. M., Eds., Plenum Press, New York, 1988, 405.

3. **Hynes, R. O.,** Molecular biology of fibronectin, in *Annual Review of Cell Biology,* Vol. 1, Palade, E., Alberts, B. M., and Spudich, J. A., Eds., Annual Reviews, Palo Alto, CA, 1985, 67.

4. **Schwarzbauer, J. E., Tamkun, J. W., Lemischka, I. R., and Hynes, R. O.,** Three different fibronectin mRNAs arise by alternative splicing within the coding region, *Cell,* 35, 421, 1983.

5. **Schwarzbauer, J. E., Paul, J. I., and Hynes, R. O.,** On the origin of species of fibronectin, *Proc. Natl. Acad. Sci. U.S.A.,* 82, 1424, 1985.

6. **Engvall, E. and Ruoslahti, E.,** Binding of soluble form of fibroblast surface protein, fibronectin, to collagen, *Int. J. Cancer,* 20, 1, 1977.

7. **Jilek, F. and Hormann, H.,** Fibronectin (cold insoluble glubulin) VI influence of heparin and hyaluronic acid on the binding of native collagen, *Hoppe-Seylers Z. Physiol. Chem.,* 360, 597, 1979.

8. **Salonen, E. M., Saksela, O., Vartio, T., Vaheri, T., Nielsen, L., and Zeuthen, J.,** Plasminogen and tissue-type plasminogen activator bind to immobilized fibronectin, *J. Biol. Chem.,* 260, 12302, 1985.

9. **Salonen, E. M., Vartio, T., Hedman, K., and Vaheri, A.,** Binding of fibronectin by the acute-phase reactant C-reactive protein, *J. Biol. Chem.,* 259, 1496, 1984.

10. **Pierschbacher, M. D., Hayman, E. G., and Ruoslahti, E.,** The cell attachment determinant in fibronectin, *J. Cell Biochem.,* 28, 115, 1985.

11. **Pytela, R., Pierschbacher, M. D., and Ruoslahti, E.,** Identification and isolation of a 140 kd cell surface glycoprotein with properties expected of a fibronectin receptor, *Cell,* 40, 191, 1985.

12. **Morrison, P. R., Edsall, J. T., and Muller, S. G.,** *J. Am. Chem. Soc.,* 70, 3103, 1948.

13. **Pierschbacher, M. D. and Ruoslahti, E.,** Cell attachment activity of fibronectin can be duplicated by small synthetic fragments of the molecule, *Nature,* 309, 30, 1984.

14. **Yamada, K. M. and Kennedy, D. W.,** Dualistic nature of adhesive protein function: fibronectin and its biologically active peptide fragments can autoinhibit fibronectin function, *J. Cell Biol.,* 99, 29, 1984.

15. **Maciag, T. et al.,** Organizational behavior of human umbilical vein endothelial cells, *J. Cell Biol.,* 94, 511, 1982.

16. **Vaheri, A., Salonen, E. M., Vartio, T., Hedman, K., and Stenman, S.,** Fibronectin and tissue injury, in *Biology and Pathology of the Vessel Wall,* Woolf, N., Ed., Praeger, Eastburn, 1983, 161.

17. **Jonkman, M. F.,** Epidermal Wound Healing between Moist and Dry, Ph.D. Thesis, University of Groeningen, 1989.

18. **Berman, M., Kenyon, K., Hayashi, K., and L'Hernault, N.,** Pathogenesis of epithelial defects and stromal ulceration, in *The Cornea: Transactions of the World Congress on the Cornea,* Vol. 3, Dwight Cavanagh, H., Ed., Raven Press, New York, 1988, 35.

19. **Saksela, O., Alitalo, K., Kiistala, U., and Vaheri, A.,** Basal lamina components in experimentally induced skin blisters, *J. Invest. Dermatol.,* 77, 283, 1981.

20. **Clark, R. A. F. et al.,** Fibronectin is produced by blood vessels in response to injury, *J. Exp. Med.,* 156, 646, 1982.

21. **Vartio, T., Seppa, H., and Vaheri, A.,** Susceptibility of soluble and matrix fibronectins to degradation by tissue proteinases, mast cell chymase and cathepsin G, *J. Biol. Chem.,* 256, 471, 1981.

22. **Vaheri, A., Salonen, E. M., and Vartio, T.,** Fibronectin in the formation and degradation of the pericelluler matrix, in *Fibrosis,* (Ciba Foundation Symposium 114), Evered, D. and Whelan, J., Eds., Pitman, London, 1985, 111.

23. **Berman, M., Manseau, E., Law, M., and Aiken, D.,** Ulceration is correlated with degradation of fibrin and fibronectin at the corneal surface, *Invest. Ophthalmol. Vis. Sci.,* 24, 1358, 1983.

24. **Kurkinen, M., Alitalo, K., Vaheri, A., Stenman, S., and Saxen, L.,** Fibronectin in the development of chick eye, *Dev. Biol.,* 69, 589, 1979.

25. **Ben Zvi, A., Rodrigues, M. M., Krachmer, J. H., and Fujikawa, L. S.,** Immunohisto-chemical characterization of extracellular matrix in the developing human cornea, *Curr. Eye Res.,* 5, 105, 1986.

26. **Cintron, C., Fujikawa, L. S., Covington, H., Foster, C. S., and Colvin, R. B.,** Fibronectin in developing rabbit cornea, *Curr. Eye Res.,* 3, 489, 1984.

27. **Suda, T., Nishida, T., Ohashi, Y., Nakagawa, S., and Manabe, R.,** Fibronectin appears at the site of corneal stromal wound in rabbits, *Curr. Eye Res.,* 1, 553, 1981/1982.

28. **Friend, J.,** Physiology of the cornea: Metabolism and biochemistry, in *The Cornea. Scientific Foundations and Clinical Practice,* Smolin, G. and Thoft, R. A., Eds., Little, Brown and Company, Boston, 1987, 16.

29. **Watanabe, K., Nakagawa, S., and Nishida, T.,** Stimulatory effects of fibronectin and EGF on migration of corneal epithelial cells, *Invest. Ophthalmol. Vis. Sci.,* 28, 205, 1987.

30. **Nishida, T., Nakagawa, S., Watanabe, K., Yamada, K., McDonald, J., Otori, T., et al.,** Pathobiology of epithelial defects: peptide (GRGDS) of fibronectin cell-binding domain inhibits corneal epithelial attachment and spreading on plasma fibronectin, *Invest. Ophthalmol. Vis. Sci. Suppl.,* 27, 53, 1987.

31. **Nishida, T. and Nakagawa, S.,** Expression of fibronectin receptors in corneal epithelial cells, in *Healing Processes of the Cornea,* (Advances in Applied Biotechnology, Vol. 1), Beuerman, R. W., Crosson, C. E., and Kaufman, H. E., Eds., Portfolio, The Woodlands, TX, 1989, 127.

32. **Nishida, T., Nakagawa, S., Awata, T., Ohashi, Y., Watanabe, K., and Manabe, R.,** Fibronectin promotes epithelial migration of cultured rabbit cornea *in situ, J. Cell. Biol.,* 97, 1653, 1983.

33. **Phan, T.-M. M., Foster, C. S., Wassan, P. J., Fujikawa, L. S., Zagachin, L. M., and Colvin, R. B.,** Role of fibronectin in healing of corneal epithelial scrape wounds, *Invest. Ophthalmol. Vis. Sci.,* 30, 377, 1989.

34. **Phan, T.-M. M., Foster, C. S., Zagachin, L. M., and Colvin, R. B.,** Role of fibronectin in the healing of superficial keratectomies *in vitro, Invest. Ophthalmol. Vis. Sci.,* 30, 386, 1989.

35. **Nishida, T., Nakagawa, S., Watanabe, K., Yamada, K. M., Otori, T., and Berman, M. B.,** A peptide from fibronectin cell-binding domain inhibits attachment of epithelial cells, *Invest. Ophthalmol. Vis. Sci.,* 29, 1820, 1988.

36. **Barnes, D., van der Bosch, J., Masui, H., Miyazaki, K., and Sato, G.,** The culture of human tumor cells in serum-free medium, *Methods Enzymol.,* 79, 368, 1981.

37. **Federgreen, W. and Stenn, K. S.,** Fibronectin (LETS) does not support epithelial cell spreading, *J. Invest. Dermatol.,* 75, 261, 1980.

38. **McKeehan, W. et al.,** in *Requirements of Vertebrate Cells In Vitro,* Waymouth, C., Ed., Cambridge University Press, Cambridge, 1981, 118.

39. **Murray, T. C., Stingl, C., Kleinman, H. K., Martin, G. R., and Katz, S. I.,** Epidermal cells adhere preferentially to type IV (basement membrane) collagen, *J. Cell Biol.,* 80, 179, 1979.

40. **Terranova, V. P., Rohrback, D. H., and Martin, G. R.,** Role of laminin in the attachment of PAM 212 (epithelial) cells to basement membrane collagen, *Cell,* 22, 719, 1980.

41. **Fujikawa, L. S., Forster, C. S., Harrist, T. J., Lanigan, J. M., and Colvin, R. B.,** Fibronectin in healing rabbit corneal wounds, *Lab. Invest.,* 45, 120, 1981.

42. **Nishida, T., Nakagawa, S., Awata, T., et al.,** Rapid penetration of purified autologous fibronectin eyedrops from patient's plasma, *Jpn. J. Ophthalmol.,* 26, 416, 1982.

43. **Ohashi, Y., Nakagawa, S., Nishida, T., Suda, T., Watanabe, K., and Manabe, R.,** Appearance of fibronectin in rabbit cornea after thermal burn, *Jpn. J. Ophthalmol.,* 27, 547, 1983.

44. **Nishida, T., Nakagawa, S., Nishibayashi, C., Tanaka, H., and Manabe, R.,** Fibronectin enhancement of corneal epithelial wound healing of rabbits *in vivo, Arch. Ophthalmol.,* 102, 455, 1984.

45. **Nishida, T., Ohashi, Y., Awata, T., and Manabe, R.,** Fibronectin: a new therpy for corneal trophic ulcer, *Arch. Ophthalmol.,* 101, 1046, 1983.

46. **Phan, T.-M. M., Foster, C. S., Shaw, C. E., and Colvin, R. B.,** Topical fibronectin in a rabbit alkali burn model of corneal ulceration, *Invest. Ophthalmol. Vis. Sci. Suppl.,* 26, 176, 1985.

47. **Newton, C., Hatchell, D. L., Klintworth, G. K., and Brown, C. F.,** Topical fibronectin and corneal epithelial wound healing in the rabbit, *Arch. Ophthalmol.,* 106, 1277, 1988.

48. **Phan, T.-M. M., Foster, C. S., Boruchoff, S. A., Zagachin, L. M., and Colvin, R. B.,** Topical fibronectin in the treatment of persistent corneal epithelial defects and trophic ulcers, *Am. J. Ophthalmol.,* 104, 494, 1987.

49. **Nishida, T., Nakagawa, S., and Manabe, R.,** Clinical evaluation of fibronectin eyedrops on epithelial disorders after herpetic keratitis, *Ophthalmology,* 92, 213, 1985.

50. **Nishida, T., Nakagawa, S., Awata, T., Tani, Y., and Manabe, R.,** Fibronectin eyedrops for recurrent traumatic erosion, *Lancet,* II, 521, 1983.

51. **Salonen, E. M., Tervo, T., Torma, E., Tarkkanen, A., and Vaheri, A.,** Plasmin in tear fluid of patients with corneal ulcers: basis for new therapy, *Acta Ophthalmol.,* 65, 3, 1987.

52. **Kono, I., Matsumoto, Y., Kono, K., Ishibashi, Y., Narushima, K., Kabashima, T., Yamane, K., Sakurai, T., and Kashiwagi, H.,** Beneficial effect of topical fibronectin in patients with keratoconjunctivitis sicca of Sjögren's syndrome, *J. Rheumatol.,* 12, 487, 1985.

53. **Spigelman, A. V., Deutsch, T. A., and Sugar, J.,** Application of homologous fibronectin to persistent human corneal epithelial defects, *Cornea,* 6, 128, 1987.

54. **Engvall, E., Ruoslahti, E., and Miller, E. J.,** Affinity of fibronectin to collagens of different types and to fibrinogen, *J. Exp. Med.,* 147, 1584, 1978.

55. **Tervo, T., Sulonen, J., Valtonen, S., Vannas, A., and Virtanen, I.,** Distribution of fibronectin in human and rabbit corneas, *Exp. Eye Res.,* 42, 399, 1986.

56. **Jensen, O. L., Gluud, B. S., and Eriksen, H. O.,** Fibronectin in tears following surgical trauma to the eye, *Acta Ophthalmol.,* 63, 346, 1985.

57. **Cameron, J. D., Hagen, S. T., Waterfield, R. R., and Furcht, L. T.,** Effects of matrix proteins on rabbit corneal epithelial cell adhesion and migration, *Curr. Eye Res.,* 7, 293, 1988.

58. **Burrill, P. H., Bernardini, J., Kleinman, H. K., and Kretchmer, N.,** Effect of serum fibronectin and laminin on adhesion of rabbit intestinal epithelial cells in culture, *J. Supramol. Struct. Cell Biochem.,* 16, 385, 1981.

59. **Grinnell, F., Feld, M., and Minter, D.,** Fibroblast adhesion to fibrinogen and fibrin substrata: requirement for cold-insoluble globulin (plasma fibronectin), *Cell,* 19, 517, 1980.

60. **Klebe, R. J.,** Isolation of collagen-dependent cell attachment factor, *Nature,* 250, 248, 1974.

61. **Norris, D. A., Clark, R. A. F., Swigart, L. M., Huff, J. C., Weston, W. L., and Howell, S. E.,** Fibronectin fragment(s) are chemotactic for human peripheral blood monocytes, *J. Immunol.,* 129, 1612, 1982.

62. **Martin, B. M., Gimbrone, M. A., Jr., Majeau, G. R., Unanue, E. R., and Cotran, R. S.,** Stimulation of human monocyte/macrophage-derived growth factor (MDGF) production by plasma fibronectin, *Am. J. Pathol.,* 111, 367, 1983.

63. **Golub, B. M., Foster, C. S., and Colvin, R. B.,** Fibronectin, laminin, type IV collagen, factor VIII and fibrin: Sequential analysis during experimental corneal neovascularization response in the guinea pig, *Invest. Ophthalmol. Vis. Sci. Suppl.,* 22, 27, 1982.

64. **Villager, B., Kelley, D. G., Engleman, W., Kuhn, C., and McDonald, J. A.,** Human alveolar macrophage fibronectin: synthesis, secretion and ultrastructural localization during gelatin-coated latex particle binding, *J. Cell Biol.,* 90, 711, 1981.

65. **Blumenstock, F. A., Laba, T. M., Weber, P., and Laffin, R.,** Biochemical and immunological characterization of human opsonic a2 SB glycoprotein: its identity with cold-insoluble globulin, *J. Biol. Chem.,* 253, 4287, 1978.

66. **Laba, T. M. and Jaffe, E.,** Plasma fibronectin (opsonic glycoprotein): its synthesis by vascular endothelial cells and role in cardiopulmonary integrity after trauma as related to reticuloendothelial function, *Am J. Med.,* 60, 577, 1980.

TEAR ENZYMES AND THE WOUND HEALING
OF THE CORNEA

Part C
Plasminogen Activators, Plasminogen Activator Inhibitors, and
Corneal Reepithelization

Quantitative studies of corneal epithelial wound healing confirmed the clinical observation that simple epithelial defects are covered rapidly by the growing adjacent epithelium. Cintron et al.[1] provided morphologic and autoradiographic evidence that large epithelial defects on rabbit corneas are covered by the movement of adjacent epithelial cells. Crosson[2] demonstrated that the leading edge of the regenerating epithelium is composed of a single layer of cells, showing signs of cellular motility (ruffled borders and filopodia). This appearance of the leading edge can be seen until complete closure of the epithelial wound occurs. Such changes in the epithelial cells are thought to be brought about by the action of intracellular contractile structures (myofilaments, actin) and special protein structures in and near the cell membrane.[3] Complete resurfacing is followed by the reformation of the anchoring hemidesmosomes that had been ruptured at the time of the injury.[4]

It has been shown both on rabbit and human corneas that the removal of epithelium by scrape[5] or by alkali burn[6] or by thermal burn[7] results in the appearance of fibrin and fibronectin on the surface of the wound and of plasminogen in the wounded cornea. It has been suggested that corneal reepithelization might involve the interaction between the epithelium and a subjacent fibrin-fibronectin matrix during resurfacing.[5,6]

Weimar[8] reported that the epithelium of injured rat corneas releases serine proteases. Berman et al.[9] provided evidence that the epithelium of ulcerating rabbit corneas releases plasminogen activators (PAs) and that the activation of plasminogen by PAs into plasmin actually takes place in the ulcerating corneas. High caseinolytic[10] and plasminogen-independent amidolytic[11] activities were detected in the tears of patients with different corneal disorders, indicating the presence and activity of the plasminogen-plasmin system in the tear fluid of these patients. Salonen and colleagues have shown that the caseinolytic activity in the tears of patients with nonhealing corneal ulcers is primarily due to the presence of plasmin.[10] We demonstrated, however, that besides plasmin a number of other proteolytic enzymes are present and act in the tears of corneal ulcer patients.[12] In further studies we found that PA levels increase in a number of corneal pathologies and that this increase usually precedes the appearance of plasmin in the tear fluid.[13,14]

The basic function of tear PAs and the generated plasmin as suggested by Storm could be the removal of fibrin from the surfaces of the cornea and conjunctiva as well as the maintenance of the patency of lacrimal puncta thereby ensuring tear outflow through the lacrimal canaliculi. This latter function would be similar to the proposed function of the plasminogen-plasmin system in the kidney glomerulus with regard to fibrin cleavage and fluid flow.[15] Pandolfi and Astrup[16] emphasized the significance of PAs in the senescence and turnover of corneal epithelial cells. Both experimental and clinical data suggest that PAs and plasmin also play a role in the healing processes of the cornea. It has been suggested that PAs and plasmin generated in the wounded cornea may assist tissue repair by preventing the excessive or permanent deposition of fibrin[16] and that the cleavage in the epithelial defect zone of the fibrin-fibronectin matrix in the normal healing process can occur in a controlled manner without destroying the basement membrane and without leading to corneal ulceration.[6]

Fibronectin on the other hand has been found to enhance the migration of epithelial cells in tissue culture[17] and *in vivo*.[18] Fibronectin eye drops have been effective in enhancing the resurfacing of the corneal epithelium in cases with persistent epithelial defect[19] and with trophic ulcer.[20] Berman et al.,[21] based on studies of growing rabbit and human corneal epithelium in tissue culture, proposed a model, in which tissue plasminogen activator (t-PA) released from the leading edge of the growing epithelium is activated *in situ* on the subjacent fibrin-fibronectin matrix, and the active t-PA activates plasminogen to plasmin. Plasmin in turn, cleaves fibronectin in a limited fashion to release the leading edge of epithelium which then advances to the next fibronectin molecule. Fibronectin-cell contact, as in other systems, can initiate the formation of intracellular F actin myofilaments via transmembrane proteins resulting in the movement of the epithelial cell in the direction of the leading edge. After epithelial resurfacing is accomplished complete resorption of the subepithelial fibrin and fibronectin is effected by plasmin and firm attachment of the epithelial cells is achieved through the formation of new hemidesmosomes.

Berman and colleagues, on the basis of fibrinolytic patterns of frozen sections and tissue cultures of ulcerating corneas, proposed that prolonged or excessive PA/plasmin production in cases of severe alkali burns of the cornea result in uncontrolled proteolysis leading to the destruction of basal membrane and Bowman's membrane, and to the secretion by keratocytes and activation of latent collagenase, and to the ulceration of the corneal stroma.[9,21,22] Hayashi et al.[23] showed on frozen sections of ulcerating rat corneas that antibodies to t-PA inhibit the fibrinolytic activity in the defect region and on the surface of the epithelium. The histologic examination of vitamin A-deficient rat corneas revealed the formation of a pseudomembrane composed of polymorphonuclear leukocytes (PMNs), cell debris, and fibrinous exudate. The frozen sections of these corneas did not show fibrinolytic activity in the defect region and still demonstrated t-PA activity along the intact epithelial surface. It has been suggested that the pseudomembrane formation which delays reepithelization in the defect region of ulcerating vitamin A-deficient corneas might have resulted from the absence of PA activity in the defect region.

Further studies demonstrated that amiloride, a selective inhibitor of urokinase PA (u-PA),[24] prevents lysis at the leading edge of the epithelium but not along the intact epithelial surface more peripherally in the frozen sections of scraped or alkali-burned rabbit corneas. The combination of amiloride and anti-t-PA antibody completely inhibited the lysis in the sections. These observations suggested a modification of the original model assuming that u-PA is produced by cells at the leading edge while other parts of the epithelial sheet produce t-PA.[25]

Ever since Storm in 1955 first detected PAs and their precursors in tears great effort has been made to find out whether tears contained t-PA or u-PA. Rijken et al.,[26] and van Haeringen and Thörig[27] found t-PA; Wang et al.,[22] Punyiczki et al.,[12] and Tözsér et al.[11] found u-PA; Hayashi et al.[23] detected both activators in normal human tears. Recent studies indicate that a marked increase in PA levels and activities can be detected in the tears of patients with different inflammatory and traumatic diseases of the cornea and the conjunctiva.[11,13,28] The great variability of normal values and differences in the detection of the two activators can be attributed to at least five different factors: (1) the use of various methods for the detection of PAs and the determination of PA activities in tears, as various procedures give different results, e.g., if fibrin(ogen) is used as a cofactor in the determinations the activity of t-PA can be increased up to 1000 times;[13] (2) the presence in tears of proactivators, which can be activated before or during the measurements, and to the presence of inhibitors and activator-inhibitor complexes that can change the results of tear PA determinations to a great extent in normal and especially in pathological cases;[13,29] (3) the use of different sampling methods, to differences in tear flow rates (dilution), and vascular permeability (transudation) that seem to change PA levels and activities both in normal and pathological cases;[13] (4) the presence of PMNs, bacteria, injured or destroyed cells in the cornea or in tear samples, and the activation of zymogens by proteolytic enzymes causing further differences in the PA levels and activities measured in the tears of patients with corneal and conjunctival inflammations;[11] and (5) that there are significant differences in the levels of PAs in the tears of humans and of different animal species even under normal circumstances.[29-31] Therefore, the results of animal experiments can be used in solving clinical problems only with special caution.

Human epithelial cells of various origin were shown to secrete u-PA in cell cultures.[32-34] In the cultures of most of these cells t-PA was also produced, suggesting that both types of PA can be secreted by the same epithelial cells, while vascular endothelial cells contain and release in response to stimuli only t-PA.[26] In experiments based on the early observations that frozen sections[16] and tissue explants[35] of normal corneas release PAs, the origin of corneal PA was found to be the epithelium,[36] and its properties were shown to be similar to that of u-PA.[37] Berman et al.[9] provided evidence that fibroblasts in ulcerating corneas also produce PAs. There exists considerable evidence showing that PAs, mainly t-PA, can be produced by the conjunctiva and the lacrimal gland as well.[27,38]

t-PA and u-PA, besides differences in molecular size and enzymatic and immunological properties, differ in their affinity for fibrin which leads to a great difference in their physiological action. t-PA activates plasminogen into plasmin on the surface of fibrin, and produces plasmin which is bound to its natural substrate (fibrin) and its action is practically restricted to fibrin cleavage. u-PA activates plasminogen into plasmin in body fluids and secretions without fibrin and produces free plasmin with a wide proteolytic activity. Therefore, t-PA is generally thought to initiate fibrinolysis while u-PA is supposed to be involved in a number of different extracellular proteolytic processes.[39,40]

u-PA was shown to be secreted also by corneal and conjunctival epithelial cells.[9,22,38] u-PA is present in normal tears, collected at low flow rates, mainly in inactive form.[29] Its active form is present in tears with epithelial defects with different origin. Especially high levels of active u-PA are detected in cases with corneal ulcers.[11,13] High proteolytic activity in the tears of patients with persistent epithelial defects and corneal ulcers can be explained by high levels of free plasmin in tears, presumably resulting from the excessive and nonregulated production of u-PA.[9,11,13]

t-PA is secreted as a single chain form (sct-PA). The single chain t-PA has high affinity for fibrin resulting from a fibrin-binding domain in the N-terminal region of the molecule. Binding to fibrin and formation of a ternary complex with fibrin and plasminogen lower the K_m for activation of plasminogen so that the activation is greatly enhanced in the presence of fibrin. Concomitant with activation of plasminogen, sct-PA is converted to two-chain t-PA (tct-PA) which is also enzymatically active and also dependent upon fibrin-binding for activity. u-PA is also thought to be secreted in a single-chain form (scu-PA). After activation of plasminogen bound to fibrin by scu-PA, the scu-PA is converted to two-chain u-PA (tcu-PA) by plasmin. Two-chain u-PA is able to activate plasminogen into plasmin without the presence of fibrin. Recent studies have indicated that cell-surface u-PA activators generate plasmin which is involved in extracellular matrix degradation in physiological processes involving tissue remodeling as well as in local and metastatic neoplastic growth.[39,41]

The early clinical and histological observations that the surface of the cornea in persistent defects and corneal ulcers is covered by cellular debris and fibrous exudate which might prevent healing suggested the use of proteolytic enzymes (trypsin, chymotrypsin) to clear up the bottom of the defect and to promote the healing process.[42,43] Our general knowledge of u-PA action and the fact that active u-PA is present in the tears of patients with epithelial defects would suggest a possible role of u-PA-generated free plasmin in the digestion of the fibrous exudate and debris in the defect region as a part of the physiological healing process. The lack of "clearing up" in the defect zone due to insufficient u-PA production and the lack of activation of plasminogen into plasmin or the excessive production of inhibitors could explain disturbances in the healing process leading to persistant epithelial defects. Excessive u-PA and plasmin production or missing inhibitors on the other hand could give an explanation for the development of stromal ulceration.

Frozen sections of alkali-burned and scraped corneas showed u-PA activity at the leading edge of the healing epithelium and t-PA activity along the surface of the regenerated epithelium. PA activity was completely inhibited by amiloride, a selective inhibitor of u-PA, while the lysis around the surrounding epithelium was unaffected. Anti-t-PA antibodies on the contrary inhibited the lysis around the epithelium and left the leading edge activity unaffected.[25] Abnormally high levels of plasmin after alkali burn of the cornea, as suggested by Berman et al.,[21] can result in the generation of a relatively high level of "uncoupled domains", which in turn competing with intact fibronectin molecules can prevent epithelial resurfacing. As described by Yamada et al.[44] in other systems and shown by Nishida et al.[45] in corneal epithelial cell cultures, synthetic peptides based the cell-binding region of fibronectin can prevent epithelial cell migration on a surface coated with intact fibronectin; therefore, the possibility that fibronectin fragments competing with intact fibronectin molecules can prevent reepithelization of the corneal surface seems to be quite plausible. It has been shown in other systems that the proper orientation of fibronectin molecules is important for the migration of cells (keratocytes or epithelial cells) in the process of wound healing.[46] On the basis of these data we propose that the process of corneal reepithelization in simple cases of epithelial defects can be described in five steps: (1) the removal of fibrous exudate consisting of fibrin, fragmented, and/or randomly positioned fibronectin molecules, plasma proteins, released intracellular proteins, and fragments of injured epithelial cells. This, besides the mechanical effect of blinking and washing effect of tears, on a molecular level could be effected by the proteolytic action of free plasmin generated by u-PA secreted by the leading edge of the growing epithelium. This could be followed by (2) the production of a layer of intact fibronectin molecules kept in the right position by appropriate attachment to the basal membrane and by the presence of fibrin molecules helping the fibronectin molecules to stay in an upright position, keeping their cell-binding domain ready to bind to the cell membrane receptors of the approaching epithelial cells. Following this would be (3) the migration of the epithelial cell. This may be followed by (4) the selective resorption of fibrin from under the regenerated epithelium effected by fibrin-bound plasmin generated by t-PA produced by the epithelial cells. The last phase of epithelial wound healing involves (5) the reformation of anchoring hemidesmosomes (Figure 1).

The actual regulatory mechanism is thought to be even more complicated involving the regulation of the secretion of single-chain u-PA and t-PA by the epithelial cells, their conversion to double-chain activators, the very probable binding of activators to cell surface receptors, and certainly the production of and interaction with PA inhibitors (PAI-1 and PAI-2) and plasmin inhibitors (α-2-antiplasmin and possibly α-2-macroglobulin). Recently, we demonstrated the presence of PAI-1 and PAI-2 in the tear fluid of patients with corneal pathologies including epithelial defects and stromal ulcers, and proposed a

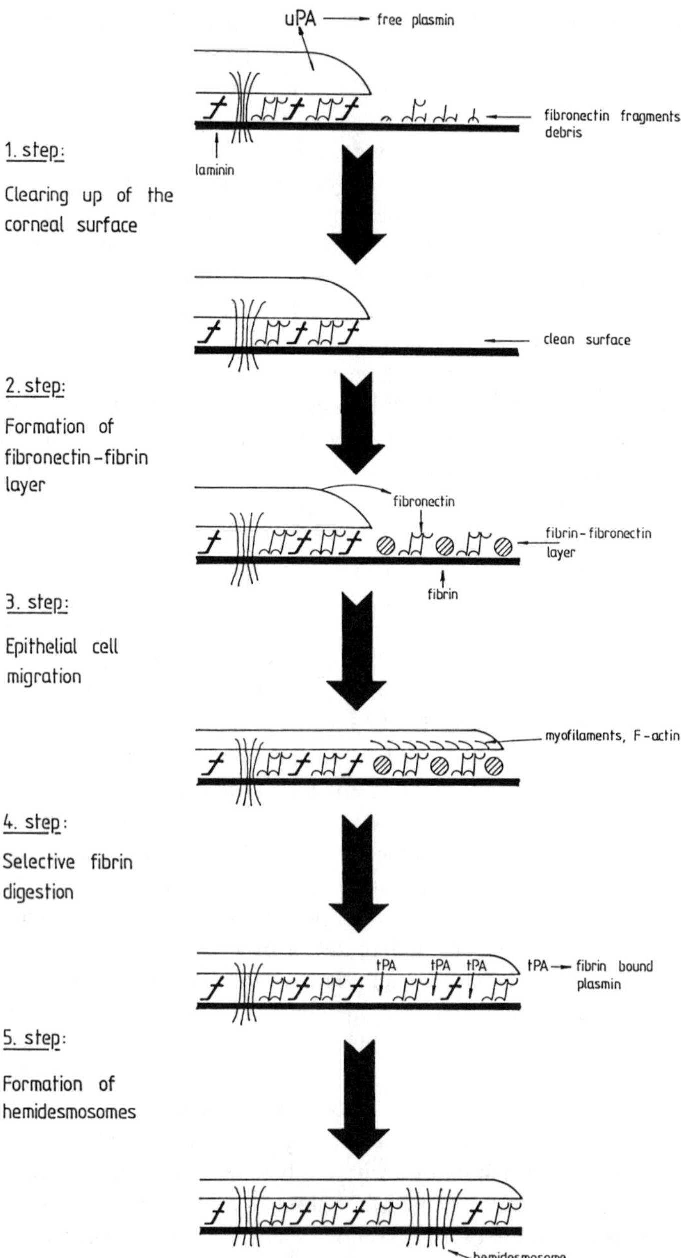

FIGURE 1. Schematic representation of a model for corneal reepithelization (proposed in five steps) in case of simple epithelial defects, and the role of the PA-plasmin system in the process of resurfacing.

mechanism that PAIs act as a first line of defense regulating the action of the released PAs.[29] More information on tear proteinase inhibitors and their possible role in the development and healing of corneal wounds and corneal ulcers is given in Chapters 7, Part D, 10, and 14.

van Setten and co-workers[47] investigated the relationship between the tear fluid level of plasmin and PA and the healing of the epithelial wound. They performed central anterior keratectomy in rabbits and determined tear plasmin and PA levels at regular intervals during the healing process and following rewounding after the epithelization was complete. Enzyme levels were determined in tear samples collected with capillary tubes using radial caseinolytic procedures. Following anterior keratectomy the plasmin concentrations increased rapidly from a mean ($\pm$SEM) of 3.9 ± 0.9 µg/ml to a mean of 37.9 ± 7.8 µg/ml ($p < 0.01$) and gradually decreased during wound healing. Rewounding also resulted in a moderate increase in tear plasmin concentrations. The PA level in the tear samples was found to decrease within the first 24 to 48 h and increase again afterward. The authors pointed out that plasmin and PAs may play an important role in the process of corneal wound healing, and emphasized that the type, depth, and nature of the wound as well as a balance between the destructive and reparative processes may affect the levels of these enzymes detected in the tear fluid.[47]

We investigated the u-PA activities in the tears of rabbits following the mechanical (scraping) or chemical (*n*-heptanol) debridement and alkali burn of the central part of the corneal epithelium. All three types of injury enhanced the PA activities in the tears. The increase in u-PA activity was highest in alkali burn, lowest for *n*-heptanol debridement. Scraping yielded an intermediate increase in u-PA activity. The maximum value in activator activity was reached at 5 h for mechanical injury and at 24 h for chemical injuries. u-PA activity values returned to the normal range by the time of reepithelization for mechanical scraping and 1 to 3 d following complete epithelial wound closure for heptanol debridement and alkali burn (Figure 2). A trend was observed between u-PA activity level and the size of the wound but the correlation was not pronounced (R = 0.538). Despite the individual differences, the tear u-PA activity-time curves showed basically the same behavior, suggesting a uniform mechanism of PA secretion and release following different injuries and observed during the healing period. The lack of high correlation between tear levels of u-PA and the area of the epithelial wound suggests that other factors such as dilution by tearing and by transudation, just like the mode of injury and enzymatic events accompanying corneal wound healing, also affect the level of u-PA in the tear fluid. We assume that at least two different processes contribute to elevated tear u-PA levels and activities: (1) a release of the activator from injured epithelial cells and (2) an active secretion of the same and possibly another (t-PA type) activator by the regenerating epithelium. Either the lack of sufficient activator and enzyme (plasmin) activity or the

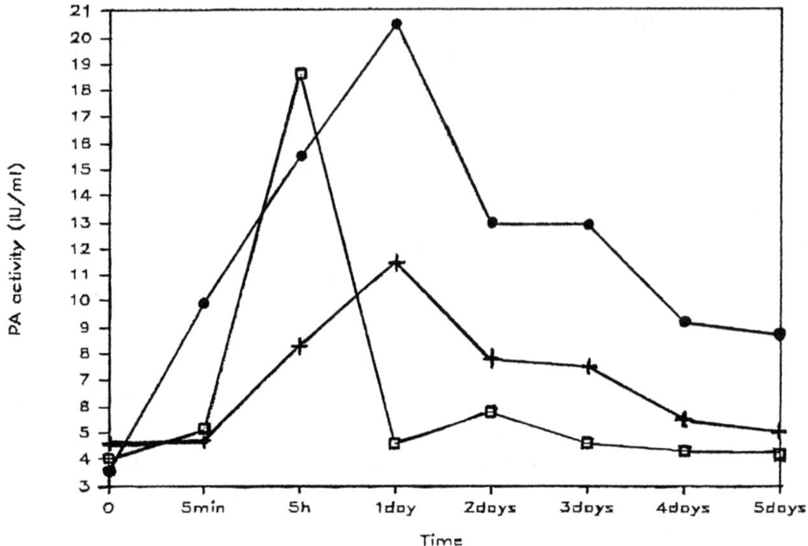

FIGURE 2. Average values of PA activity determinations performed by radial caseinolysis in the absence of fibrin in the tears of five rabbits prior to (0) and at different time intervals subsequent to alkali burn, mechanical scrape, and *n*-heptanol debridement of the corneal epithelium. ● Alkali burn, + *n*-heptanol debridement, □ mechanical scrape. (Data taken from Berta, A., Holly, F. J., Tözsér, J., and Holly, T. F., *Int. Ophthalmol.*, 15, 363, 1991.)

prolonged and uncontrolled secretion of these enzymes may lead to abnormal epithelial healing and may lead to persistent epithelial defects and may be the starting point of destructive processes leading to corneal ulceration.[14]

REFERENCES

1. **Cintron, C., Kublin, C. L., and Covington, H.,** Quantitative studies of corneal epithelial wound healing in rabbits, *Curr. Eye Res.*, 1, 507, 1981/1982.
2. **Crosson, C. E.,** Cellular changes following epithelial abrasion, in *Healing Processes in the Cornea,* (Advances in Applied Biotechnology Series, Vol. 1), Beuerman, R. W., Crosson, C. E., and Kaufman, H. E., Eds., Gulf, Houston, 1989, 3.
3. **Nakagawa, S., Nishida, T., and Manabe, R.,** Actin organization in migrating corneal epithelium of rabbits *in situ, Exp. Eye Res.,* 41, 335, 1985.
4. **Gipson, I. K., Spurr-Michaud, S. J., and Tisdale, A. S.,** Anchoring fibrils form a complex network in human and rabbit cornea, *Invest. Ophthalmol. Vis. Sci.,* 28, 212, 1987.
5. **Fujikawa, L. S., Forster, C. S., Harrist, T. J., Lanigan, J. M., and Colvin, R. B.,** Fibronectin in healing rabbit corneal wounds, *Lab. Invest.,* 45, 120, 1981.
6. **Berman, M., Manseau, E., Law, M., and Aiken, D.,** Ulceration is correlated with degradation of fibrin and fibronectin at the corneal surface, *Invest. Ophthalmol. Vis. Sci.,* 24, 1358, 1983.

7. **Ohashi, Y., Nakagawa, S., Nishida, T., Suda, T., Watanabe, K., and Manabe, R.,** Appearance of fibronectin in rabbit cornea after thermal burn, *Jpn. J. Ophthalmol.*, 27, 547, 1983.

8. **Weimar, V.,** Polymorphonuclear invasion of wounded corneas, *J. Exp. Med.*, 105, 141, 1957.

9. **Berman, M., Leary, R., and Gage, J.,** Evidence for a role of the plasminogen activator-plasmin system in corneal ulceration, *Invest. Ophthalmol. Vis. Sci.*, 19, 1204, 1980.

10. **Salonen, E. M., Tervo, T., Törma, E., Tarkkanen, A., and Vaheri, A.,** Plasmin in tear fluid of patients with corneal ulcers: basis for new therapy, *Acta Ophthalmol.*, 66, 3, 1987.

11. **Tözsér, J., Berta, A., and Punyiczki, M.,** Plasminogen activator activity and plasminogen independent amidolytic activity in tear fluid from healthy persons and patients with anterior segment inflammation, *Clin. Chim. Acta*, 183, 323, 1989.

12. **Punyiczki, M., Berta, A., and Tözsér, J.,** Study of neutral proteinases of human tear samples using gels containing substrates, *Clin. Chem. Enzym. Comns.*, 1, 115, 1988.

13. **Berta, A., Tözsér, J., and Holly, F.,** The determination of plasminogen activator activity in normal and pathological human tears, and the significance of tear plasminogen activators in the inflammations of the cornea and the conjunctiva, *Acta Ophthalmol.*, 68, 508, 1990.

14. **Berta, A., Holly, F. J., Tözsér, J., and Holly, T. F.,** Tear plasminogen activators — indicators of epithelial cell destruction. The effects of scraping, n-heptanol treatment, and alkali burn of the cornea on the plasminogen activator activity of rabbit tears, *Int. Ophthalmol.*, 15, 363, 1991.

15. **Storm, O.,** Fibrinolytic activity in human tears, *Scand. J. Clin. Lab. Invest.*, 7, 55, 1955.

16. **Pandolfi, M. and Astrup, T.,** A histochemical study of the fibrinolytic activity. Cornea, conjunctiva, and lacrimal gland, *Arch. Ophthalmol.*, 77, 258, 1967.

17. **Nishida, T., Nakagawa, S., Awata, T., Tani, Y., and Manabe, R.,** Fibronectin eyedrops for recurrent traumatic erosion, *Lancet*, II, 521, 1983.

18. **Nishida, T., Nakagawa, S., Nishibayashi, C., Tanaka, H., and Manabe, R.,** Fibronectin enhancement of corneal epithelial wound healing of rabbits *in vivo*, *Arch. Ophthalmol.*, 102, 455, 1984.

19. **Nishida, T., Ohashi, Y., Awata, T., and Manabe, R.,** Fibronectin: a new therpy for corneal trophic ulcer, *Arch. Ophthalmol.*, 101, 1046, 1983.

20. **Nishida, T., Nakagawa, S., Awata, T., Ohashi, Y., Watanabe, K., and Manabe, R.,** Fibronectin promotes epithelial migration of cultured rabbit cornea *in situ*, *J. Cell. Biol.*, 97, 1653, 1983.

21. **Berman, M., Kenyon, K., Hayashi, K., and L'Hernault, N.,** The pathogenesis of epithelial defects and stromal ulceration, in *The Cornea: Transactions of the World Congress on the Cornea*, Vol. 3, Cavanagh, H. D., Ed., Raven Press, New York, 1988, 35.

22. **Wang, H.-M., Berman, M., and Law, M.,** Latent and active plasminogen activator in corneal ulceration, *Invest. Ophthalmol. Vis. Sci.*, 26, 511, 1985.

23. **Hayashi, K., Frangieh, G., Kenyon, K. R., Berman, M., and Wolf, G.,** Plasminogen activator activity in vitamin A-deficient rat corneas, *Invest. Ophthalmol. Vis. Sci.*, 29, 1810, 1988.

24. **Vassali, J.-D. and Belin, D.,** Amiloride selectively inhibits the urokinase-type activator, *F.E.B.S. (Federation of European Biochemical Societies) Letters*, 214, 187, 1987.

25. **Berman, M.,** The pathogenesis of corneal defects, *Acta Ophthamol.*, 192(suppl.), 55, 1989.

26. **Rijken, D. C., Wijngaards, G., and Welbergen, J.,** Relationship between tissue plasminogen activator and activators in blood and vascular wall, *Thromb. Res.*, 18, 815, 1980.

27. **van Haeringen, N. J. and Thörig, L.,** Enzymology of tear fluid, *Protides Biol. Fluids Proc. Colloq.*, 32, 399, 1985.

28. **Barlati, S., Marchina, E., Quaranta, C. A., Vigasio, F., and Semerano, F.,** Analysis of fibronectin, plasminogen activators, and plasminogen in tear fluid as markers of corneal damage and repair, *Exp. Eye Res.*, 51, 1, 1990.

29. **Tözsér, J. and Berta, A.,** Plasminogen activator inhibitors in human tears, *Acta Ophthalmol.,* 69, 426, 1991.

30. **Thörig, L., van Agtmaal, E. J., Glasius, E., Tan, K. L., and van Haeringen, N. J.,** Comparisosn of tears and lacrimal gland fluid in the rabbit and guinea pig, *Curr. Eye Res.,* 4, 913, 1985.

31. **Thörig, L., van Haeringen, N. J., and Wijngaards, G.,** Comparison of enzymes of tears, lacrimal gland fluid, and lacrimal gland tissue in the rat, *Exp. Eye Res.,* 38, 605, 1984.

32. **Bernik, M. B.,** Increased plasminogen activator (urokinase) in tissue culture after fibrin deposition, *J. Clin. Invest.,* 52, 823, 1973.

33. **Bernik, M. B., Wijngaards, G., and Rijken, D. C.,** Production by human tissues in culture of immunologically distinct, multiple molecular weight forms of plasminogen activators, *Ann. N.Y. Acad. Sci.,* 370, 592, 1981.

34. **Vetterlein, D., Bell, T. E., Young, P. L., and Roblin, R.,** Immunological quantitation and immunoabsorption of urokinase-like plasminogen activators secreted by human cells, *J. Biol. Chem.,* 255, 3665, 1980.

35. **Pandolfi, M., Astedt, B., and Dyster-Aas, K.,** Release of fibrinolytic enzymes from human cornea, *Acta Ophthalmol.,* 50, 199, 1972.

36. **Lanz, E., Dyster-Aas, K., and Pandolfi, M.,** *In vitro* release of fibrinolytic activators from cornea, *Graefe's Arch. Clin. Exp. Ophthalmol.,* 206, 157, 1978.

37. **Pandolfi, M. and Lantz, E.,** Partial purification and charactarization of keratokinase, the fibrinolytic activator of the cornea, *Exp. Eye Res.,* 29, 563, 1979.

38. **Lantz, E. and Andersson, A.,** Release of fibrinolytic activators from the cornea and conjunctiva, *Graefe's Arch. Clin. Exp. Ophthalmol.,* 219, 263, 1982.

39. **Ogston, D.,** *Antifibrinolytic Drugs. Chemistry, Pharmacology and Clinical Usage,* John Wiley & Sons, Chichester, 1984.

40. **Saksela, O.,** Plasminogen activation and regulation of pericellular proteolysis, *Biochim. Biophys. Acta,* 823, 35, 1985.

41. **Haber, E., Quertermous, T., Matsueda, G. R., and Runge, M. S.,** Innovative approaches to plasminogen activator therapy, *Science,* 243, 51, 1989.

42. **Illig, K. M.,** Die Behandlung rezidivierender Hornhauterosionen mit verdauender Fermenten, *Klin. Mbl. Augenheilk.,* 176, 839, 1980.

43. **Rigó, Gy.,** A disjunctio epithelialis gyógyítása tripszinnel. (The treatment of epithelial dysjunction with trypsin), *Szemészet,* 122, 87, 1985.

44. **Yamada, K., Akiyama, S., Hasegawa, T., Hasegawa, E., Humphries, M., Kennedy, D., et al.,** Recent advances in research on fibronectin and other cell attachment proteins, *J. Cell Biochem.,* 28, 79, 1985.

45. **Nishida, T., Nakagawa, S., Watanabe, K., et al.,** A peptide from fibronectin cell binding domain inhibits attachment of epithelial cells, *Invest. Ophthalmol. Vis. Sci.,* 29, 1820, 1988.

46. **McDonald, J. A.,** Fibronectin: a primitive matrix, in *The Molecular and Cellular Biology of Wound Repair,* Clark, R. A. F. and Henson, P. M., Eds., Plenum Press, New York, 1988, 405.

47. **van Setten, G.-B., Salonen, E.-M., Vaheri, A., Beuerman, R. W., Hietanen, J., Tarkkanen, A., and Tervo, T.,** Plasminogen and plasminogen activator activities in tear fluid during corneal wound healing after anterior keratectomy, *Curr. Eye Res.,* 8, 1293, 1989.

Chapter 11

TEAR ENZYMES AND THE WOUND HEALING OF THE CORNEA

Part D
Collagens, Collagenases, Collagenase Inhibitors, and
Corneal Wound Healing

I. COLLAGENS IN WOUND HEALING

Wound healing, in general, can be divided into four phases: (1) inflammatory (infiltrative) phase, (2) neomatrix formation (synthesis, deposition, and remodeling), (3) wound contraction, and (4) re-epithelization. These phases or processes are not distinct in time but rather overlap during the course of repair; different types of collagen and collagen fragments seem to play an important role in each of them.[1]

1. The inflammatory phase of wound healing is initiated almost immediately following tissue injury by the activation of tissue complement and/ or the activation of the coagulation cascade. This leads to the invasion of inflammatory cells into the wound area through the chemotactic action of complement factor C5a, platelet factor 4, fibrin degradation products, and other chemotactic agents. Fibrillar collagens (types I and III) were shown to play a role in this early phase of wound healing since they promote platelet aggregation.[2] The binding of platelets to collagen structures results in the release of glycoproteins, fibronectin, and thrombospodin. The latter is responsible for the subsequent aggregation of platelets to one another.[3] This collagen-induced aggregation of platelets, together with the activation of the coagulation cascade, is responsible for the formation of the fibrin/thrombus plug forming the first physical connection between the edges of the wound. Collagen fragments generated by collagenases of granulocyte and macrophage origin were shown to have a chemotactic effect on fibroblasts,[4] and thus may mediate repair by promoting connective tissue cell migration and proliferation leading to later phases of wound repair.

2. The migration of fibroblasts into the wounded area is the starting point of the second phase of repair which is characterized by the synthesis and deposition into the defect zone of collagen and noncollagen extracellular matrix molecules. This process leads to the formation of a fibrotic or granulotic tissue. This phase usually begins 3 to 5 d after wounding and lasts for 10 to 12 d. During this period the synthesis of collagen types I and III continues.[5,6] This is followed by a longer period of remodeling characterized by the formation of fibrils from the deposited collagen molecules and that of crosslinks between different macromolecules of

213

the matrix, leading to the formation of scar tissue. The fibers gradually become oriented in the scar resulting in an increase in tensile strength.[7]

3. The mechanism of wound contraction is not yet fully understood. It is believed that connections are formed between the newly synthesized type I and type III collagen molecules and special contractile cells called myofibroblasts, located in the extracellular matrix surrounding the wound. Wound contraction is different from scar contracture. Wound contraction is a relatively early event in the wound healing process. This phase involves an interaction between myofibroblasts and the newly formed extracellular matrix molecules. Scar contracture is a later event associated primarily with scar remodeling.[8,9]

4. Reepithelization occurs on surfaces covered by or composed of a matrix of collagen molecules (type I, III, or IV) and noncollagenous proteins such as fibronectin, laminin, and fibrin. Resurfacing may occur on intact basement membranes, on bare surfaces of differentiated connective tissues, or on the surface of a granulation (fibrotic) tissue. The provisional matrix provided by the granulation tissue and the serum-derived components is quite different from the basement membrane, containing collagen type IV, on which epithelial cells normally reside. This results in totally different cellular behavior characterized by lateral cell movement onto the wound area and cell proliferation, rather than the vertical movement and terminal differentiation characteristic of epithelial cells, when they rest on an intact basement membrane.[10]

II. THE INTERACTIONS OF COLLAGENS WITH OTHER EXTRACELLULAR MATRIX PROTEINS AND CELL MEMBRANES

Cells involved in the wound healing require attachment to a surface or substratum to grow or to divide. Collagens are the major proteins of the matrix surrounding these cells. In tissue culture, collagen enhances growth or differentiation of various cell types including fibroblasts and epithelial cells. The cells do not bind directly to the collagen of the matrix. Rather, extracellular glycoproteins mediate this attachment. Epithelial cells, however, may also attach directly to type IV collagen, which is the major constituent of basement membranes. Fibronectins and laminin are the attachment proteins whose interactions with collagen have been studied most extensively. Whereas fibronectin binds to all types of collagen, laminin binds preferentially to type IV collagen. Apart from promoting the attachment of epithelial cells and fibroblasts to the surroundig matrix, fibronectin-collagen and laminin-collagen interactions also play a role in cellular differentiation. The attachment by laminin seems to be effected by membrane proteins and focal contacts between laminin and cytoplasmic microfilaments in a way very similar to that of fibronectin.[11,12]

Fibronectin forms immobilized fibrillar arrays across the surface of fibroblasts and epithelial cells, and interconnects cells with each other or with collagen structures. Fibronectin possesses specific high-affinity binding sites for both the cell surface and the collagen molecule. In cell cultures, fibronectin is found at or close to the focal contacts, where cells stick to the collagen substrate and at points of cell-to-cell adhesion. These are the points where the cytoplasmic actin myofilaments insert into the plasma membrane. Vinculin and actin are also found in these regions. By electron microscopy the extracellular fibronectin fibers appear to have close contact with actin fibers within the cell. The membrane junction for these fibers is termed fibronexus.[13] Fibronectin receptor, a 140-kDa cell surface glycoprotein may anchor both fibronectin and the intracellular fibers to the opposite sides of the plasma membrane.[14] Such interactions are thought to form at focal contacts between cells, attachment proteins, and collagen structures in the wound healing process, too.[1]

III. COLLAGENS AND COLLAGEN SYNTHESIS IN CORNEAL WOUND HEALING

In corneal tissue, collagen constitutes 70% of the dry weight. Collagen type I comprises about 80 to 90% of total amount of collagens in the cornea.[15] This collagen is primarily located in the extracellular matrix of the corneal stroma. Type III collagen production was detected in the cornea in association with stromal wound healing.[16] Most of the collagen in the basement membrane of the corneal epithelium, as in other epithelial basement membranes throughout the human body, is type IV collagen.

The collagen molecules in the corneal stroma are organized into fibers, which measure 24 to 30 nm in diameter and have a 64-nm longitudinal banding periodicity. The collagen fibers, which are enmeshed in a glycosaminoglycan matrix, form lamellar sheets that lie parallel to the corneal surface. Parallel fibers in one lamella are at oblique angles to those in an adjacent lamella. Each lamella is 1.5 to 2.5 μm thick and continuous across the cornea. An unusual feature of the collagen fibrils of the cornea is their diameter of 24 to 30 nm, a size found elsewhere in the adult human body only in the aortic valve and the vitreous humor, all other collagen having a fibril size of 40 to 280 nm.[9] In embryonic tissues all collagen has a small fibril size (diameter) but in the cornea no further enlargement occurs during maturation. The size of the fibrils and their regular spacing in the extracellular matrix is essential for the transparency of the cornea.[17] The collagen fibrils in the sclera and those in the corneal wound (scar) tissue reach larger diameters more typical of mature collagen.[18]

The healing process of deep or penetrating corneal wounds consists of similar steps as described for wound healing in general. A fibrin plug is formed in the wound and the surrounding tissues are infiltrated first by inflammatory cells, then due to action of chemotactic factors by keratocytes. Collagen (predominantly type I, but also type III and AB) is secreted by keratocytes showing signs of fibroblastic conversion. A remodeling occurs in the fibrotic tissue which is characterized by the formation of fibrils from the newly synthesized collagen molecules and by the appearance of collagen-collagen and collagen-glycosaminoglycan cross-links and interactions. The structural macromolecules and cross-links produced during wound healing in the cornea resemble in many respects those found in the developing embryonic cornea. Collagen fibrils that are produced vary markedly in diameter from 20 to 120 nm and never become organized into the lattice-like array with the elements less than 200 nm apart characteristic of normal lamellar architecture. Stromal scar tissue differs from normal cornea in glycosaminoglycan content and composition, too. It is thought that the macromolecules of the interfibrillar matrix are important determinants of collagen fibril size and spatial arrangement.

IV. COLLAGENASES AND COLLAGENASE INHIBITORS IN THE HEALING OF CORNEAL WOUNDS

Collagenases are generally defined as a class of proteinases having the unique ability to cleave the triple helix of collagen at neutral pH under nondenaturing conditions. Other proteinases such as elastase, trypsin, chymotrypsin, and others are unable to cleave native collagens in spite of the fact that they readily attack the same substrates when the triple helical structure is disrupted by thermal denaturation. Similarly, cathepsins are unable to digest native collagen at neutral pH but have the capacity to degrade the molecule when the triple helix is deformed in an acidic environment.[19]

The role of collagenolytic activity in wound healing in general is not yet fully elucidated. Nevertheless, it seems likely that both interstitial collagenase and the tissue collagenase inhibitor must play an active role in dermal wound repair.[20] Collagen fragments generated by collagenolytic enzymes of inflammatory cells were shown to play a role in the chemotactic recruitment of fibroblasts in the wound healing process.[21] Epithelial cells were suggested to burrow between the eschar and granulation tissue by expressing collagenase and other hydrolases dissolving the collagenous matrix in front of them as they move.[22] Soluble factors secreted by the regenerating epithelium may be capable of modulating the expression of collagenases and other factors responsible for controlling collagen turnover in dermal wound repair. Recently collagenase responsible for degrading collagen type IV was shown to be released by migrating epidermal cells[10] and by dermal fibroblasts.[23] Whether such a mechanism and such proteinases also play a role in remodeling the surface of the basement membrane, or that of the bare stromal surface in corneal wound healing, are problems that need further investigation.

Ever since the first observation that corneal epithelial cells release collagenolytic enzymes, especially if they originate from ulcerating corneas, the idea that collagenase plays an important role in corneal ulceration has been very popular. The idea to treat corneal ulcers by collagenase inhibitors occurred to the first investigators. This was followed by 2 decades of experimental and clinical investigation. Various investigators were concerned about finding ways corneal ulcers can be treated; the process of corneal wound healing can be promoted by the use of collagenase inhibitor eyedrops. The results were rather contradictory. The first reports showed the beneficial effect of the use of inhibitors in the treatment of corneal ulcers, persistent epithelial defects, and different types of keratitis. Other reports indicated that collagenase inhibitors do not necessarily promote the healing of corneal ulcers; some of these preparations may even hinder reepithelization and stromal healing due to toxic effects. This is why the use of collagenase inhibitors has become less popular among ophthalmologists.

Experimental data are accumulating proving that proteinases (collagenase, plasmin, plasminogen activators, etc.) play important roles in the healing of corneal defects by digesting the debris within the defect zone and by promoting the sliding of the regenerating epithelial cells on the denuded surface. It has been observed that excessive or prolonged proteinase action (or the lack of physiological inhibitors?) may hinder the healing process, but the total absence (or complete inhibition) of these proteolytic enzymes also may prevent epithelial and stromal repair. The diversity of the role of proteinases in the healing processes and the complexity of the regulatory mechanisms explain the unpredictable outcome of the use of collagenase inhibitors for the promotion of the healing of corneal ulcers. Beneficial effect can be expected only if inhibitors are used in selected cases when the existing high collagenase activity hinders the wound healing process. Another possibility for the successful use of collagenase inhibitors is the prevention of ulcer formation. Various collagenase inhibitors were shown to prevent ulceration in different forms of ocular surface disease. It is also possible to stop the enzymatic degradation of corneal tissue in case of an already existing ulcer and by doing so indirectly promote the repair mechanisms.[24]

REFERENCES

1. **McPherson, J. M. and Piez, K. A.,** Collagen in dermal wound repair, in *The Molecular and Cellular Biology of Wound Repair,* Clark, R. A. F. and Henson, P. M., Eds., Plenum Press, New York, 1988, 471.
2. **Shosan, S.,** Wound healing, in *International Review of Connective Tissue Research,* Vol. 9, Hall, D. A. and Jackson, D. S., Eds., Academic Press, New York, 1981, 1.
3. **Jaffe, E. A., Leung, L. L. K., Nachman, R. L., Levin, R. I., and Mosher, D. F.,** Thrombospondin is the endogenous lectin of human platelets, *Nature,* 295, 246, 1982.
4. **Postlethwaite, E. E., Seyer, J. M., and Kang, A. H.,** Chemotactic attraction of human fibroblasts to type I, II and III collagens and collagen derived peptides, *Proc. Natl. Acad. Sci. U.S.A.,* 75, 871, 1978.
5. **Madden, J. W. and Peacock, E. E.,** Studies on the biology of collagen during wound healing. I. Rate of collagen synthesis and deposition in cutaneous wounds of the rat, *Surgery,* 64, 288, 1968.

6. **Ross, R. and Benditt, E. P.,** Wound healing and collagen formation. I. Sequential changes in components of guinea pig skin wounds observed in the electron microscope, *J. Biophys. Biochem. Cytol.,* 11, 677, 1961.

7. **Heughan, C. and Hunt, T.,** Some aspects of wound healing reasearch: a review, *Can. J. Surg.,* 18, 118, 1975.

8. **Erlich, H. P.,** The role of connective tissue matrix in hypertrophic scar contracture, in *Soft and Hard Tissue Repair,* Hunt, T. K., Heppenstall, R. B., Pines, E., and Rovee, D., Eds., Praeger, New York, 1984, 533.

9. **Gabbiani, G., Hirschel, B. J., Ryan, G. B., Statkov, P. R., and Majno, G.,** Granulation tissue as a contractile organ. A study of structure and function, *J. Exp. Med.,* 135, 719, 1972.

10. **Woodley, D. T., Liotta, L. A., and Brondage, R.,** Adult human epidermal cells migrating on nonviable matrix produce a type IV collagenase, *Clin. Res.,* 30, 266, 1982.

11. **Timpl, R.,** Laminin — a glycoprotein from basement membranes, *J. Biol. Chem.,* 254, 9933, 1979.

12. **Yamada, K., Akiyama, S., Hasegawa, T., Hasegawa, E., Humphries, M., Kennedy, D., Nagata, K., Urishihara, H., Olden, K., and Chen, W.,** Recent advances in the research on fibronectin and other attachment proteins, *J. Cell Biochem.,* 28, 79, 1985.

13. **Singer, I. L.,** The fibronexus: a transmembrane association of fibronectin containing fibers and bundles of 5 nm microfilaments in hamster and human fibroblasts, *Cell,* 16, 675, 1979.

14. **Pytela, R., Pierschbacher, M. D., and Rouslahti, E.,** Identification and isolation of a 140 kd cell surface glycoprotein with properties expected of a fibronectin receptor, *Cell,* 40, 191, 1985.

15. **Nishida, T., Nakagawa, S., Ohashi, Y., Awata, T., and Manabe R.,** Fibronectin in corneal wound healing: appearance in cultured rabbit cornea, *Jpn. J. Ophthalmol.,* 26, 410, 1982.

16. **Newsome, D. A., Foidart, J. M., Hassell, J. R., et al.,** Detection of specific collagen types in normal and keratoconus corneas, *Invest. Ophthalmol. Vis. Sci.,* 20, 738, 1981.

17. **Maurice, D. M.,** The structure and transparency of the cornea, *J. Physiol.,* 136, 263, 1957.

18. **Schwarz, W.,** *Z. Zellforsch. Mikrosk. Anat.,* 38, 78, 1953.

19. **Burleigh, M. C., Barrett, A. J., and Lazarus, G. S.,** A lysosomal enzyme that degrades native collagen, *Biochem. J.,* 137, 387, 1974.

20. **Mignatio, P., Welgus, H. G., and Rifkin, D. B.,** Role of degradative enzymes in wound healing, in *The Molecular and Cellular Biology of Wound Repair,* Clark, R. A. F. and Henson, P. M., Eds., Plenum Press, New York, 1988, 497.

21. **Woolley, D. E.,** Mammalian collagenases, in *Extracellular Matrix Biochemistry,* Piez, K. A. and Reddi, A. H., Eds., Elsevier, New York, 1984, 119.

22. **Harris, E. D. and Krane, S. M.,** Collagenases, *N. Engl. J. Med.,* 291, 652, 1974.

23. **Salo, T., Turpeenniemi-Hujanen, T., and Tryggvason, K.,** Tumor-promoting phorbol esters and cell proliferation stimulate secretion of basement membrane (type IV) collagen-degrading metalloproteinase by human fibroblasts, *J. Biol. Chem.,* 260, 8526, 1985.

24. **Berta, A., Holly, F. J., and Tözsér, J.,** Proteinase-inhibitors in the diagnosis and treatment of ocular surface diseases, in *Proc. Symp. Pharmacotherapy of the Eye,* Springer-Verlag, Heidelberg, accepted for publication.

Chapter 12

TEAR ENZYMES IN HERPETIC KERATITIS AND OTHER VIRAL INFECTIONS OF THE CORNEA

I. EPIDEMIOLOGY, PATHOGENESIS, AND CLINICAL MANIFESTATIONS OF HERPETIC KERATITIS

Herpes simplex virus (HSV) is a ubiquitous neurotrop virus. It is the most frequent causative agent of human viral infections. Primary infection usually occurs in early childhood. The virus has an ability to survive in the sensory ganglia. The virus persists in a latent form and is triggered by a variety of aspecific mechanisms (stress, fever, or other stimuli) often produces recurrent disease. More than 80% of adults in the U.S. have been infected with and have detectable anibodies to HSV.[1]

Herpetic keratitis, caused primarily by HSV type 1 (oral), and occasionally by HSV type 2 (genital), is the leading cause of corneal blindness in the U.S., with 300,000 to 500,000 cases reported annually.[2,3] Herpetic dermatitis and herpes labialis (cold sore) are also caused by HSV type 1. These diseases afflict approximately one third of the world population. Herpes genitalis is the second most common venereal disease in the U.S. with over 100,000 cases reported annually.[73] Disseminated infections and herpetic encephalitis are rare but usually very serious diseases with high mortality rate.[5,6] Patients suffering from primary or secondary immunodeficiencies, leukemics, AIDS patients, and those receiving systemic immunosuppressive therapy are at high risk of severe systemic and ocular herpetic infections.[7] Herpes viruses have been related to oncogenesis: HSV type 1 has been associated with nasopharyngeal tumors, and HSV type 2 with cervical carcinoma.[8]

After entering through one of the mucous mebranes (mouth, nose, or eye) or starting with dermal lesions of the surrounding areas, the virus reaches the trigeminal and the cervical ganglia. The virus persists in the ganglia in a latent state, from time to time reactivates, and via nerve fibers gets back to the periphery (skin or mucous membranes) and produces recurrent disease.[9] Rabbit experiments revealed that if the eye was infected with a particular herpes virus strain the strain colonized the cervical and trigeminal ganglia on both sides.[10] The same virus (one particular strain) seems to persist in the ganglia during the whole life of the living being (animal or human). Once a ganglion is infected this seems to prevent the superinfection with another strain, even if this latter virus is more virulent than the original strain.[11]

Based on the observations that total body irradiation or immunosuppression or the use of corticosteroids worsen herpetic disease, it was generally thought that differences in the severity and forms of herpetic disease among various individuals are caused by differences in immunologic resistance. It was generally accepted that variations in the diseases are basically host related, and severe disease develops in hosts who, even though seem to be otherwise healthy, have lower resistance to a particular virus strain. Substantial experimental data and clinical observations suggest that this may not be the right explanation. Certain strains were found to cause stromal disease (interstitial or disciform keratitis) in a higher percentage of the cases than others. The same strains produce large amounts of highly antigenic glycoproteins. The occurrence of disciform keratitis, a kind of virus-induced autoimmune reaction, is to a large extent determined by the characteristics of the infecting virus (by the antigens it produces). Similarly, the patterns of the dentritic lesion were shown to depend on the virus causing the infection and did not seem to depend on the immunologic status of the infected animal.[12] Experimental animals infected by different strains showed different recurrence rates, basically determined by the strain and not the immunological resistance of the animals.[13] Corticosteroids, whose effect on virus infections is thought to be closely related to immunosuppression, were found to worsen herpetic infection in tissue culture where no immunologic elements were present.[14] The *in vivo* effects of corticosteroids on herpetic keratitis were also found to be different depending the genome of the infective agent.[15] The site of the herpetic eruption may also be determined by the virus. Different strains grow preferentially at different temperatures (some at body core temperature, others at lower values near the temperature of the corneal surface); strains that prefer cooler temperatures are more likely to cause corneal diseases.[16] These findings do not rule out the possible role of immunological factors in the development of the disease, but reveal that differences in recurrence rates and clinical patterns are determined to a large extent by the virus genome, and infections by different strains may lead to different lesions as for severity and clinical forms in otherwise healthy hosts.[1]

The clinical forms of ocular herpes can be classified into three main groups: (1) neonatal (connatal, congenital) ocular herpes, (2) primary ocular herpes, and (3) recurrent ocular herpes.

1. Apart from a single case (two children, dizygotic twins) reported by Hutchinson and colleagues[17] where the supposed mechanism was transplacental infection with HSV-1, all other neonatal herpetic infections reported were either caused by exposure to an HSV-2 infected birth canal or were the results of an ascending HSV-2 infection during the late prenatal period. Ocular herpes in infants may include conjunctivitis, superficial keratitis, stromal reaction, chorioretinitis, and cataract.[18] In most cases ocular involvement was associated with vesicular eruptions of the skin. Maternal transplacental antibodies do not seem to protect the infants from connatal ocular disease, though this passive immunity seems to prevent systemic infections and life-threatening visceral involvement in the majority of the cases.[19]

2. Primary ocular involvement usually occurs between 6 months and 6 years. Newborns are protected by antiherpes antibodies of maternal origin that were acquired through the placenta from the blood of the

mother. This protection, which may prevent infection with the herpes virus, lasts for about 6 months. By this time the passively received antibodies are eliminated and the infant becomes susceptible to herpes virus infections. More than 80% of the population become virus carriers by the age of 6 years.[20] Primary ocular herpes usually appears as acute follicular conjunctivitis, or acute keratoconjunctivitis, often associated with vesicular eruption of the periocular skin or the lid margins, and the enlargement of the preauricular lymph nodes. Corneal involvement in primary infections is limited to the epithelial layer, may appear as typical dendritic lesion, or in atypical forms as diffuse punctate superficial keratitis.[3]

3.   The most frequent form of recurrent ocular herpes is herpetic epithelial keratitis. Epithelial keratitis may present as punctate keratitis, as small bullous epithelial eruptions, as dentritic, or as geographic epithelial lesions. Epithelial lesions stain with fluorescein and have a tendency to progress in a branching linear fashion. Epithelial cells at the edges of the herpetic lesion contain actively replicating virus and stain with rose bengal. Faint superficial scars may develop under the epithelial lesion especially if it is chronic and the treatment was inadequate or had been delayed.[1] The disease may involve the corneal stroma causing herpetic stromal keratitis (formerly called herpetic interstitial keratitis) or deep herpetic ulcer. These forms are believed to be caused by direct viral invasion of the corneal stroma, and are basically different from disciform keratitis (disciform edema). Disciform keratitis is a virus-induced immunological reaction of the cornea. It is characterized by a disc-shaped edema and cellular infiltration of the central corneal stroma.[21,22] In disciform keratitis there are no clinical signs of epithelial involvement, but keratic precipitates and other symptoms of a mild iritis are often present. No intact virus can be detected in the stroma, but aqueous humor samples obtained by paracenthesis may be positive for herpes virus.[23] This form of ocular herpes responds well to corticosteroids. In long-standing cases the so-called metaherpetic ulcer may develop. This is a chronic trophic ulcer characterized by a gray, thickened border formed by heaped-up epithelium unable to move across and adhere to the necrotic ulcer bed.[3]

II. THE PROPAGATION OF THE VIRUS IN OCULAR TISSUES AND THE MECHANISM OF TISSUE DESTRUCTION IN HERPETIC CORNEAL DISEASE

In the eye corneal nerves are thought to be the source of reactivated herpes virus, which may be shed in the tears and may cause corneal disease.[1] Another mechanism for the virus to reach the epithelial cells is via the nerve endings located in the middle and superficial layers of the corneal epithelium, a direct transport without the suggested role of the tear film. The localized nature of the epithelial lesion in ocular herpes (dentritic keratitis) may support this latter mechanism, while in the case of other non-neurotrop viruses causing dissemi-nated lesions on the surface of the cornea (superficial punctate keratitis) a transport through the tear film is more probable. The possible role of hydrolytic enzymes in the propagation of the virus along the nerve axons, and in the development of the characteristic epithelial vesicules, and in the development of dentritic lesions is rather speculative, though such mechanisms have been demonstrated in other viral diseases.

The epithelial lesion in epithelial herpetic keratitis typically forms a dentric figure and progresses in a branching linear fashion. Because this branching figure closely resembles the arborizing intraepithelial nerve fiber network, it has been postulated that the structural pattern of the dentritic fugure corre-sponds to the anatomic pattern of the nerve filaments in the cornea. This assumption is based entirely on the clinical appearance of the lesion rather than on any supporting experimental data.[24] Sugiura and Kondo,[25] after the resec-tion of the trigeminal nerve, and Baum,[24] after penetrating keratoplasty, dem-onstrated in rabbit experiments that following the inoculation of herpes virus the characteristic dentritic pattern developed in the corneas also in the absence of sensory nerves. These results suggest that the dentritic epithelial lesion of acute herpes simplex keratitis is not associated with the anatomic pattern of corneal epithelial nerves. There is no histological proof for the existence of the suggested polygonal cells either, whose dedritic processes also were proposed to be the anatomical basis of herpetic epithelial lesions.[26] It has been observed that, following the rubbing of the cornea through closed eyelids, after the instillation of a drop of fluorescein, a mosaic-like pattern of polygons appear in the epithelium.[27,28] The dimensions of the branchings of these polygons and those of the dendritic figure are approximately the same, and the pattern of both may be determined by the same but yet unknown structural characteristics of the corneal epithelium.[24]

In severe cases of herpetic keratitis the herpes virus invades the corneal stroma and infects stromal keratocytes.[29] Necrotizing stromal keratitis is char-acterized by extensive tissue damage. Besides the direct effect of the virus, antibody- and cell-mediated immune mechanisms have been implicated as causes of the damage.[30] As in cases with bacterial ulcers in the formation of herpetic ulcers hydrolases deriving from polymorphonuclear leukocytes (PMNs), from macrophages, and possibly from infected, necrotizing corneal cells effect matrix damage. Direct immune mechanisms (cytotoxic antibodies, macroph-ages, T lymphocytes, natural killer cells, etc.) also play a role in the destruction of keratocytes.[30]

Campbell et al.[31] injected herpes virus in the corneal stroma of presensitized rabbits. They demonstrated, besides cellular infiltration and vascularization of the corneal stroma, accompanying iritis and conjunctivitis, that the virus-injected corneas produced collagenolytic enzymes (latent and active collag-enase). The levels of total (latent plus active) collagenase detected in the culture media of the herpes-infected corneas were high, similar to those observed previously culturing alkali-burned rabbit corneas. Herpes-infected corneas, unlike alkali-burned corneas, showed a high proportion of the enzyme in latent, inactive form. The authors suggested that collagenase and probably other proteolytic enzymes deriving from keratinocytes and PMNs invading the infected cornea, once activated, are important in effecting stromal damage in herpetic keratitis. Different factors leading to their activation, however, need further investigation.

The incidence of primary herpetic epithelial disease is comparable in men and women but men demonstrate statistically more stromal damage than do women.[32] Stromal disease in women appears to worsen just before the onset of menstruation, at a time when progesterone levels drop rapidly (Berman, 1980). Based on these observations it has been suggested that sexual steroid hormones have a regulatory effect on stromal destruction in herpetic corneal ulceration. Berman and co-workers using rabbits sensitized with herpes simplex virus demonstrated that subconjunctival medroxyprogesterone beneficially affected the clinical course of the stromal disease. The inflammatory cell infiltrate of the stroma diminished, and the level of collagenase released by the hormone-treated cornea, when cultured *in vitro,* was greatly reduced in comparison with nontreated corneas. The exact mechanism how medroxyprogesterone suppressed stromal infiltration and ulceration is not known. A corticosteroid-like effect is very probable as, besides the beneficial antiinflammatory effect, the epithelial herpetic disease was exacerbated by medroxyprogesterone treatment like it is experienced when corticosteroids are used in the active phase of ocular herpetic infections. Though medroxyprogesterone has certain advantages over the use of corticosteroids (does not suppress collagen synthesis, does not cause steroid cataract or steroid glaucoma), this hormone has not become a part of the therapeutic arsenal of practicing ophthalmologists.[33]

Alterations in the physical and clinical characteristics of the preocular tear film in patients with herpetic keratitis were reported by different authors. The decreased stability of the tear film (<10 s break-up time) is probably brought about by changes in the smoothness of the epithelial surface caused by vesicles, epithelial striae, and the dendritic lesion itself.[34,35] Epithelial lesions and decreased tear film stability may form a vicious circle in which localized drying out of the surface may worsen the condition of the epithelium. These physico-chemical changes are probably accompanied by alterations in the composition of tears[36,37] and can only partly be compensated by an increased (reflex) tear secretion.[38,39]

Tervo et al.[40] measured radial caseinolytic (plasmin) activity in the tears of patients with different lesions of the superficial cornea, including 11 eyes with superficial punctate keratitis, 3 eyes with dentritic keratitis, and 25 eyes with corneal ulcers; 72.0% of tear samples from patients with corneal ulcer, 45.5% of samples from superficial punctate keratitis patients, and 66.7% from detritic keratitis patients were found to contain active plasmin. The ranges of enzyme levels in tears were 0.5 to 21.0, 1.0 to 12.0, and 4.0 to 10.0 µg/ml in the case of ulcers, punctate keratitis, and dentritic keratitis, respectively. The authors pointed out that plasmin was also detected in the tear fluid of eyes showing neither bacterial growth nor leukocyte discharge, and suggested that the appearance of plasmin in the tears of patients with various lesions of the corneal epithelium may be associated with the production and release of tissue-type plasminogen activator (t-PA) from the affected epithelial cells.

We measured the PA activity in the tears of patients with different inflammatory diseases of the cornea including herpetic epithelial keratitis and corneal ulcers. We demonstrated elevated activator activities in the tears of these patients both with an amidolytic method using a chromogenic synthetic substrate[41] and with plasminogen-containing casein plates.[37] The range of PA activities in pathological tears was 0.1 to 2.0 IU/ml (normal value: 0.03 ± 0.02 IU/ml).[37] The detected activator activity was shown to be almost exclusively of urokinase (u-PA) type, and showed good correlation with the severity of the disease.[41,42] In another study we detected plasminogen activator inhibitor (PAI) activity in the tears of patients with corneal epithelial diseases including herpetic (dentritic) keratitis.[43] PAIs are not normal components of the normal tear fluid but can be detected in tears of patients with different inflammatory and traumatic lesions of the cornea. Immunoblotting revealed the presence of two different types of PAI, PAI-1 and PAI-2, in the tears of patients, and proved that most of the PAIs are present in the form of activator-inhibitor complexes. Unlike PAI-1, which is present in the plasma, PAI-2 is thought to be produced by epithelial cells (including those of the human cornea). This may be supported by the observations that PAI-2 was detected at low tear flow rates, and PAI-1 was present in tear samples that showed signs of increased vascular permeability.[43]

Based on these experimental data we propose a multilevel regulatory system controlling the action of proteinases present in the tears of patients suffering from herpetic keratitis. Affected epithelial cells release u-PA. The first level of regulation is the inhibition of the activator by specific activator-inhibitors PAI-2 produced by the epithelium and PAI-1 deriving from the blood. In the case of excessive u-PA production, or caused by the lack of sufficient activator-inhibitor production, the PAI capacity of the tears is used up and the activator activity in the tears increases. This results in the conversion of plasminogen into plasmin. The activity of plasmin is inhibited by plasmin inhibitors (α-2-antiplasmin, α-2-macroglobulin, and α-1-antitrypsin) deriving from the blood through the walls of the blood vessels of the inflamed conjunctiva. The presence of these inhibitors and inhibitor-enzyme complexes in pathological human tears was also demonstrated.[44-47] If this second level of inhibition should prove to be insufficient, the excessive plasmin levels and the activation of procollagenase into active collagenase may result in the destruction of the basal membrane, and the formation of a stromal ulcer.[33,48] This latter process is probably also effected by similar proteolytic enzymes (regulated by activators and inhibitors of the same type) deriving from leukocytes and infected stromal keratinocytes located in or invading the corneal stroma. The third level of

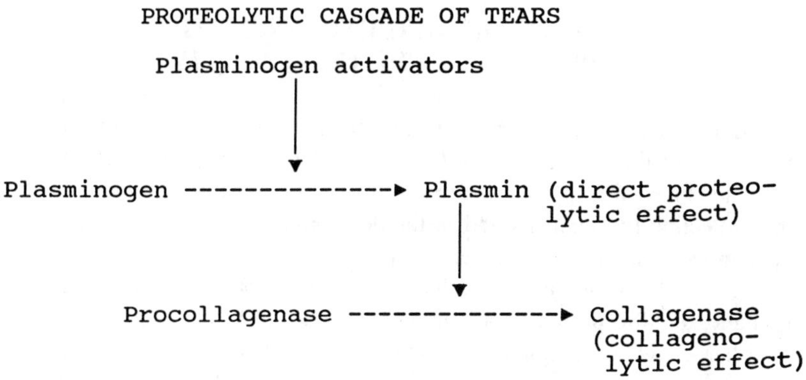

PROTEOLYTIC CASCADE OF TEARS

Plasminogen activators

Plasminogen --------------► Plasmin (direct proteo-
                                   lytic effect)

Procollagenase --------------► Collagenase
                                   (collageno-
                                   lytic effect)

FIGURE 1. The schematic representation of the proteolytic cascade of tears.

TABLE 1
Inhibitors of Tear Proteases — Therapeutic Possibilities

Enzymes	Natural inhibitors	Artificial inhibitors
Collagenases	α-2-Macroglobulin	EDTA Na-EDTA (Ca^{++},Zn^{++}) Ca-ETDA (Zn^{++}), L-cysteine, N-acetyl-L-cysteine, specific antibodies
Plasmin	α-1-Antiplasmin, α-2-macroglobulin, α-1-antitrypsin	Aprotinin (Trasylol), SBTI, specific antibodies
Plasminogen activators	PAI-1, 2, 3, 4	Amiloride, synthetic thrombin inhibitors, specific antibodies

Note: Specific antibodies are natural substances but are not present in humans and in animals under physiological conditions. They are produced in animals for diagnostic purposes. That is why they are listed among "artificial inhibitors".

regulation is the inhibition of collagenase action by physiological collagenase inhibitors.[49] The three levels of proteinase action are shown in Figure 1, and their inhibitors in Table 1.

The determination of enzymes, activators, and inhibitors in tears may provide useful information concerning the progress of tissue destruction in herpetic keratitis. PA levels show good correlation with the degree of epithelial involvement. An increase in their activity is a sign that the first line of defense (PAIs) is either not functioning properly or is used up. The appearance of active plasmin in the tears of herpetic keratitis patients may indicate that the progression of the disease and the ulceration of the stroma has to be anticipated. In the case of an already existing ulcer, enzyme activity levels may reflect the activity of the disease. These tear enzyme activity measurements may be used in deciding whether inhibitor therapy might be necessary in the case of a patient suffering from one particular form and stage of this multifaceted disease.[49]

III. IMMUNOLOGY OF HERPETIC INFECTIONS OF THE CORNEA AND THE CONJUNCTIVA

Both animals and man respond to ocular HSV infection with the same immunological and inflammatory reactions as to other viral infections. Cell-mediated immunity appears to be more important than humoral factors both in the protecting mechanisms and in the development of severe complications, beginning with stromal damage.

Primary ocular infection results in the production of circulating antiviral immunoglobulins primarily belonging to IgG but also to IgA and IgM classes. Anti-herpes IgG is thought to play an important role in inhibiting extracellular virus spread and preventing viremia.[50] Ocular infection leads to the appearance of specific secretory IgA in tears. This antibody is capable of neutralizing herpes virus, but anti-herpes antibody is present in tears only for 90 to 180 d after primary infection. It has been suggested that IgG molecules in tears may interact with HSV and thereby may prevent its complete neutralization by anti-viral IgA.[51] Stevens and Cook[52] demonstrated that anti-herpes IgG is able to inhibit the reactivation of the latent virus in neurons in tissue culture, but there is no direct evidence that the same event also occurs *in vivo*.

Cellular immunological reactions are considered to be the most important in ocular HSV infections; the details of these reactions, however, are not yet fully known. Lymphocytes recognizing herpes virus antigens in regional lymph nodes following ocular HSV infection were demonstrated by lymphocyte transformation tests. Similar but less-intensive response of lymphocytes from the blood, spleen, and distant lymph nodes was also detected. This has lead to the conclusion that, though signs of a systemic immune response are present, ocular herpetic infections are primarily controlled by local immune reactions.[53] The participation of humoral factors and mediators was also demonstrated. The production of lymphokines and migration inhibition factor was induced by intracorneal injection of HSV in animals and was followed by the appearance of sensitized lymphocytes. Another example for the combined action of hu-moral and cellular factors in ocular herpes is the presence of the antibody-dependent cell-mediated cytotoxicity reaction.[50] Lymphocytes of patients in-fected with HSV were shown to produce interferon, which in turn proved to be an effective inhibitor of the virus at least in *in vitro* experiments.[54]

Macrophages may also play a protective role phagocyting virus particles and removing virus-infected cells. Macrophages themselves are probably not susceptible to herpes infection. In experimental systems human macrophages were not found to replicate HSV virus, and thereby may play a role in restrict-ing herpetic infection.[55] Other cellular elements, the natural kiler cells, were shown to be very important in HSV infections. Neutral killer cells are non-lymphocyte, non-macrophage cells probably of neutrophilic origin. The pro-tection provided by the natural killer cells is nonspecific, is greatly augmented by interferon, and does not require prior contact with the herpes virus to be active. The natural killer cell system is thought to be one of the most ancient protective mechanisms in the development of human immunity. The possible role of this mechanism in herpetic infections is supported by the observation that many patients with recurrent herpes showed low natural killer cell activity.[56]

IV. ANTIVIRAL AND ANTIINFLAMMATORY THERAPY AND THE HEALING OF HERPETIC KERATITIS

Of all viral diseases primarily affecting the eye the most important are those caused by DNA viruses: herpes simplex virus, varicella zoster virus, adenovirus, and vaccinia virus. Of these four viruses only herpes and vaccinia virus are highly sensitive to antiviral therapy, varicella virus only moderately responsive, and adenovirus is resistant to therapy with currently available antiviral agents.[57]

Any antiviral drug can treat only those forms and phases of viral keratitis, including different forms and stages of ocular HSV infection, that are associated with viral replication. It is generally accepted that epithelial (dentritic) keratitis is caused by replicating virus, stromal necrotizing keratitis (interstitial herpetic keratitis) is caused primarily by replicating virus and a significant immunologic component, and disciform keratitis (disciform edema) is caused primarily by immune reactions that can be suppressed by corticosteroids.

The presently used antiviral drugs (effective in the treatment of herpes and vaccinia viruses) are pirimidine-base analogues. They incorporate into the virus DNA or inhibit DNA polymerase and block replication of viral DNA or the production of virus proteins. These antiviral agents are iodoxuridine (IDU), vidarabine (ara-A), trifluorothymidine (TFT), and acyclovir (ACV).

IDU was the first antimetabolite sucessfully used in the treatment of herpetic keratitis.[58] It is an effective antiviral agent, and has the ability to antagonize the worsening effect of corticosteroids on herpetic keratitis,[59] but also has significant toxic effects. Its toxicity, like that of most other antiviral drugs, is related to the fact that besides incorporating into the viral DNA they adversely affect the DNA synthesis of normal, noninfected, and especially that of regenerating cells. Toxicity explains the unfavorable effect of this drug on corneal wound healing in topical use and also explains why it cannot be administered systemically.[60] IDU has become less popular in the past few years because of its requirement for frequent application and because more potent antiviral drugs have been developed.[1]

ara-A was the second antiviral agent. Because of its insolubility it must be used in ointment form.[61] Its toxicity is not less than that of IDU, as it inhibits DNA polymerase in normal cells, too.[62] ara-A has been shown to be effective systemically against herpetic encephalitis and other herpetic diseases.[63]

Trifluridine was developed as an anticancer drug, but proved to be more effective as an antiviral agent.[64] It is now the drug of choice for the topical treatment of herpetic epithelial keratitis. Because trifluridine is not very selective in sparing host DNA and inhibiting viral DNA and is rapidly degraded in the bloodstream it cannot be used systemically. It is more effective than IDU in antagonizing the corticosteroid effect, and given topically can be used as antiviral coverage.[1]

The youngest but most promising antiviral drug, ACV, has been extensively studied and widely used in the past few years. This drug is specifically activated to its monophosphate form by thymidine kinase. It is highly effective both topically and systemically, and has minimal or no toxicity.[62] ACV is active against both HSV-1 and HSV-2, and is comparable in efficacy to trifluridine for topical use.[65] When given topically ACV penetrates the cornea and at least in this respect is superior to other topical antiviral agents.[66]

Although antiviral agents alone are effective in the treatment of herpetic eye diseases caused by actively replicating virus (dentritic keratitis, geographic ulcer), they are not effective in forms in which immunologic components play a major role. The use of corticosteroids is indicated in severe cases with stromal involvement (disciform and interstitial keratitis) and they appear to reduce scarring and vascularization, relieve pain, and restore vision. Steroids must be used for approximately 3 months, and antiviral coverage is highly advisable to avoid the reactivation of herpetic epithelial keratitis. If corticosteroids are discontinued prematurely a relapse is likely to occur, characterized by an increased number of inflammatory cells in the corneal stroma and a rise in the levels of hydrolytic enzymes in tears causing further tissue destruction.[1]

Corticosteroids can be given topically, systemically, and subconjunctivally. To achieve therapeutic concentration in the corneal stroma and in the anterior chamber by topical administration, high concentrations and frequent application are required causing an increased risk of the relapse of the epithelial disease. In contrast, by systemic administration sufficient tissue and aqueous humor concentrations are accompanied by relatively low corticosteroid levels at the epithelium and, as a consequence, there should be fewer epithelial complications. In spite of this, most ophthalmologists prefer the topical use fearing the general side effects of long-standing systemic corticosteroid therapy. The third possibility, subconjunctival corticosteroid injections, may be a very good form of treatment. This may combine the effectiveness of systemic administration without the unfavorable effects of corticosteroid eyedrops. Subconjunctival steroid injections are not very popular among ophthalmologists in the U.S., because they fear that such injections cause discomfort to the patient, may lead to subconjunctival hemorrhage, and in case of depot formulations it is difficult to terminate the therapeutic effect. In our experience subconjunctival corticosteroids are extremely useful in the treatment of immunological reaction associated with ocular herpes. The discomfort can be minimized by the use of topical anesthetics, and the possible complications are negligible as compared to those of the systemic steroid therapy.

The use of other drugs (interferons, growth factors, and proteinase inhibitors) in the treatment of herpetic keratitis is in experimental phase. It has been shown that interferons have an effect against herpes simplex virus and can be used in the treatment of dendritic keratitis.[67] In controlled trials, however, it appeared to be inferior in efficacy to the other antiviral drugs. Interferon did not prove to be effective in the prevention of herpetic recurrences either.[68,69] They can, however, potentiate the effect of the presently available antiviral agents and have a potential to be used as a form of adjunct therapy.[70,71]

Epidermal growth factor (EGF) was found to have no positive effect in human herpetic epithelial keratitis.[72] In rabbit experiments, however, the healing of herpetic epithelial lesions was faster when the antiviral therapy was supplemented by the use of EGF-containing eye drops. It has been suggested that EGF may counteract the adverse effects of antiviral drugs on cell division and on epithelial regeneration.[73]

The use of proteinase inhibitors in the treatment of nonhealing ulcers and of persistent epithelial defects including those associated with herpetic infection has been suggested.[33,46,49,74] No controlled study has been published so far proving the efficacy of either collagenase inhibitors or aprotinin in the treatment of herpes-induced stromal ulcers or in the prevention of ulcer formation in herpetic disease. Based on tear protease activity determinations performed on patients with herpetic keratitis, we feel, however, that such tests may help in selecting the appropriate cases and determining the optimal time interval when inhibitor eyedrops are more likely to be effective in preventing or limiting tissue destruction. Specific inhibitors of proteolytic enzymes, if their use is based on the detection of excessive proteinase activity in tears, has a good potential to become a form of useful therapy, in conjunction with antiviral and antiinflammatory drugs, in the control of tissue damage in necrotizing stromal keratitis, and in other ulcerating forms of herpetic corneal disease.

REFERENCES

1. **Kaufman, H. E. and Rayfield, M. A.,** Viral conjunctivitis and keratitis, in *The Cornea,* Kaufman, H. E., McDonald, M. B., Barron, B. A., and Waltman, S. R., Eds., Churchill Livingstone, New York, 1988, 299.
2. **Bell, D. M., Holman, R. C., and Pavan-Langston, D.,** Herpes simplex keratitis: epidemiologic aspects, *Ann. Ophthalmol.,* 14, 421, 1982.
3. **Pavan-Langston, D.,** Viral diseases: herpetic diseases, in *The Cornea. Scientific Foundations and Clinical Practice,* Smolin, G. and Thoft, R. A., Eds., Little, Brown and Company, Boston, 1983, 178.
4. Workshop on the Treatment and Prevention of Herpes Simplex Infection, *J. Infect. Dis.,* 127, 117, 1973.
5. **Fraser, N., Lawrence, W., Wroblewska, Z., et al.,** Herpes simplex type I DNA in human brain tissue, *Proc. Natl. Acad. Sci. U.S.A.,* 78, 6461, 1981.
6. **Nahmias, A., Alford, C., and Korones, S.,** Infection of the newborn with *Herpesvirus hominis, Adv. Pediatr.,* 17, 185, 1970.
7. **Whitley, R. et al.,** Adenine arabinoside therapy of herpes zoster in the immunosuppressed, *N. Engl. J. Med.,* 294, 1193, 1976.
8. **Rapp, F. and Reed, C.,** Experimental evidence for the oncogenic potential of herpes simplex virus, *Cancer Res.,* 36, 800 (1976).
9. **Stevens, J. G., Nesburn, A. B., and Cook, M. L.,** Latent herpes simplex virus from trigeminal ganglia of rabbits with recurrent eye infection, *Nature,* 235, 216, 1972.
10. **Lonsdale, D. M., Moria Brown, S., Subak-Sharpe, J. H., et al.,** The polypeptide and the DNA restriction enzyme profiles of spontaneous isolates of herpes simplex virus type 1 from explants of human trigeminal, superior cervical, and vagus ganglia, *J. Gen. Virol.,* 43, 151, 1979.

11. **Centifanto-Fitzgerald, Y. M., Varnell, E. D., and Kaufman, H. E.,** Initial herpes simplex virus type 1 infection prevents ganglionic superinfection by other strains, *Infect. Immun.,* 35, 1125, 1982.

12. **Centifanto-Fitzgerald, Y. M., Yamaguchi, T., Kaufman, H. E., et al.,** Ocular disease pattern induced by herpes simplex virus is genetically determined by a specific region of viral DNA, *J. Exp. Med.,* 155, 475, 1982.

13. **Gerdes, J. C. and Smith, D. S.,** Recurrence phenotypes and establishment of latency following rabbit keratitis produced by multiple herpes simplex virus strains, *J. Gen. Virol.,* 64, 2441, 1983.

14. **Nishiyama, Y. and Rapp, F.,** Regulation of persistent infection with herpes simplex virus *in vitro* by hydrocortisone, *J. Virol.,* 31, 841, 1979.

15. **Kaufman, H. E., Varnell, E. D., Centifanto, Y. M., and Kissling, G. E.,** Effect of the herpes simplex virus genome on the response of infection to corticosteroids, *Am. J. Ophthalmol.,* 100, 114, 1985.

16. **Centifanto-Fitzgerald, Y. M.,** Recurrent and non-recurrent HSV-1 strains: effect of temperature, in *Herpetic Eye Diseases,* Maugdal, P. C. and Missoten, L., Eds., Dr. W. Junk, Dordrecht, The Netherlands, 1985, 27.

17. **Hutchinson, D., Smith, R., and Haughton, P.,** Congenital herpetic keratitis, *Arch. Ophthalmol.,* 93, 70, 1975.

18. **Cibis, A. and Bunde, R.,** Herpes simplex virus-induced congenital cataracts, *Arch. Ophthalmol.,* 85, 220, 1971.

19. **Hagler, W., Walters, P., and Nahmians, A.,** Ocular involvement in neonatal herpes simplex virus infection, *Arch. Ophthalmol.,* 82, 169, 1969.

20. **Smith, J., Pentherer, J., and MacCallum, F.,** The incidence of *Herpesvirus hominis* antibody in the population, *J. Hyg. Cambr.,* 65, 395, 1967.

21. **Dawson, C., Togni, B., and Moore, R.,** Structural changes in chronic herpetic keratitis, *Arch. Ophthalmol.,* 79, 740, 1968.

22. **Meyers, R. and Chitjian, P.,** Immunology of herpesvirus infections: immunity to herpes simplex virus in eye infections, *Surv. Ophthalmol.,* 21, 194, 1976.

23. **Kaufman, H. E., Kanai, A., and Ellison, E. D.,** Herpetic iritis: demonstration of virus in the anterior chamber by fluorescent antibody techniques and electron microscopy, *Am. J. Ophthalmol.,* 71, 465, 1971.

24. **Baum, J.,** Morphogenesis of the dedritic figure in herpes simplex keratitis, *Am. J. Ophthalmol.,* 70, 222, 1970.

25. **Sugiura, S. and Kondo, E.,** Development of dendritic keratitis on the cornea of the rabbit after previous resection of the trigeminal nerve, *Acta Soc. Ophthalmol. Jpn.,* 65, 502, 1961.

26. **Sugiura, S., Wakui, K., and Kondo, E.,** Polygonal cell system of the corneal epithelium its comparative anatomy and embryology, *Jpn. J. Ophthalmol.,* 6, 220, 1962.

27. **Schweitzer, N. M. J.,** A fluorescein colored polygonal pattern in the human cornea the reflectographic "Furchenbild" of Fischer, *Arch. Ophthalmol.,* 77, 548, 1967.

28. **Zander, E. and Waddell, G.,** Reaction of corneal nerve fibers to injury, *Br. J. Ophthalmol.,* 35, 61, 1951.

29. **Dawson, C. and Togni, B.,** *Surv. Ophthalmol.,* 21, 121, 1976.

30. **Meyers, R.,** Immunology of herpes simplex virus infection, *Int. Ophthalmol. Clin.,* 15(4), 37, 1975.

31. **Campbell, R., Pavan-Langston, D., Lass, J., Berman, M., Gage, J., and Albert, D.,** Collagenase levels in a new model of experimental herpetic interstitial keratitis, *Arch. Ophthalmol.,* 98, 919, 1980.

32. **Cavanagh, H.,** Herpetic ocular disease therapy of persistent epithelial defects, *Int. Ophthalmol. Clin.,* 15(4), 67, 1975.

33. **Berman, M. B.,** Collagenase and corneal ulceration, in *Collagenase in Normal and Pathological Connective Tissues,* Woolley, D. E. and Evanson, J. M., Eds., John Wiley & Sons, Chichester, 1980, 141.

34. **Brodrick, J. D., Dark, A. J., and Peace, G. W.,** Finger print striae of cornea following herpes simplex keratitis, *Ann. Ophthalmol.,* 8, 481, 1976.

35. **Norn, M. S.,** Dendritic (herpetic) keratitis III. Follow-up examination of the tear film. Fischer-schweitzer's fluorescein pattern and vital staining, *Acta Ophthalmol.,* 48, 227, 1970.

36. **Avisar, R., Menaché, R., Shaked, P., and Savir, H.,** Lysozyme content of tears in some external infections, *Am. J. Ophthalmol. Vis. Sci.,* 92, 555, 1981.

37. **Berta, A., Tözsér, J., and Holly, F. J.,** Determination of plasminogen activator activities in normal and pathological human tears. The significance of tear plasminogen activators in the inflammatory and traumatic lesions of the cornea and the conjunctiva, *Acta Ophthalmol.,* 68, 508, 1990.

38. **de Koning, E. W. J. and van Bijsterveld, O. P.,** Schirmer test values and lysozyme content of tears in acute dendritic keratitis, *Invest. Ophthalmol. Vis. Sci.,* 25, 55, 1984.

39. **Moudgil, S. S., Singh, M., Parmar, I. P. S., and Khurana, A. K.,** Tear film flow and stability in herpes simplex keratitis and chronic blepharitis, *Acta Opthalmol.,* 64, 509, 1986.

40. **Tervo, T., Salonen, E.-M., Vaheri, A., Immonen, I., van Setten, G.-B., Himberg, J. J., and Tarkkanen, A.,** Elevation of tear fluid plasmin in corneal disease, *Acta Ophthalmol.,* 66, 393, 1988.

41. **Tözsér, J., Berta, A., and Punyiczki, M.,** Plasminogen activator activity and plasminogen independent amidolytic activity in tear fluid from healthy persons and patients with anterior segment inflammation, *Clin. Chim. Acta,* 183, 323, 1989.

42. **Tözsér, J. and Berta, A.,** Urokinase-type plasminogen activator in rabbit tears. Comparison with human tears, *Exp. Eye. Res.,* 51, 33, 1990.

43. **Tözsér, J. and Berta, A.,** Plasminogen activator inhibitors in human tears, *Acta Ophthalmol.,* 69, 426, 1991.

44. **Berman, M. B., Barber, J. C., Talamo, R. C., and Langley, C. E.,** Corneal ulceration and the serum antiproteases. I. Alpha-1-antitrypsin, *Invest. Ophthalmol.,* 12, 759, 1973.

45. **Berta, A., Vámosi, P., and Tözsér, J.,** Eine neue Methode zur semiquantitativen Alpha-1-Antitrypsinbestimmung in der menschlichen Tranenflüssigkeit, *Fortschr. Ophthalmol.,* 86, 773, 1989.

46. **Prause, J. U.,** Cellular and biochemical mechanisms involved in the degradation and healing of the cornea, *Acta Ophthalmol.,* 168(suppl.), 7, 1984.

47. **Punyiczki, M., Berta, A., and Tözsér, J.,** Study of neutral proteinases of human tear samples using gels comtaining substrates, *Clin. Chem. Enzym. Comns.,* 1, 115, 1988.

48. **Berta, A., Punyiczki, M., and Tözsér, J.,** Collagenase in normal and pathological human tears determined by microtiter plate method, in *Ophthalmology Today,* Ferraz de Oliveira, L. N., Ed., Elsevier Science, Amsterdam, 1988, 503.

49. **Berta, A., Holly, F. J., and Tözsér, J.,** Proteinase-inhibitors in *Diagnosis and Treatment of Ocular Surface Diseases,* Proc. Symp. Pharmacotherapy of the Eye, Springer-Verlag, Heidelberg, accepted for publication.

50. **Nesburn, A. B.,** Immunological aspects of ocular herpes simplex disease, in *Immunology of the Eye. Workshop III: Immunologic Aspects of Ocular Diseases: Infection, Inflammation and Allergy,* Suran, A., Gery, I., and Nussenblatt, R. B., Eds., Information Retrieval, Washington, DC, 1981, 21.

51. **Centifanto, Y. M. and Kaufman, H. E.,** The development of tear antibodies and herpes simplex keratitis, in *The Secretory Immunologic System,* Dayton, D. H., Jr., Small, P. A., Jr., Chanock, R. M., Kaufman, H. E., Tomas, T. B., et al., Eds., U.S. Department of Health and Human Services, Bethesda, MD, 1969, 331.

52. **Stevens, J. G. and Cook, M. L.,** Maintainance of latent herpetic infection: an apparent role for antiviral IgG, *J. Immunol.,* 113, 1685, 1974.

53. **Jakobs, R. P., Aurelian, L., and Cole, G. A.,** Cell mediated immune response to herpes simplex virus: type-specific lymphoproliferative responses in lymph nodes draining the site of primary infection, *J. Immunol.,* 116, 1520, 1976.

54. **Rasmussen, L. E. et al.,** Lymphocyte interferon production and transformation after herpes simplex infection in humans, *J. Immunol.,* 112, 728, 1974.
55. **Stevens, J. G. and Cook, M. L.,** Restriction of herpes simplex virus by macrophages. An analysis of cell-virus interaction, *J. Exp. Med.,* 133, 19, 1971.
56. **Bloomfield, S. E. and Lopez, C.,** Herpes infections in immunosuppressed host, citation by **Nesburn, A. B.,** Immunological aspects of ocular herpes simplex disease, in *Immunology of the Eye. Workshop III: Immunologic Aspects of Ocular Diseases: Infection, Inflammation and Allergy,* Suran, A., Gery, I., and Nussenblatt, R. B., Eds., Information Retrieval, Washington, DC, 1981, 21.
57. **Pavan-Langston, D.,** Ocular antiviral therapy, *Int. Ophthalmol. Clin.,* 20(3), 149, 1980.
58. **Kaufman, H. E., Martola, E.-L., and Dohlman, C. H.,** The use of 5-iodo-2′-deoxyuridine (IDU) in the treatment of herpes simplex keratitis, *Arch. Ophthalmol.,* 68, 235, 1962.
59. **Kaufman, H. E., Martola, E.-L., and Dohlman, C. H.,** Herpes simplex treatment with IDU and corticosteroids, *Arch. Ophthalmol.,* 69, 468, 1963.
60. **Pavan-Langston, D.,** Diagnosis and management of ocular herpes simplex, *Int. Ophthalmol. Clin.,* 15(4), 19, 1975.
61. **Pavan-Langston, D. and Dohlman, C. H.,** A double-blind clinical study of adenine arabinoside therapy of viral keratoconjunctivitis, *Am. J. Ophthalmol.,* 74, 81, 1972.
62. **Pavan-Langston, D.,** Ocular viral diseases, in *Antiviral Agents and Viral Diseases of Man,* Pavan-Langston, D., Buchanan, R., and Galasso, G., Eds., Raven Press, New York, 1979, 254.
63. **Whitley, R. J., Soong, S. J., Dolin, R., et al.,** Adenine arabinoside therapy of biopsy-proved herpes simplex encephalitis, *N. Engl. J. Med.,* 297, 289, 1977.
64. **Kaufman, H. E. and Heidelberger, C.,** Therapeutic antiviral action of 5-trifluoromethyl-2′-deoxyuridine, *Science,* 145, 585, 1964.
65. **La Lau, C., Oosterhuis, J. A., Versteeg, J., et al.,** Multicenter trial of acyclovir and trifluorothymidine in herpetic keratitis, *Am. J. Med.* (Acyclovir Symposium), 305, 1982.
66. **Poirier, R. H., Kingham, J. D., deMiranda, P., and Annel, M.,** Intraocular antiviral penetration, *Arch. Ophthalmol.,* 100, 1964, 1982.
67. **Jones, B. R., Coster, D. J., Falcon, M. G., and Cantell, K.,** Topical therapy of ulcerative herpetic keratitis with human interferon, *Lancet,* 2, 128, 1976.
68. **Kaufman, H. E., Meyer, R. F., Laibson, P. R., et al.,** Human leukocyte interferon for the prevention of recurrences of herpetic keratitis, *J. Infect. Dis.,* 133, A165, 1976.
69. **Shuster, J. J., Kaufman, H. E., and Nesburn, A. B.,** Statistical analysis of the rate of recurrence of herpesvirus ocular epithelial disease, *Am. J. Ophthalmol.,* 91, 328, 1981.
70. **de Koning, E. W. J., van Bijsterveld, O. P., and Cantell, K.,** Combination therapy for dentritic keratitis with human leukocyte interferon and trifluorothymidine, *Br. J. Ophthalmol.,* 66, 509, 1982.
71. **Sundmacher, R., Cantell, K., and Neumann-Haefelin, D.,** Combination therapy of dentritic keratitis with trifluorothymidine and interferon, *Lancet,* II, 687, 1978.
72. **Daniele, L., Frati, L., Fiore, C., and Santoni, G.,** The effect of epidermal growth factor (EGF) on the corneal epithelium in humans, *Graefe's Arch. Klin. Exp. Ophthalmol.,* 210, 159, 1979.
73. **Assouline, M., Montefiore, G., and Pouliquen, Y.,** Fibroblast growth factor effects on corneal wound healing in the rabbit, in *Healing Processes in the Cornea,* (Advances in Applied Biotechnology Series, Vol. 1), Beuerman, R. W., Crosson, C. E., and Kaufman, H. E., Eds., Gulf, Houston, 1989, 79.
74. **Salonen, E.-M., Tervo, T., Törma, E., Tarkkanen, A., and Vaheri, A.,** Plasmin in tear fluid of patients with corneal ulcers: basis for new therapy, *Acta Ophthalmol.,* 65, 3, 1987.

Chapter 13

TEAR ENZYMES IN ALLERGIC AND IMMUNOLOGICAL DISEASES OF THE EYE

I. PATHOLOGICAL BASIS OF IMMUNOLOGICAL DISEASES (HYPERSENSITIVITY REACTIONS)

The human body is defended from microorganisms and foreign (non-self) molecules, primarily of protein nature, by specific reactions called immunological reactions. Molecules, or special structures on macromolecules, recognized by an organism as non-self (consequently harmful) are called antigens. The immune response of the body to antigens is generally helpful, leading to the elimination of the antigen. When the immune response is excessive or inappropriate, these reactions may as well be harmful, leading to tissue damage in the host. The harmful immunological mechanisms are called hypersensitivity reactions. Gell et al.[1] described four types of hypersensitivity reactions (Table 1), all of which can occur in the eye.

Type I — Immediate hypersensitivity. Type I hypersensitivity reaction is initiated by the binding of an antigen to IgE molecules attached by their Fc portion to the cell membrane of basophil leukocytes or mast cells. This contact is the starting point of a sequence of events leading to the release from the cells of a number of biologically active substances (histamine, prostaglandins, thromboxane, slow reacting substance, chemotactic factors, etc.).[2] Cyclic AMP and possibly cyclic GMP play a role in modulating this process.[3] Vasoactive amines such as leukotrienes or SRS-A (slow reacting substance of anaphylaxis) and histamine cause a flow of fluids through the walls of blood vessels and contract the smooth muscles. ECF-A (eosinophilic chemotactic factor of anaphylaxis) attracts eosinophil leukocytes. The effect on the host depends on the site and the extent of the damage.[4]

Type II — Antibody- and cell-mediated cytotoxic reactions. These reactions are mediated by specific antibodies of isotypes IgG1, IgG3, or IgM and complement. Macrophages and polymorphonuclear leukocytes and their intracellular enzymes are also involved in the tissue destruction in this type of hypersensitivity reaction. Cytotoxic antibodies bind to the cell membrane of target cells. This is followed by complement fixation, resulting in cell membrane lysis. In addition, phagocytes, attracted by chemotactic factors, bind to the complement and/or the Fc receptors of the membrane-bound antibodies and the attacked and partly destroyed cells. Lymphocytes (killer cells) may also cause direct damage to the involved cells through antibody-dependent cell-mediated cytotoxic reactions.[5]

Type III — Immune complex diseases. Complexes of antigens with complement-fixing antibodies of isotypes IgG1, IgG2, IgG3 or IgM interact with the first component of the complement cascade and activate the whole cascade.[6] The activation of the complement cascade can lead to tissue injury both directly

TABLE 1
Types of Hypersensitivity Reaction
and their Mediators

I   Immediate — mediated by IgE
II   Cytotoxic — mediated by IgG, IgM, and complement
III   Immune complex — mediated IgG, IgM, and complement
IV   Delayed type hypersensitivity — mediated by mononuclear cells

or through the liberation of hydrolytic enzymes of the emigrating leukocytes. The injection of antigen into the skin of a previously sensitized host, possessing circulating antibodies, following a 2- to 8-h latency period leads to the development of erythema and edema of the skin, called Arthus reaction. This reaction results from the binding of the injected antigen locally in the tissue to antibodies and fixing of complement.[7] Not only localized immunocomplexes but also soluble circulating immunocomplexes can activate complement-mediated tissue-destructive reactions. Wherever immune complexes are deposited, determined by their size, shape, and chemical nature, the complement cascade is activated and tissue damage is likely to occur.[8,9]

Type IV — Delayed-type hypersensitivity. Type IV hypersensitivity reaction is effected by cells and not by antibodies. The effector cells of delayed hypersensitivity are antigen-specific T lymphocytes. The specificity and the reactivity of these lymphocytes is dependent on antigen presentation from macrophages. Once this cell-mediated immune memory is established and the lymphocytes are exposed to the specific sensitizing antigen, they exhibit direct cytotoxicity on cells bearing the offending antigens and release lymphokines.[5] Lymphokines are humoral factors produced by lymphocytes with a variety of different actions. These include direct cytotoxic effect, chemotactic effects on macrophages and lymphocytes including killer cells, and the potentiation of phagocytosis and cell-mediated cytotoxic reactions. Delayed hypersensitivity is known to play a role in certain infections (e.g., those caused by mycobacteria), autoimmune reactions, contact dermatitis, and tumor immunity.[4]

II. OCULAR ALLERGY AND IMMUNOLOGICAL
DISORDERS AT THE SURFACE OF THE EYE

The ocular surface is exposed to allergic stimuli and to antigens, and this exposure often leads to allergic and other abnormal immunological reactions. These diseases have three characteristics common in other allergic and immunologic diseases: (1) chronicity, (2) the presence of inflammatory reactions, and (3) a tendency for recurrence.[10]

The unique location and structural and chemical characteristics of the cornea and the conjunctiva make these tissues susceptible to involvement in all types of immunologic reactions.[11] Mucous membrane-associated lymphoid elements which are important in antigen processing are present in the conjunctiva. Lymphocytes and immunoglobulins (IgA, IgG, and IgM) are present in the conjunctiva and in lesser quantities in the cornea.[12] The presence of collagen in the cornea and that of blood vessels in the limbus predisposes these tissues for the involvement in systemic collagen and vascular diseases.[13]

The exposure of the ocular surface to environmental antigens promotes the development of type I, immediate hypersensitivity reactions. This is the pathogenetic mechanism for allergic conjunctivitis, vernal conjunctivitis, atopic keratoconjunctivitis, and contact lens-associated giant papillary conjunctivitis. Type II hypersensitivity, antibody- or cell-mediated cytotoxic reactions form the pathological basis of cicatriceal pemphigoid, pemphigus vulgaris, dermatitis herpetiformis, corneal immune rings, and Mooren's ulcer. Type III hypersensitivity reaction, caused by the formation of immune complexes and the activation of the complement cascade are the mechanisms leading to the development of bullous pemphigoid, Stevens-Johnson syndrome, Lyell's disease, Reiter's syndrome, Mooren's ulcer, and different forms of systemic vasculitis. Type IV hypersensitivity is mediated by cellular immunological reactions. Such mechanisms are involved to some extent in the pathogenesis of disciform keratitis, allograft rejections, phyctenulosis, vernal conjunctivitis, giant papillary conjunctivitis, and atopic keratoconjunctivitis.[14]

III. TEAR ENZYMES AND THE MEDIATORS OF OCULAR INFLAMMATION

Histamine is the most studied of the mediators involved in the development of the symptoms of ocular allergy. Histamine was found to be present in normal tears at low concentrations.[15] Elevated tear levels of histamine were detected in allergic conjunctivitis and in vernal conjunctivitis.[16] The elevation in the tear levels of histamine could be suppressed by the topical use of 4% chromoglycate.[17] Other mediators, prostaglandins and leukotrienes, were also demonstrated in the tears of allergic patients challenged with specific allergens.[18,19]

Histamine is present in the granules of mast cells and basophil leukocytes, where it is pharmacologically inactive and reversibly bound to heparin. The release mechanism is an active, calcium-dependent exocytic process, which begins with the binding of a bivalent antigen with two IgE molecules already attached to the cell membrane of mast cells or basophils.[20,21] This attachment activates the enzyme adenyl cyclase resulting in the conversion of ATP to cyclic AMP. Cyclic AMP then activates cytoplasmic cyclic AMP-dependent protein kinase. The action of this enzyme through the phosphorylation of various intracellular proteins leads to the aggregation of tubulin into intracellular microtubules, which participate in the degranulation and the exocytosis of vasoactive amines.[22] Once released, histamine is either quickly metabolized to methyl histamine by methyltransferase or it can be converted to imidazole acetic acid by diamine oxidase, also called histaminase. The action of histamine is determined by the rate of release and the function of the inactivating mechanisms.[23,24]

In allergic conjunctivitis histamine and other mediators are released from mast cells, leukocytes and histiocytes, located in the lymphoid elements of the subconjunctiva.[25] The possible role of the lacrimal glands in the production of vasoactive amines in allergic patients was also suggested.[26] Mast cells, lymphocytes, plasma cells, monocytes, and basophilic and eosinophilic leukocytes were detected in conjunctival scrapings in the acute and subacute phases of hay fever conjunctivitis, vernal conjunctivitis, atopic keratoconjunctivitis, and contact lens-associated giant papillary conjunctivitis.[27-29] Various studies demostrated the presence and activity of allergen-specific IgE antibodies present in the tears and in the conjunctiva of patients suffering from the same allergic conjunctival diseases.[30-34] In such patients the histamine concentration in tears can reach values greater than 100 ng/ml compared with normal values of 5 to 15 ng/ml.[16,17,26]

Histamine causes similar changes (capillary dilatation, increased vascular permeability, smooth muscle contraction) in the eye as it does in other parts of the human body. Abelson and Allansmith[25] found that 10 μl of 50 ng/ml histamine phosphate solution (corresponding to the concentration of 240 nmol/l) caused conjunctival redness and increased vascular permeability in 50% of the subjects studied. Kari et al.[26] studied the time sequence of the appearance of histamine in the tears of allergic patients following the challenge with a specific allergen. They determined the tear histamine levels of allergic patients in a symptom-free period and at regular intervals following the performance of a conjunctival challange test in one eye. Tear histamine levels found in the non-atopic controls did not differ from the values obtained from the allergic patients during the symptom-free period. The clinical signs (redness of the conjunctiva, increased tearing, itching) developed within 5 min following the exposure of the allergen. This rapid reaction to allergen was also reflected in histamine levels in tears, which reached its peak value 5 min after the allergic challenge, and did not change in an interval of 30 min. Even though the clinical signs of allergic reaction were seen in all subjects challenged, no increase in tear histamine levels was observed at 5 min in 36% of the patients. This may indicate that either the histamine response was over by this time or mechanisms other than histamine release were also acting in these cases. The histamine levels followed the same course of elevation in the contralateral (nonchallenged) eye of the responsive patients, possibly mediated by some unknown neural reflex mechanism.[26]

Tryptase, a neutral endopeptidase, is also secreted by activated mast cells of allergic patients. Tryptase levels in various body fluids have been used as an indicator of mast cell activation. Butrus and colleagues[35] determined tryptase levels in the tears of patients with asymptomatic and symptomatic seasonal allergic conjunctivitis, vernal conjunctivitis, and contact lens-associated giant papillary conjunctivitis. They also studied the release of tryptase into the tear fluid after provoking the conjunctiva with different allergens, chemical stimuli, and rubbing. Tryptase levels were elevated in the tears of patients with active ocular allergy, and also increased after provoking the conjunctiva with allergens in atopic subjects, as well as a response to exposure to certain chemicals and rubbing in non-atopic subjects. Based on their findings they suggest that tryptase levels in tears may be used as an indicator of mast cell involvement in ocular allergic diseases and may be used to evaluate and compare the effects of different mast cell stabilizing agents.[35]

The complement cascade can be activated by various immunologic and nonimmunologic stimuli to initiate inflammatory reactions in the conjunctiva and the cornea. The activation process may follow two distinct routes called the classical pathway and alternate pathway of complement activation. Each of these pathways involves the sequential activation of several enzymes and proteins. The majority of the steps in the activation cascade involves proenzyme enzyme conversions. Nonenzyme proteins of the cascade act as cofactors which accelerate enzymatic reactions or assemble macromolecules which interact with cells. Practically all enzymes, proteins, protein fragments, and peptides produced during the activation of the complement cascade have potent biological activity and are critical constituents of inflammatory processes.[36]

The classic pathway of complement activation builds membrane-bound complexes capable of enzymatic cell lysis via the specific sequence of complement components C1, C4, C2, C3, C4, C5, C6, C7, C8, C9 or produces biologically active cleavage products via the specific sequence C3b, C3a, C5a, and C5,6,7. The alternate pathway involves C3 and five other serum proteins, bypassing the initial C1, C4, and C2 components of the classic pathway. Both pathways cleave the pivotal C3 component, activate the terminal enzyme sequence C5 through C9, and produce the same biologic reactions. The alternate complement pathway can be directly activated by bacterial and fungal polysaccharides and endotoxins, whereas the classic pathway is activated only by aggregates of immunoglobulins or antigen-antibody complexes.[5,36-38]

The complement system seems to be less active in tears than in serum. Yamamoto and Allansmith[38] determined complement activity in tears in hemolytic assays up to dilutions of 1:4, whereas serum was active in the same system up to dilutions of 1:32. Elevated levels or activities of different complement fractions were measured in the tears of patients with different inflammatory or immunological diseases. Chandler et al.[39] determined the tear levels of different complement components in tears. Tears from normal subjects contained undetectable or low concentrations of C4 and no detectable C3 or C1. C4 complement fraction was present at a level 8 times higher than normal in the tears of a patient with staphylococcal blepharoconjunctivitis and 32 times higher than normal in the tears of a patient with Mooren's ulcer. C1 or C3 fraction was not detectable in the pathological tear samples either. In another study Bluestone et al.[40] detected equally low levels of B1C complement fraction both in normal tears and in the tears of patients with various ocular diseases. Mondino and Zaidman[41] performed hemolytic assays for C1, C4, C3, and C5 on tear samples from seven normal subjects and ten patients with corneal ulcers. Absent or low hemolytic activities of C1, C4, and C3 were found in normal tears. C5 was detected in the tears of four of seven normal subjects. Tear samples from corneal ulcer patients showed elevated levels of C1, C4, C3, and C5, but wide ranges in values were found. It was suggested that complement in tears may contribute to host defense in microbial corneal ulcers.

Chandler et al.[39] reported a patient with corneal graft rejection who had normal level of C4 fraction and no detectable C3 and C1 complement in the tear fluid. We studied the immunoglobulin and complement levels in a series of experiments in the tears of patients with corneal graft rejection and detected elevated C4 complement level in the tears of a patient who did not receive systemic immunosuppressive therapy.[42] Both immunoglobulin (IgG, IgA, IgM) and C4 level elevations in tears were found to precede the onset of the clinical symptoms of corneal graft rejection and were suggested to be used to predict transplant rejection in the postoperative period following corneal transplantation.[43] In an unpublished study, however, we demostrated that systemic immunosuppressive treatment or subconjunctival corticosteroid injections suppress the elevation of immunoglobulins and complement in tears, even in cases in which the treatment is not effective in suppressing the clinical symptoms of corneal graft rejection.

Kijlstra and Veerhuis[44] used a sensitive microtechnique to assay the classic complement pathway in stimulated normal tears. No complement activity was detected; however, the tears of all 11 normal individuals tested contained substances that strongly interfered with the complement-mediated lysis of antibody-coated red blood cells. In a further study Kijlstra et al.[45] demonstrated the same type of complement inhibitor activity in the tears of various species. Gel filtration and sodium dodecyl sulfate-polyacrylamide gel electrophoresis suggested that a small molecular weight (approximately 20,000 Da) tear protein is involved in the anticomplementary activity of human tears. Studies concerning the mechanism of inhibition showed that the complement cascade was inhibited after the EAC14 intermediate had been formed.[45] Detailed analysis revealed that the inhibition of complement activity is associated with lactoferrin, a constant component of human tears. The addition of human tears or that of purified lactoferrin to an *in vitro* system (in which after incubating immune complexes with fresh normal human serum the deposition of complement components was measured with an ELISA technique) resulted in the inhibition of the classical pathway, the deposition of C3 and C5, while the deposition of C4 was unaffected. High concentrations of normal tears also inhibited alternative pathway C3 deposition on immune complexes, whereas purified lactoferrin did not affect this pathway. The inhibition of complement activation by tears was not due to masking of the immune complexes or the already deposited C3. Further experiments with purified lactoferrin demonstrated that lactoferrin did not bind to the complexes, either before or during complement activation. The saturation of lactoferrin with iron or copper resulted in a decrease in complement inhibition. C4 deposition was not affected by lactoferrin, which suggests that the inhibition of the classical pathway was due to an effect on the classical C3 convertase.[46] The observation that lactoferrin inhibited the classical, but not the alternate, pathway C3 convertase suggests that the effect is not mediated through competition for metallic ions, but rather by protein-protein interactions. These findings indicate a possible antiinflammatory role of tear lactoferrin in modulating the activation of the complement system.[47]

Salonen et al.[48] demonstrated the appearance of proteolytic activity, related to the activation of plasmin from the inactive plasminogen, in the tears of allergic patients following topical exposure to the specific allergen. Plasmin was identified using zymographic analysis and blockage by antiplasmin antibodies. The plasmin appeared in the tears of the patients within 3 to 5 min after the challenge with the allergen. The plasmin level in the tear samples was very high ranging from 2.2 to 28.6 µg/ml. No plasmin was detected in the tears of non-atopic controls, or in atopic persons when the eyes were provoked with test materials to which they were not sensitized.

We studied the plasminogen activator (PA) activity in the tears of patients suffering from various corneal and conjunctival diseases of immunological origin, including conjunctival pemphigoid and Mooren's ulcer clinical type I, and type II.[49] PA activities showed good correlation with the severity of the diseases. High activator activities were measured in the tears in the active phases and in the clinically symptomatic forms of the diseases. The activity values gradually normalized in the healing phases as the clinical symptoms disappeared. We think that the results of tear PA activity determinations reflect the degree of epithelial involvement in immunological and nonimmunological inflammations of the cornea and the conjunctiva, as the urokinase-type activator (measured in the system we used without the addition of fibrin) is produced by or released from the affected epithelial cells.[50] Urokinase-type PA (u-PA) and plasmin, the effector enzymes of extra- and pericellular proteolysis, probably also play a role in the invasion of imunocompetent cells into the otherwise compact structure of the cornea.[49]

In another study we determined total collagenolytic activity of the tears in patients suffering from Mooren's ulcer.[51] Tear enzyme activity tests were performed on six patients at regular intervals during the course of the disease. No collagenolytic activity was present in the tears of the patients when the disease was inactive and the eyes were quiet. High collagenolytic activities were measured in the active phases of Mooren's ulcer, also characterized by progressive stromal ulceration. The activity values were higher in patients with clinical type I (a more severe and rapidly progressing form of Mooren's ulcer) and were only moderately elevated in the less-active and slowly progressing clinical type II. In one of the patients, when all other therapeutic measures failed, we covered the cornea with a conjunctival flap in the active phase of the disease. The collagenolytic activity in the tears of this patient was extremely high when the conjunctival flap was performed. The conjunctiva, covering the active ulcer, was digested by proteolytic enzymes within 2 d after surgery. We successfully performed a penetrating keratoplasty, however, on another patient with severe Mooren's ulcer clinical type I, in an inactive phase, without detectable collagenolytic activity in the tears. We compared the diagnostic value of tear immunoglobulin and tear collagenolytic enzyme activity determinations in the course of the disease.[52] While immunoglobulin (especially IgG) levels were usually high in the tears of the patients in all forms and stages with considerable irritation (including simple conjunctivitis and postsurgical irritation), collagenolytic activity was detectable only when the ulcer was progressing. Immunosuppressive and cytostatic therapy easily suppressed the elevation of immunoglobulins in the tears even in cases when the therapy did not alter the progression of Mooren's ulcer. Collagenolytic activity on the other hand normalized only in cases when the therapy successfully arrested the progression of the disease or when there was a spontaneous remission. Therefore we think that collagenolytic activity measurements can be used to differentiate between the active and the inactive phases and between the two clinical forms of Mooren's ulcer. Such information may be useful in choosing the optimal time for surgical intervention (surgery is more successful in the inactive phase when conjunctival flaps or corneal grafts are not in danger of possible enzymatic digestive attack), in evaluating the therapeutic effects of conservative treatment (normalizing enzyme activities indicate responsiveness to therapy), and to a certain extent tear enzyme activities may inform on the prognosis, too (very high and long-standing collagenolytic activities in tears may be considered as signs of poor prognosis for Mooren's ulcer patients).[43,51]

REFERENCES

1. **Gell, P. G. H., Coombs, R. A., and Lackman, P., Eds.,** *Clinical Aspects of Immunology,* Section 4., Blackwell Scientific, Oxford, 1974.
2. **Johansson, S. G. O., Bennich, H. H., and Berg, T.,** The clinical significance of IgE, in *Progress in Clinical Immunology I,* Schwartz, R. S., Ed., Grune and Stratton, New York, 1972.
3. **Lichtenstein, L. M.,** Allergy, in *Clinical Immunology,* Vol. 1, Bach, F. H. and Good, R. A., Eds., Academic Press, New York, 1972.
4. **Tomar, R. H.,** Hypersensitivity reactions, in *Clinical Diagnosis and Management by Laboratory Methods,* Henry, J. B., Ed., W.B. Saunders, Philadelphia, 1984, 955.
5. **Foster, C. S.,** Basic ocular immunology, in *The Cornea,* Kaufman, H. E., McDonald, M. B., Barron, B. A., and Waltman, S. R., Eds., Churchill Livingstone, New York, 1988, 85.
6. **Rahi, A. H. S. and Garner, A.,** *Immunopathology of the Eye,* Blackwell Scientific, Oxford, 1976.
7. **Friedlander, M. H.,** *Allergy and Immunology of the Eye,* Harper & Row, New York, 1979.
8. **Berta, A., Kálmán, K., Patvaros, I., and Kelenhegyi, Cs.,** Changes in the tear protein composition in Graves' disease, *Orbit,* 5, 97, 1986.
9. **Cohrane, C. G. and Koffler, D.,** Immune complex disease in experimental animals and man, *Adv. Immunol.,* 16, 185, 1973.
10. **Allansmith, M. R.,** Ocular allergy, in *Immunology of the Eye. Workshop III: Immunologic Aspects of Ocular Diseases: Infection, Inflammation and Allergy,* Suran, A., Gery, I., and Nussenblatt, R. B., Eds., Information Retrieval, Washington, DC, 1981, 347.

11. **Smolin, G. and O'Connor, G. R.,** *Ocular Immunology,* Lea & Febiger, Philadelphia, 1981.

12. **Allansmith, M. R. and McClellan, B.,** Immunoglobulins in the human cornea, *Am. J. Ophthalmol.,* 80, 123, 1975.

13. **James, D. G., Graham, E., and Hamblin, A.,** Immunology of multisystem ocular disease, *Surv. Ophthalmol.,* 30, 155, 1985.

14. **Robin, J. B. and Dugel, R.,** Immunologic disorders of the cornea and the conjunctiva, in *The Cornea,* Kaufman, H. E., McDonald, M. B., Barron, B. A., and Waltman, S. R., Eds., Churchill Livingstone, New York, 1988, 511.

15. **Abelson, M. B., Soter, N. A., Simon, M. A., Dolman, J., and Allansmith, M. R.,** Histamine in human tears, *Am. J. Ophthalmol.,* 83, 417, 1977.

16. **Abelson, M. B., Baird, R. S., and Allansmith, M. R.,** Tear histamine levels in vernal conjunctivitis and other ocular inflammations, *Ophthalmology,* 87, 812, 1980.

17. **Della Dora, F.,** Histamine levels of tears during allergic conjunctivitis: physiopathology and therapy with disodium chromoglycate, *Chibret Int. J. Ophthalmol.,* 3(1), 16, 1985.

18. **Bisgaard, H., Ford-Hutchinson, A. W., Charleson, S., et al.,** Detection of leukotrine C4-like immunoreactivity in tear fluid from subjects challanged with specific allergen, *Prostaglandins,* 27, 369, 1984.

19. **Dhir, S. P., Garg, S. K., Shamra, Y. R., and Lath, N. K.,** Prostaglandins in human tears, *Am. J. Ophthalmol.,* 87, 103, 1979.

20. **Siraganian, R. P., Hook, W. W., and Levine, B. B.,** Specific *in vitro* release from basophyls by bivalent haptens: evidence for activation by simple bridging of membrane bound antibody, *Immunochemistry,* 12, 149, 1975.

21. **Stanworth, D. R.,** The role of antibody in immunological cell triggering processes, *Hematologica,* 8, 299, 1974.

22. **Winslow, C. M. and Auisten, K. F.,** Enzymatic regulation of mast cell activation and secretion by adenylate cyclase and cyclic AMP-dependent protein kinases, *Fed. Proc.,* 41, 22, 1982.

23. **Beaven, M. A.,** Histamine, *N. Engl. J. Med.,* 294, 320, 1976.

24. **Toselli, C. and Miglior, M.,** *Oftalmologica Clinica,* Moduzzi Editore, Bologna, 1979, 209.

25. **Abelson, M. B. and Allansmith, M. R.,** Histamine and the eye, in *International Symposium on the Immunology and Immunopathology of the Eye,* Siverstein, A. M. and O'Connor, G. R., Eds., Masson, New York, 1979, 362.

26. **Kari, O., Salo, O. P., Halmepuro, L., and Suvilehto, K.,** Tear histamine during allergic conjunctivitis challenge, *Graefe's Arch. Clin. Exp. Ophthalmol.,* 223, 60, 1985.

27. **Abelson, M. B., Madiwale, N., and Weston, J. H.,** Conjunctival eosinophils in allergic ocular disease, *Arch. Ophthalmol.,* 101, 555, 1983.

28. **Allansmith, M. R., Baird, R. S., and Block, K.,** Ocular anaphylaxis — the sequence of mast cell alteration during 24 hours of observation, *Invest. Ophthalmol. Vis. Sci.,* 22, 211, 1982.

29. **Udell, I. U., Allansmith, M. R., Acherman, S. J., et al.,** Eosinophil granule major basic protein and Charlot-Leyden crystal protein in human tears, *Am. J. Ophthalmol.,* 92, 824, 1981.

30. **Aalders-Deenstra, V. and Brunzeel, P. L. B.,** Measurement of total IgE antibody levels in lacrimal fluid of patients suffering from atopic and nonatopic eye disorders — evidence for local IgE production in atopic eye disorders, *Br. J. Ophthalmol.,* 69, 380, 1985.

31. **Allansmith, M. R., Hahn, G. S., and Simon, M. A.,** Tissue, tear and serum IgE concentrations in vernal conjunctivitis, *Am. J. Ophthalmol.,* 81, 506, 1976.

32. **Ballow, M. and Mendelson, L.,** Specific immunoglobulin E antibodies in tear secretions of patients with vernal conjunctivitis, *J. Allerg. Clin. Immunol.,* 66, 112, 1980.

33. **Brauninger, G. E. and Centifanto, Y. M.,** Immunoglobulin E in human tears, *Am. J. Ophthalmol.,* 72, 558, 1971.

34. **Kari, O., Salo, O. P., Bjorkstein, F., and Backman, A.,** Allergic conjunctivitis — total and specific IgE in the tear fluid, *Acta Ophthalmol.*, 63, 97, 1985.

35. **Butrus, S. I., Ochsner, K. I., Abelson, M. B., and Schwartz, L. B.,** The level of tryptase in human tears: an indicator of activation of conjunctival mast cells, *Ophthalmology*, 97, 1678, 1990.

36. **Kaplan, A. P.,** The complement, coagulation and kinin-forming cascades in inflammation. in *Immunology of the Eye. Workshop III: Immunologic Aspects of Ocular Diseases: Infection, Inflammation and Allergy,* Suran, A., Gery, I., and Nussenblatt, R. B., Eds., Information Retrieval, Washington, DC, 1981, 383.

37. **Allansmith, M. R.,** *The Eye and Immunology,* C.V. Mosby, St.Louis, 1982.

38. **Yamamoto, G. K. and Allansmith, M. R.,** Complement in tears from normal humans, *Am. J. Ophthalmol.*, 88, 758, 1979.

39. **Chandler, J. W., Leder, R., Kaufman, H. E., and Caldwell, J. R.,** Quantitative determinations of complement components and immunoglobulins in tears and aqueous humor, *Invest. Ophthalmol.*, 13, 151, 1974.

40. **Bluestone, R., Easty, D. L., Goldberg, L. S., Jones, B. R., and Pettit, T. H.,** Lacrimal immunoglobulins and complement quantified by counter-immunoelectrophoresis, *Br. J. Ophthalmol.*, 59, 279, 1975.

41. **Mondino, B. J. and Zaidman, G. W.,** Hemolytic complement in tears, *Ophthalmic Res.*, 15, 208, 1983.

42. **Berta, A.,** Könnyvizsgálatok jelentôsége a keratoplasztikát követô immunológiai reakció elôjelzésében, (Significance of tear protein tests in the prediction of corneal graft rejection), *Szemeszet,* 122, 172, 1985.

43. **Berta, A., Zajácz, M., and Jaross, N.,** A könnyvizsgálatok jelentôsége az ulcus rodens corneae (Mooren fekély) diagnosztikájában, [Significance of tear protein tests in the diagnosis of ulcus rodens corneae (Mooren's ulcer)], *Szemeszet,* 124, 220, 1987.

44. **Kijlstra, A. and Veerhuis, R.,** The effect of an anticomplementary factor on normal human tears, *Am. J. Ophthalmol.*, 92, 24, 1981.

45. **Kijlstra, A., Jeurissen, S. H., and Koning, K. M.,** Complement inhibitors in tears of different species, *Exp. Eye Res.*, 38, 57, 1984.

46. **Kijlstra, A. and Jeurissen, S. H.,** Modulation of classical C3 convertase of complement by tear lactoferrin, *Immunology*, 47, 263, 1982.

47. **Kievits, F. and Kijlstra, A.,** Inhibition of C3 deposition on solid-phase bound immune complexes by lactoferrin, *Immunology*, 54, 449, 1985.

48. **Salonen, E.-M., Lauharanta, J., Sim, P.-S., Stephens, R., and Vaheri, A.,** Rapid appearance of plasmin in the tear fluid after ocular allergen exposure, *Clin. Exp. Immunol.*, 73, 146, 1988.

49. **Berta, A., Tözsér, J., and Holly, F. J.,** Determination of plasminogen activator activities in normal and pathological human tears. The significance of tear plasminogen activators in the inflammatory and traumatic lesions of the cornea and the conjunctiva, *Acta Ophthalmol.*, 68, 508, 1990.

50. **Tözsér, J., Berta, A., and Punyiczki, M.,** Plasminogen activator activity and plasminogen independent amidolytic activity in tear fluid from healthy persons and patients with anterior segment inflammation, *Clin. Chim. Acta,* 183, 323, 1989.

51. **Zajácz, M. and Berta, A.,** Proteolytische Enzymaktivitat und chirurgusche eingriffe bei ulcus Mooren, *Klin. Mbl. Augenheilk.*, 187, 401, 1985.

52. **Berta, A.,** Diagnosztikus könnyfehérje-meghatározások, (Diagnostic Tear Protein Determinations), Ph.D. thesis, Hungarian Academy of Sciences, Budapest, 1987.

ENZYMES AND ENZYME-INHIBITORS IN KERATOCONJUNCTIVITIS SICCA AND OTHER OCULAR SURFACE DISEASES

I. OCULAR SURFACE DISEASE

The expression "ocular surface disease", in general, means all the disorders (progressive diseases and pathological states) that change the structural, physical, or chemical characteristics of the outermost epithelial layer of the cornea and the conjunctiva, and by doing so interfere with the formation of a sufficiently stable preocular tear film.

These pathological changes can be macroscopical, such as pterygium, symblepharon, or deforming scars of the cornea or conjunctiva or they can be microscopical, such as small epithelial lesions and different forms of epitheliopathy. Lesions may affect only the superficial epithelial layer(s) or the whole epithelium. The defect may even be deeper, penetrating the basal membrane, the Bowman layer, and involve a part of the corneal stroma. The affliction may be temporary (traumatic abrasion), long lasting (persistent epithelial defect, nonhealing corneal ulcer), or recurring (recurrent erosion).

Ocular surface disease is often seen in dry eyes, where the pathological changes of the epithelium are caused by inappropriate wetting of the epithelial surface by the tear fluid. Other possible causes include injuries (mechanical, chemical), inflammations (herpetic or adenoviral keratitis, Thygeson keratitis), and degenerative (band-shaped keratopathy, spheroid degeneration, epithelial dystrophies). Contact lens wear and the use of preservatives in eye drops and contact lens solutions are also frequent causes of ocular surface disease.

There is a bilateral correlation between tear film stability and ocular surface damage. On the one hand unstable tear film, often seen in various forms of the dry eye syndrome, sooner or later leads to ocular surface damage (ocular surface disease developing in dry eyes is called keratoconjunctivitis sicca). Alterations of the ocular surface, on the other hand, adversely affect tear film stability, and by doing so set up a vicious cycle, further worsening the condition of the eye.[1]

II. OCULAR SURFACE DISEASE AND CORNEAL ULCERATION

It is known from clinical experience that long-standing epithelial lesions often lead to corneal ulceration. Most patients with noninfected ulcers have a history of persistent or recurrent epithelial defects. The work of Berman and colleagues provided evidence that an excessive and unregulated production of plasminogen activators (PAs) from the injured epithelial cells, followed by the activation of plasmin and collagenase, leads to the enzymatic degradation of collagen and interfibrillary matrix of the corneal stroma.[2] The same enzymes and activators were found to play an important role in corneal wound healing, re-epithelization, neovascularization, and different inflammatory processes of the cornea and conjunctiva.[3] The presence of proteinase inhibitors in wounded and ulcerating corneas has been demonstrated and a regulatory mechanism to control proteolysis during the healing process has been proposed.[4]

It was soon realized that the tears of patients with ulcerating corneas contain proteolytic enzymes: collagenases, plasmin, PAs, and the determination of tear proteinases for diagnostic purposes was suggested. The idea to use natural (serum) or synthetic inhibitors of these proteinases in the treatment of corneal ulcers was born at that time.[5,6] The lack of correlation between the detection of proteinases and the results of clinical trials to treat corneal ulcers with proteinase inhibitors indicated the importance of regulatory mechanisms.[3,7]

III. PROTEINASE INHIBITORS IN TEARS

The first observations concerning the proteinase inhibitor content of human tears were those of Erickson in 1965, who reported the presence of antitrypsin activity in human tears and nasal secretions. Kueppers determined the antitrypsin activity in stimulated normal tears to be 25 ($\pm$13) μg/ml, and that in unstimulated tears approximately 10 μg/ml higher. He suggested that the physiological role of tear proteinase inhibitor(s) is the protection of the cornea from "naturally occurring" proteinases.[8]

Early attempts to determine tear trypsin-inhibitor capacity for the detection of α-1-antitrypsin deficiency and for heterozygote screening were not successful, because other inhibitors are also present and contribute to the trypsin-inhibitor capacity of human tears.[9] Patients with α-1-antitrypsin deficiency usually do not have greater predilection for corneal ulceration than do patients with normal serum α-1-antitrypsin levels. The total trypsin-inhibitor activity in the tear fluid of these patients is usually normal. We detected, however, the absence of trypsin inhibitor activity in the tears of a patient with congenital α-1-antitrypsin deficiency who was suffering from recurrent corneal ulcers on both eyes.[10] Another observation supporting that the lack of trypsin inhibitors might play a role in the development of ulcers was that of Tabbara et al.,[11] who reported of a patient with α-1-antitrypsin deficiency suffering from chronic ulcerative conjunctivitis. The possible genetic background for certain types of anterior uveitis and the role of α-1-antitrypsin in acute anterior uveitis and rheumatic diseases was discussed in detail by Saari et al.[12,13]

Immunological studies have shown the presence of different plasma proteinase inhibitors α-1-antitrypsin, α-1-antichymotrypsin, and inter-α-trypsin inhibitor in normal tears.[14,15] The level of α-1-antitrypsin in the tears of patients with corneal ulceration was shown to be elevated.[4] High tear concentrations of the inhibitor α-1-antitrypsin were also found in bacterial keratitis and conjunctivitis.[16] Since serum albumin levels were also elevated in tears with high inhibitor concentrations, it has been suggested that the increased α-1-antitrypsin levels observed in inflamed eyes were caused by leakage of proteins from blood vessels of the conjunctiva.[4]

A thiol proteinase inhibitor which inhibits papain was detected in normal human tears.[17] Its level in tears of healthy individuals was compared with that of patients with different inflammatory eye diseases. Activities lower than normal were found in some patients with bacterial conjunctivitis and blepharitis. Higher than normal activities were found in patients with herpes simplex and allergic conjunctivitis.[18]

Attempts to detect inhibitors of the fibrinolytic system in normal tears were unsuccessful.[19] However, plasminogen activator inhibitor (PAI) activities[20] and plasmin inhibitor activities[21] were demonstrated in the tears of patients with keratoconjunctivitis sicca and other inflammatory diseases of the cornea and the conjunctiva.

Along with elevated PAI activities we detected plasmin inhibitor activities in most of the tears from patients with keratoconjunctivitis sicca. PA activities were usually also increased in the tears of the patients. Both PA and PAI activities showed correlation with the degree of epithelial damage (evaluated on the basis of the number and size of epithelial defects). Plasmin activities were demonstrated only in only about 25% of the cases, with no obvious correlation with the degree of the severity of the epithelial involvement. Most interesting was the finding that relatively high serum albumin levels were measured in the samples having no plasmin activity while the samples with plasmin activity contained serum albumin at significantly lower concentration (Berta et al., unpublished data).

IV. DIAGNOSTIC IMPLICATIONS

Ocular surface diseases do affect the levels of proteinase inhibitors in tears. Proteinase inhibitor activity measurements (enzyme activity tests) usually yield values different from the results of concentration determinations obtained with immunological methods. The latter measure the total (active + inactive) inhibitor levels in the tear fluid. In ocular surface diseases the total inhibitor levels are usually higher than the active inhibitor levels indicating that a significant portion of inhibitors are present as enzyme-inhibitor complexes. In most cases there is an inverse relationship between the inhibitor levels in the tears and the tear flow rates at which the samples were collected. Most proteinase inhibitors either derive from the conjunctival vessels through transudation, are released from the injured epithelium, or are secreted by the regenerating epithelial cells of the cornea and the conjunctiva. In these instances the excessive tearing causes dilution resulting in lower or even undetectable inhibitor levels.[22]

Changes in tear inhibitor levels observed in various ocular surface diseases suggest that these compounds may play a regulatory role in the proteolytic processes that accompany the formation and healing of ocular surface defects.

The application of inhibitor determinations for diagnostic purposes is presently not a simple task. Their levels or activities have to be evaluated in conjunction with those of proteinases. In addition to dilution by reflex tearing, permeability changes of the conjunctival vessel walls also have to be considered.[20,23]

Proteinases and their activators and inhibitors are known to play important roles in inflammatory processes and in the healing of such afflictions. As suggested by Anderson and Leopold[18] the initial reaction to inflammation may result in an increase in proteolytic enzymes. Complex formation of inhibitors with these proteinases would mask most of their activity so that low levels of active inhibitors would be found in the tears from eyes in active stages of inflammation. However, as the process proceeds toward regression or healing, the protease activity would decrease and the level of inhibitor activity would increase. As it has been shown for skin inflammation, inappropriate levels of inhibitors may to lead to abnormal inhibitory responses also in inflammations of the cornea and conjunctiva.[24]

Various inhibitors were shown to behave in different ways when measured in the tears of patients with ocular surface diseases, and thus may have different diagnostical implications. The level of α-1-antitrypsin is about 30 times higher in plasma than in tears. Its level in tears is usually elevated in keratoconjunctivitis sicca. This change is supposed to be the result of leakage of serum proteins from conjunctival vessels due to irritation and inflammation caused by the dry eye state. Type I PAI can appear in tears from blood plasma in a similar way. Type II PAI is normally absent in blood (it can be detected only in pregnancy). It is known to be secreted, however, by different epithelial cells including those of the human cornea. The appearance of type II PAI in the tears of patients with ocular surface diseases is a sign of the activation of regulatory processes that intend to control proteolyses triggered by the damage of the surface epithelial cells. Permeability changes result in transudation in the appearance of plasma inhibitors in the tears of patients with ocular surface disease. This process can be looked upon as a physiological protective mechanism preventing excessive damage to the ocular surface by activated proteinases.[25] The determination of plasma inhibitors may yield information on how this protective mechanism works. Besides their presence and inhibitory action may alter the results if enzyme activity determinations are performed in the tears of patients with ocular surface diseases.

V. INHIBITOR THERAPY

Therapeutic approaches to external eye diseases, as suggested by Kenyon in 1982,[26] may be considered on three levels:

1.   The determination of etiology and introduction of primary therapy.
2.   Promotion of the healing process.
3.   Limitation of ulceration (the prevention of ulcer formation or progression).

1. In most cases of ocular surface disease the specific etiology (dry eyes, infectious or inflammatory diseases, degenerations, structural deformities) can be determined and a therapy using antimicrobial, anti-inflammatory agents or surgical procedures to cure the primary disease can be initiated. Apart from rare cases in which the lack of proteinase inhibitors may be a causative factor, and missing inhibitors may have to be replenished, the use of proteinase inhibitors usually cannot be looked upon as primary (causative) therapy.

2. Pharmaceutical agents that may be employed for the promotion of wound healing or reepithelization include fibronectin, epithelial growth factor, serum or plasma proteins, vitamins, and synthetic polymers.[27] Experimental data are accumulating which suggest that proteinases play an important role in the healing of corneal defects by digesting the debris within the defect zone and by promoting the sliding of the regenerating epithelial cells on the denuded surface. It has been observed that excessive or prolonged proteinase action (lack of inhibitors?) may hinder the healing process, but the total absence (or complete inhibition) of these proteolytic enzymes may also prevent reepithelization. The diversity of the roles of proteinases in the healing processes and the complexity of the regulatory mechanisms explain the unpredictable outcome of the use of inhibitors for the promotion of epithelial wound healing. Beneficial effect can only be expected in cases where unfavorably high proteinase activity hinders the wound healing process. Tear proteinase activity determinations can be very helpful in this respect.

3. Prevention of ulceration (a complication that may occur in cases severe and long-standing ocular surface disorders) and the inhibition of the enzymatic degradation of corneal tissue in case of an already-existing ulcer (by doing so indirectly promote the repair mechanisms) are presently the only acceptable initiatives for enzyme-inhibitor therapy.

Autologous serum in form of eyedrops or subconjunctival injections has been successfully used to treat severe chemical burns in order to prevent conjunctival necrosis and corneal ulceration.[28,29] The beneficial effect of serum-based artificial tears in keratoconjunctivitis sicca has been reported.[30] Drops made from serum or plasma are also used to administer fibronectin which is believed to promote the healing of epithelial defects. Accumulation of data on the inhibitory action of plasma inhibitors on proteinases involved in the degradation of the cornea may explain some of the early observations concerning the beneficial effect of serum-based eyedrops, and may justify attempts to resurrect topical autologous serum therapy in ocular surface diseases.

The first synthetic inhibitors were intended to block the action of collagenases (of corneal, leukocyte, or bacterial origin) that were detected in ulcerating corneas. Though several inhibitors (Na_2EDTA, CaEDTA, cysteine, *N*-acetylcysteine, penicillamine) were shown to inhibit collagenases *in vitro,* the same inhibitors used in the form of eyedrops were not as effective therapeutically as had been expected.[3,7] At the time of these trials some basic aspects of proteinase action and inhibitor therapy were not clear. Most authors used the inhibitor eyedrops without determining the activity of proteinase(s) in the tears of the patients, and judged the effectiveness of the therapy only evaluating whether the ulcer healed or not. Different forms and stages of corneal ulcers were treated the same way; inhibitors in inappropriate forms and concentrations (too high, therefore toxic; or too low, therefore ineffective) were tested and were found to be useless and even harmful. That is why very few ophthalmologists use collagenase inhibitor eyedrops and no such formulation is commercially available nowadays.

A new possibility for treating nonhealing epithelial lesions and corneal ulcers with enzyme inhibitor was suggested by Salonen et al.,[31] who detected high plasmin activities in the tears of patients with various corneal lesions and reported cases of chronic ulcers successfully treated with aprotinin (plasmin inhibitor) eyedrops. Though no well-controlled clinical studies have been done to evaluate the effectiveness of aprotinin therapy, several ophthalmologists started to use aprotinin eyedrops to treat corneal ulcer patients. After the early positive impressions and opinions, doubts arose questioning the effectiveness of aprotinin in curing all, otherwise noncurable, corneal ulcers.

The possibility to use aprotinin eyedrops to prevent ulceration after severe alkali burn was suggested by Cejkova et al.[32] Unpublished data mentioned in a recent publication of van Setten et al.[33] suggests that aprotinin therapy is only effective in cases where plasmin activity can be detected in tears. Reim and co-workers[34] demonstrated that topical inhibitor therapy (in combination with fibronectin, chloramphenicol, and prednisone) did not effect the healing rate of ulcerating alkali-burned rabbit corneas. This therapy, however, proved to be effective in the prevention of corneal ulcers and hindering the progression of already existing ulcers both on humans and in rabbit experiments.

Our results concerning PA activities detected in the tears of patients with different traumatic and inflammatory corneal lesions provided further data that the starting point of the pathological process in ocular surface disease is the release of urokinase-type PA from the affected epithelial cells.[35,36] The excessive PA production, if not controlled by the action of PAIs, results in the activation of plasmin and in really severe cases that of collagenase. These three proteinases form a three level cascade (see Figure 1 in Chapter 12). The members of this proteolytic cascade are the enzymes that are responsible for the variety of enzymatic mechanisms known as extracellular proteolysis. The regulatory mechanisms involved in the physiological control of this cascade can also be used for therapeutic purposes[22] (see Table 1 in Chapter 12).

There are basically two approaches, both of which can provide for a selective and probably more effective inhibitor therapy to meet the specific needs of a patient suffering from a certain form or stage of ocular surface disease.

1. Different proteinases, activators, and inhibitors can be determined (or monitored) in the tears of patients and specific intervention can be designed intending to block all abnormally active enzymes and replace the missing elements. To determine several enzymes, activators, and inhibitors in the tears of patients, sophisticated methods and expensive instruments have to be used. Moreover, the tears obtained at one sitting are usually insufficient to perform multiple determinations.

2. Another approach, which probably has more of a chance to be used in clinical practice is the use of simpler and quicker diagnostic methods (such as caseinolytic plates or simple spectrophotometric tests) to detect proteolytic activity in the patients' tears and test the available inhibitors or inhibitor combinations in a similar system to find out which drug or drug combination would be the most effective in blocking the "overall effect" of proteinases, activators, and inhibitors being present in the tears of a particular patient at a given time.

REFERENCES

1. **Holly, F. J.,** Vitamins and polymers in the treatment of ocular surface disease, *Contact Lens Spectrum,* May, 37, 1990.
2. **Berman, M., Leary, R., and Gage, J.,** Evidence for a role of the plasminogen activator-plasmin system in corneal ulceration, *Invest. Ophthalmol. Vis. Sci.,* 19, 1204, 1980.
3. **Berman, M.,** Collagenase and corneal ulceration, in *Collagenase in Normal and Pathological Connective Tissues,* Woolley, D. E. and Evanson, J. M., Eds., John Wiley & Sons, New York, 1980, 141.
4. **Berman, M. B., Barber, J. C., Talamo, R. C., and Langley, C. E.,** Corneal ulceration and serum antiproteases, *Invest. Ophthalmol.,* 12, 759, 1973.
5. **Brown, S. I., Akiya, S., and Weller, C. A.,** Prevention of the ulcers of the alkali burned cornea. Preliminary studies with collagenase inhibitors, *Arch. Ophthalmol.,* 82, 95, 1969.
6. **Itoi, M., Gnadinger, M. C., Slansky, H. H., and Dohlman, C. H.,** Prévention d'ulcéres du stroma de la corneé grace á l'utilisation d'unsel de calcium d'EDTA, *Arch. Ophthal.,* 29, 389, 1969.
7. **Berman, M. B.,** Collagenase inhibitors: rationale for their use in treating corneal ulceration, *Int. Ophthalmol. Clin.,* 15, 49, 1975.
8. **Kueppers, F.,** Proteinase inhibitor in human tears, *Biochim. Biophys. Acta,* 229, 845, 1971.
9. **Rennert, O. M., Kaiser, D., Sollberger, H., and Joller-Jemelka, S.,** Antiprotease activity in tears and nasal secretions, *Humangenetik,* 23, 73, 1974.
10. **Berta, A., Vámosi, P., and Tözsér, J.,** Eine neue Methode zur semiquantitativen Bestimmung von Alpha-1-Antitrypsin Inhalt der menschlichen Tranenflüssigkeit, *Fortschr. Augenheilk.,* 86, 773, 1989.
11. **Tabbara, K. F., Ostler, H. B., Biswell, R., and Daniels, T. E.,** Chronic ulcerative conjunctivitis in a patient with alpha-1-antitrypsin deficiency, *Ann. Ophthalmol.,* 14, 280, 1982.

12. **Saari, K. M., Kaarela, K., Korpela, T., Laippala, P., Frants, R. R., and Erickson, A. W.,** Alpha-1-antitrypsin in acute anterior uveitis and rheumatic diseases, *Acta Ophthalmol.,* 64, 522, 1986.

13. **Saari, K. M., Solja, J., Hackli, J., Seppanen, S., Fülikainen, A., Koskimies, S., Erickson, A., and Frants, R.,** Genetic background of acute anterior uveitis, *Am. J. Ophthalmol.,* 91, 711, 1981.

14. **Zirm, M., Schmut, O., and Hofmann, H.,** Quantitative Bestimmung der Antiproteinasen in der menschlichen Tranenflüssigkeit, *Graefe's Arch. Clin. Exp. Ophthalmol.,* 198, 89, 1976.

15. **Zirm, M.,** Die Bedeutung der Proteinaseinhibitoren in der Tranenflüssigkeit, *Klin. Mbl. Augenheilk.,* 177, 759, 1980.

16. **Zirm, M. and Ritzinger, I.,** Der diagnostische und prognostische Wert einer Alpha-1-Antitrypsinbestimmung in der Tranenflüssigkeit, *Klin. Mbl. Augenheilk.,* 173, 221, 1978.

17. **Anderson, J. A.,** Some characteristics of an SH-protease inhibitor found in human tears, *Invest. Ophthalmol. Vis. Sci. Suppl.,* 197, 1979.

18. **Anderson, J. A. and Leopold, I. H.,** Antiproteolytic activities found in human tears, *Ophthalmology,* 88, 82, 1981.

19. **Hayashi, K. and Sueishi, K.,** Fibrinolytic activity and species of plasminogen activator in human tears, *Exp. Eye Res.,* 46, 131, 1988.

20. **Berta, A. and Tözsér, J.,** Plasminogen Activator and Plasminogen Activator Inhibitor Determinations in the Tears of Patients with Keratoconjunctivitis Sicca and Sjögren's Syndrome, 26th Int. Congress of Ophthalmology, Free Papers, Posters & Videos, Singapore, March 18–24, 1990, 180.

21. **Tözsér, J.,** Neutral Proteinases and Proteinase-Inhibitors in Human Tears, Ph.D. Thesis, University of Debrecen, Hungary, 1988.

22. **Berta, A., Tözsér, J., and Holly, F. J.,** Multilevel inhibition of proteases in tears. New concept for the treatment of corneal ulcers, *Proc. 6th Int. Symp. Lacrimal System,* Kugler, Amsterdam, accepted.

23. **Berta, A., Holly, F. J., Tözsér J., and Holly, T. F.,** Tear plasminogen activators — indicators of epithelial cell destruction. The effects of scraping, n-heptanol treatment, and alkali burn of the cornea on the plasminogen activator activity of rabbit tears, *Int. Ophthalmol.,* 15, 363, 1991.

24. **Davis, W. L. and Anderson, J. A.,** Effects of ocular inflammation on tear proteins. I. Thiol protease inhibitors, *J. Ocular Pharm.,* 2, 31, 1986.

25. **Tözsér, J. and Berta, A.,** Plasminogen activator inhibitors in human tears, *Acta Ophthalmol.,* 69, 426, 1991.

26. **Kenyon, K. R.,** Decision-making in the therapy of external eye disease: noninfected corneal ulcers, *Ophthalmology,* 89, 44, 1982.

27. **Holly, F. J.,** Ocular surface update: diagnosis and treatment, manuscript.

28. **Alberth, B.,** *Surgical Treatment of Caustic Injuries of the Eye,* Akadémia, Budapest, 1968.

29. **Sallai, S., Valu, L., Fehér, J., and Podhorányi, G.,** On subconjunctival autochemotherapy of experimental lime burns, *Dtsch. Gesundheitsw.,* 22, 1994, 1967.

30. **Fox, R. I., Chan, R., Michelson, J. B., Belmont, J. B., and Michelson, P. E.,** Beneficial effect of artificial tears made with autologous serum in patients with keratoconjunctivitis sicca, *Arthritis Rheum.,* 27, 459, 1984.

31. **Salonen, E. M., Tervo, T., Torma, E., Tarkannen, A., and Vaheri, A.,** Plasmin in tear fluid of patients with corneal ulcers: basis of new therapy, *Acta Ophthalmol.,* 66, 3, 1987.

32. **Cejkova, J., Lojda, Z., Salonen, E.-M., and Vaheri, A.,** Histochemical study of alkali-burned rabbit anterior eye segment in which severe lesions were prevented by aprotinin treatment, *Histochemistry,* 92, 441, 1989.

33. **van Setten, G.-B., Salonen, E.-M., Vaheri, A., Beuerman, R.W., Hietanen, J., Tarkkanen, A., and Tervo, T.,** Plasminogen and plasminogen activator activities in tear fluid during corneal wound healing after anterior keratectomy, *Curr. Eye Res.,* 8, 1293, 1989.

34. **Reim, M., Becker, J., Borchers, H., Kohnen, S., Knott, H., Kranz, B., Leber, M., Lund, G., Marchand, M., Salla, S., and Steiger, C.,** Fibronectin, Plasmin, and Proteinase Inhibitor in Experimental and Human Severe Eye Burns, 9th Int. Congress of Eye Research, Abstracts, Helsinki, Finland, July 29–August 4, 1990, 111.
35. **Berta, A., Tözsér, J., and Holly, F. J.,** Determination of plasminogen activator activities in normal and pathological human tears. The significance of tear plasminogen activators in the inflammatory and traumatic lesions of the cornea and the conjunctiva, *Acta Ophthalmol.,* 68, 508, 1990.
36. **Tözsér, J., Berta, A., and Punyiczki, M.,** Plasminogen activator activity and plasminogen independent amidolytic activity in tear fluid from healthy persons and patients with anterior segment inflammation, *Clin. Chim. Acta,* 183, 323, 1989.

Chapter 15

TEAR ENZYMES AND CONTACT LENSES

Part A
Corneal Physiology in Contact Lens Wear

The first contact lenses were made of ground or blown glass and were poorly tolerated by the patients. At that time little was known about the physiology of the cornea. Fitting of the lenses became more successful when plexiglas was introduced in the contact lens praxis. At first scleral lenses were made of plexiglas. The edge of these lenses was in close connection with the bulbar conjunctiva and the underlying sclera, trapping some of the precorneal tear fluid under the lens, preventing the fluid and gas exchange with the rest of the tear film and the air. It has been soon recognized that the corneal changes seen after wearing these contact lenses are caused by the fact that the cornea has no access to the air, its oxygen supply.

Scleral lenses were soon replaced by smaller, thinner, and more flexible corneal lenses. These lenses were made of PMMA and were designed to cover only a part of the corneal surface. This type of lens usually floats on a thin layer of tears, and has the ability to move on the surface of the cornea on each blink. This movement is supposed to provide an exchange of oxygen between the tear fluid under the contact lens and the rest of the tear film. It has been observed, however, that after prolonged wearing of these lenses signs of corneal anoxia still develop.

A new generation of soft contact lenses has been developed, with higher water content, with a quality of being, at least to a certain extent, permeable to air. These lenses were better tolerated by patients and were possible to be worn for a longer period of time. The use of soft lenses brought some disadvantages in optical effectiveness and in the necessity to use chemical disinfectants and preservatives. The advantages of easy fitting, better comfort, and extended wear still made these lenses very popular.

Further development was reached by the introduction of gas-permeable hard contact lenses that intend to combine the advantages of hard lenses in easy handling and good optical properties with the better tolerability of soft lenses. A continuous effort is being made, however, to find better materials to make contact lenses and to find out more about the physiology and the metabolic needs of the cornea to secure better lens tolerance, longer duration of possible lens wear, and to reduce the number and severity of the still-existing complications.

Some of the corneal changes caused by prolonged wear of contact lenses, such as epithelial lesions and the swelling of different layers of the cornea, have been recognized to be secondary to anoxia. Epithelial edema caused by extensive wear or poorly fitted lenses is a recognized indicator of cellular dysfunction caused by the fact that these cells are deprived of oxygen necessary for their normal metabolism. Recently Hamano and Kaufman,[1] Friend,[2] and Klyce and Beuerman[3] reviewed the literature on the physiology and the metabolism of the cornea with respect to changes caused by contact lens wear. The subject is in the focus of general interest shown by the large number of publications in ophthalmological, optometrical, and contact lens journals.

The first observations suggesting changes in corneal metabolism caused by inappropriate contact lens wear were the studies showing that the thickness of the cornea increases during contact lens wear.[4] Very similar thickening was shown to occur when the cornea was deprived of oxygen by exposing its surface to nitrogen gas.[5] The normal corneal thickness was found to show a diurnal variation.[6] Corneal thickening observed after the wearing of soft contact lenses for 8 to 17 h, being less than 5% of the initial thickness of the cornea,[7-9] can be considered to be within the limits of normal variation. Hard contact lenses, on the other hand, were found to cause up to 20% increase in corneal thickness when worn for a similar period of time, indicating significant swelling of the cornea secondary to anoxia.[8]

The most important point in corneal physiology concerning contact lens wear is the fact that a contact lens covering the corneal surface may prevent oxygen reaching the cornea. Corneal epithelial cells were shown to need continuous oxygen supply from the atmosphere to function properly. Fatt and Hill[10] used polarography to study oxygen activity in corneal epithelium, Polse and Mandell[11] investigated gas exchange at the level of the epithelium in anoxia, and provided details concerning the oxygen need and oxygen uptake of the cornea.

Hamano et al.[12,13] studied the effect of contact lens wear on the oxygen tension in the corneal stroma and in the anterior chamber using an enamel-coated copper electrode according to the method of Yagi et al.[14] They provided further evidence that oxygen from the air really penetrates the cornea and even gets into the anterior chamber of the eye. A decrease in the electrical potentials reflecting the oxygen tension in the corneal stroma and the anterior chamber was observed when a contact lens was put on the cornea. This change started from the moment the lens was applied, and was greater with hard (non-gas permeable) lenses and with large lenses covering the whole surface of the cornea. The decrease was gradual in the anterior chamber and more rapid in the corneal stroma. No change in the potentials and the oxygen tension was observed when oxygen was blown on the surface of the contact lens (non-permeable hard lens), but the tension rapidly increased first in the stroma then in the anterior chamber when the lens was removed.

The use of polyurethane-coated platinium electrode covered with an oxygen-permeable membrane made possible the direct measurement of oxygen tensions on epithelial surfaces and in different layers of the cornea.[15] Using this method Hamano and colleagues demonstrated on enucleated rabbit eyes that 10 min of PMMA lens wear reduced the oxygen tension to 0 both on the surface of the epithelium and in the corneal stroma. In *in vivo* experiments it has been shown both in rabbits and humans that the oxygen tension correlated with the gas permeability (Dk values) of the lenses. The more permeable, thinner, and smaller the lenses were, the higher was the oxygen tension measured on the ocular surfaces.[16,17]

Different electrical potentials measured in living cells and living tissues can be used as indicators of normal or disturbed function. Major disturbances in the metabolic status of the living cells are usually accompanied by electrophysiological changes. Electrophysiological methods are widely used in experimental and clinical ophthalmology, most often to detect the dysfunction of the retina or the optic nerve.

It has been shown by different authors that there exists a difference in electrical potentials between the epithelium and the endothelium of the cornea. This difference, called the transcorneal potential, was measured on enucleated rabbit eyes to be –10 to –40 mV, the epithelium being negative and the endothelium being positive.[18-20] Hamano and Kikkawa[21] have demonstrated on rabbit eyes that the transcorneal potential decreased after 10 h wear of a hard contact lens from –30 mV to near 0, and gradually returned to normal level in 4 to 5 h when the lens was removed. The change in electrical potential was observed only in the area covered by the contact lens while the rest of the cornea retained normal transcorneal potential. A comparison was made between the transcorneal potential-lowering effect of hard and soft contact lenses. The decrease in potential with soft contact lens was only 5 mV compared to the 30-mV change caused by the hard contact lens with the same duration of wear (10 h) in the other eye of the same rabbit. The recovery of transcorneal potential was good after 6 h and poor after 17 h of hard contact lens wear. The transcorneal potential returned to the normal level in 1 h in the first and only after 5 h in the second case. There was not such a difference in soft contact lens; the transcorneal potential returned to normal both after 6 h and 17 h in 60 min.

Very similar results were obtained for the intracellular potential measured by a glass microelectrode in the epithelial cells of rabbit corneas following contact lens wear. The normal intracellular potential of these cells is known to be –20 to –40 mV caused by differences of K^+ and Cl^- concentrations in the intracellular and extracellular space.[22,23] The decrease in intracellular potential values in case of soft lenses was detected only after 8 h wear and was found to be as little as 5 mV, while the decrease in case of hard lenses occurred much earlier and was higher (20 to 30 mV) after 17 h. This potential change was also observed only at the sites of the cornea that were covered by the contact lens.[24]

The electrophysiological findings indicate characteristic changes in the function of corneal cells brought about by contact lens wear. These changes seem to be reversible with both types of lenses, though the disturbance is more severe and the recovery takes longer in case of hard (non-permeable) lenses.

Most of the energy used by the epithelial cells is produced by the step-by-step enzymatic breakdown of glucose, a process called glycolysis. Glucose is used by all layers of the cornea and mostly derives from the aqueous humour. Glucose enters the cornea, together with water and anorganic ions, through diffusion. Water and most of the salts are then turned back by an active transport mechanism, called the endothelial pump function. There are basically two ways to produce energy from glucose, depending whether the cells have enough oxygen or are deprived of their outer oxygen supplies. The two metabolic processes (the aerobic and the anaerobic glycolysis) follow basically the same route until they reach pyruvate, a tricarbonic acid. In aerobic conditions pyruvate is bound to acetyl-Co-A and enters the tricarboxylic acid cycle producing much more energy for the cell than it is possible without oxygen. Under anaerobic circumstances pyruvate is turned into lactate (lactic acid) by lactate dehydrogenase (LDH) producing only one tenth of the energy from the same amount of glucose as could be produced through aerobic glycolysis. Glucose is stored in the epithelial cells in the form of glycogen, a polysaccharide made up of glucose monomers. Extensive energy production results in the enzymatic breakdown (glycogenolysis) of glycogen granules stored in the cytoplasm of the cells.[25-27]

It has been shown that in aerobic conditions lactic acid accumulates in the cornea.[27,30] This is accompanied by an increased lactic acid-based activity of LDH and a decrease in the activity of enzymes of the tricarboxylic acid cycle in the epithelium of the cornea.[29,30] The excessive use of glucose due to the ineffective energy production under anaerobic circumstances sooner or later results in a reduction of the glycogen content of the epithelial cells.[31]

The glycogen content of corneal epithelial cells has been shown by histological and biochemical methods to decrease considerably when hard contact lenses were worn.[31,32] In a comparative study Hamano et al.[33] found that the corneal epithelial cells had lost their glycogen granules following 17 h of hard contact lens wear. The loss of glycogen was first observed in the basal cells of the epithelium, and the granules were missing from all layers of the corneal epithelium after 8 h. Following the wear of soft contact lenses with the same duration (17 h) the epithelial cells still contained less amounts of glycogen granules. In a long-run experiment it has been shown that when covered with high-water-content soft lens for 1 month, the cornea still continues to show some evidence of glycogen in the form of stored granules. The histological findings were also confirmed by biochemical data. Hamano et al.[8] measured glycogen levels in the corneal epithelial cells with the glucose-oxidase method. The glycogen levels were shown to decrease in correlation with the duration of lens wear.

Thoft and Friend[32] demonstrated a considerable decrease in the ATP content of epithelial cells after 6 h wear of hard contact lenses. A decrease in the glycogen content and an increase in the lactic acid level of the cornea following contact lens wear were found by Burns et al.[34] They also demonstrated that these changes were less pronounced in cases with silicon rubber lenses as compared to hard and soft contact lenses. Further data concerning the increase of LDH activity of the epithelial cells were provided by Ehlers[35] and Hamano et al.[8] The LDH activity increased when hard contact lenses had been worn for 24 h. High LDH activities were demonstrated in the epithelial cells also when gas-permeable silicon rubber lenses were worn.

It has been demonstrated that any reduction in the energy production of the epithelial cells results in a decrease in the mitotic activity of the basal layer. Generally, a large number of mitotic configurations can be seen in the basal cell layer on histological examination. Sugiura and colleagues have found on rabbit eyes a significant decrease in the number of mitotic configurations after 6 h of PMMA contact lens wear. The number of mitotic figures after 6 h of HEMA lens wear was 95% of that of the control eyes. Prolonged wear, however, caused a significant decrease in the number of mitotic configurations (74% after 24 h, 47% after 48 h) even in the case of HEMA lenses. The effect of PMMA lenses on epithelial cell mitosis was similar to that of the exposure of the cornea to nitrogen gas for the same period of time. The metabolic activity shown by the accumulation of lactic acid in the cornea was similarly depressed in both cases. With gas-permeable hard contact lenses the degree of mitotic suppression was found to be inversely related to the Dk value of the lenses.[36,37]

On the basis of these experimental data it seems to be well established that contact lens wear effects the energy-producing mechanisms of corneal cells by turning the metabolic pathways toward anaerobic mechanisms. The basic cause of this switchover is the fact that contact lenses block the way of oxygen that would otherwise diffuse into the cornea through precorneal tear film and the epithelium. The use of a less-effective energy-producing metabolic pathway results in the exhaustion of the glycogen reserves, in electrophysiological changes, and in the reduction of mitotic activity of the epithelial cells. These changes increase in intensity with the duration of lens wear and are more pronounced in the case of non-permeable hard contact lenses. By the use of soft lenses and gas-permeable hard contact lenses, the degree of corneal hypoxia can be reduced considerably, but signs of metabolic disturbances are still detectable with the currently used gas-permeable lenses if they are worn for a long time.

Contact lens wear was found to affect the function of the sensory nerves in the cornea. Touch and pain, as well as the sensation of warmth and cold, are mediated by the bare nerve endings located in the corneal epithelium. Corneal sensitivity can be measured by the force of various nylon threads (or Frey's hairs) still sensed when touching the cornea. The sensitivity is expressed in milligrams (mg); its normal value in the central cornea is 15 mg, at the limbus 20 mg, and the conjunctiva varies between 72 and 200 mg.[38] Different sensitivity values can be measured at various parts of the cornea; the horizontal meridian is the most sensitive part; the temporal half appears to be more sensitive than the nasal part. Corneal sensitivity has been found to decrease with age.[39,40] Byron and Wesely, using Frey's hairs, reported a decrease in corneal sensitivity after 4 h of contact lens wear.[41] Boberg-Ans found that the average corneal sensitivity fell from 25 to 72 mg.[42] Corneal hypersensitivity cannot be considered a contraindication for the fitting of contact lenses. A contact lens patient with reduced corneal sensitivity, on the other hand, needs closer observation than the normal patient to make sure that undetected abrasions do not occur.[43] Corneal swelling caused by contact lens wear was one of the proposed causes for corneal hyposensitivity. Polse has found that corneal swelling resulting from exposing the cornea either to an oxygen-free environment or to a hypotonic solution did not change corneal sensitivity.[44] It has been suggested that the change in corneal touch threshold accompanying contact lens wear is not caused by metabolic disturbance but rather by sensory adaptation as a consequence of continuous mechanical stimulation.[43] Based on the analysis of experimental data and on clinical observations, Millidot and O'Leary suggested that decreased corneal sensitivity associated with contact lens wear was hypoxia.[40] Polse provided further data supporting the role of corneal hypoxia. They observed that the long-term wear of PMMA lenses resulted in a significant decrease in corneal sensitivity, which could be reversed when the PMMA lens was replaced by an oxygen-permeable contact lens.[45,46]

Contact lens wearers occasionally feel hot when they wear their lenses for a period of time. This feeling has been suggested to be associated with a mechanism by which the lens does not allow a proper tear circulation between the lens and the cornea. It has been presumed that the plastic lens could act as an insulator and may raise the temperature of the cornea. There is some doubt whether heat sensation is actually caused by a rise in corneal temperature.[47] Hill and Leighton did not find a significant difference in central corneal temperature before and after contact lens was worn by rabbits, as long as the lids remained open.[47] When the lids were closed, however, the temperature in the cornea rose by more than 4°C. A temperature rise was detected in association with blepharospasm and forced blinks. Corneal temperature was shown to be correlated also by palpebral slit width. An association has been found between corneal temperature and corneal metabolism.[48] It has been suggested that incorrect fit of the contact lens interferes with tear exchange behind the contact lens which unfavorably affects corneal metabolism. Such a disturbance may cause conjunctival hyperemia, which in turn may raise the temperature in the surrounding tissues (cornea and sclera) producing the sensation of heat in warm receptors.[43]

REFERENCES

1. **Hamano, H. and Kaufman, H. E.,** *The Physiology of the Cornea and Contact Lens Applications,* Churchill Livingstone, New York, 1987.
2. **Friend, J.,** Physiology of the cornea: metabolism and biochemistry, in *The Cornea. Scientific Foundations and Clinical Practice,* Smolin, G. and Thoft, R. A., Eds., Little, Brown and Company, Boston, 1987.
3. **Klyce, S. D. and Beuerman, R. W.,** Structure and function of the cornea, in *The Cornea,* Kaufman, H. E., McDonald, M. B., Barron, B. A., and Waltman, S. R., Eds., Churchill Livingstone, New York, 1988.
4. **Smelser, G. K. and Chen, D. K.,** Physiological change in cornea induced by contact lenses, *Arch. Ophthalmol.,* 53, 676, 1955.
5. **Mandell, R. B. and Polse, A. K.,** Corneal thickness changes as a contact lens fitting index. Experimental results and a proposed model, *Am. J. Optom. Arch. Am. Acad. Optom.,* 46, 479, 1969.

6. **Kikkawa, Y.,** Diurnal variation in corneal thickness, *Exp. Eye Res.,* 15, 1, 1973.

7. **Bailey, L. I. and Carney, G. L.,** Corneal changes from hydrophylic contact lenses, *Am. J. Optom. Arch. Am. Acad. Optom.,* 50, 299, 1973.

8. **Hamano, H., Hori, M., and Hirayama, K.,** The effects of hard and soft contact lenses on rabbit cornea, *J. Jpn. C. L. Soc.,* 14, 29, 1972a.

9. **Schauer, K. R., Hughes, C. C., and Jarrell, R. L.,** Corneal thickness change associated with hard and soft contact lens wear, *Optom. Weekly,* 5, 29, 1973.

10. **Fatt, I. and Hill, M. R.,** Oxygen tension under a contact lens during blinking. A comparison of theory and experimental observation, *Am. J. Optom. Arch. Acad. Optom.,* 47, 50, 1970.

11. **Polse, K. A. and Mandell, B. R.,** Critical oxygen tension at the corneal surface, *Arch. Ophthalmol.,* 84, 505, 1970.

12. **Hamano, H., Hirayama, K., and Hori, M.,** Variation of oxygen tension within the cornea, *J. Jpn. C. L. Soc.,* 15, 7, 1973.

13. **Hamano, H., Hirayama, K., and Hori, M.,** Variation of oxygen tension in the aqueous humor under the contact lens, *J. Jpn. C. L. Soc.,* 17, 65, 1975.

14. **Yagi, S., Kojima, K., Mochiki, F., and Tanaka, T.,** Enameled Cu wires as oxygen electrode in place of Pt wires, *J. Iwate Med. Assoc.,* 15, 193, 1963.

15. **Hamano, H., Mikami, M., Mohri, H., and Mitsunaga, S.,** Measurement of oxygen partial pressure at the rabbit cornea under various types of contact lenses, *J. Jpn. C. L. Soc.,* 26, 295, 1984.

16. **Hamano, H., Mikami, M., Mohri, H., Mitsunaga, S., and Kotani, S.,** Measurement of oxygen tension in anterior ocular segments by a platinum microelectrode. I. Preliminary experiments in *in vitro* systems, *J. Jpn. C. L. Soc.,* 28, 47, 1986a.

17. **Hamano, H., Mikami, M., Mitsunaga, S., and Kotani, S.,** Measurement of oxygen tension in anterior ocular segments by a platinum microelectrode. II. *In vivo* measurement of oxygen tension on rabbit and human cornea under various gas permeable hard contact lenses, *J. Jpn. C. L. Soc.,* 28, 51, 1986b.

18. **Donn, A., Maurice, D. M., and Mills, N. L.,** Studies on the living cornea *in vitro, Arch. Ophthalmol.,* 62, 741, 1959.

19. **Lambert, B. and Donn, A.,** Effect of quabain on active transport of sodium in the cornea, *Arch. Ophthalmol.,* 72, 525, 1964.

20. **Modrell, R. W. and Potts, A. M.,** The influence of medium composition, pH, and temperature in the transcorneal potential, *Am. J. Ophthalmol.,* 48, 834, 1959.

21. **Hamano, H. and Kikkawa, Y.,** Corneal potential as influenced by the contact lens, *J. Jpn. C. L. Soc.,* 11, 145, 1969.

22. **Akaike, N. and Hori, M.,** Effect of anions and cations on membrane potential of rabbit corneal epithelium, *Am. J. Physiol.,* 219, 1811, 1970.

23. **Kikkawa, Y.,** The intracellular potential of the corneal epithelium, *Exp. Eye Res.,* 3, 132, 1964.

24. **Hamano, H., Komatsu, S., Hori, M., and Akaike, N.,** Effect of contact lens on the resting potential of the rabbit corneal epithelium, *J. Jpn. C. L. Soc.,* 11, 162, 1969.

25. **Kinoshita, J. H.,** Some aspects of the carbohydrate metabolism of the cornea, *Invest. Ophthalmol.,* 1, 178, 1962.

26. **Kinoshita, J. H. and Masurat, T.,** Aerobic pathways of glucose metabolism in bovine corneal epithelium, *Am. J. Ophthalmol.,* 48, 47, 1952.

27. **Kishida, K. and Ootori, T.,** Biochemical and histochemical studies on rabbit corneal glycogen following contact lens wear, *Acta Soc. Jpn.,* 75, 1387, 1971.

28. **Kuhlman, R. E. and Resnick, R. A.,** The oxidation of C14 labelled glucose and lactate by the rabbit cornea, *Arch. Biochem. Biophys.,* 85, 29, 1959.

29. **Lowther, E. G. and Hill, M. R.,** Recovery of the corneal epithelium after a period of anoxia, *Am. J. Optom. Arch. Am. Acad. Optom.,* 50, 234, 1973.

30. **Rengstorff, H. R., Hill, M. R., Petrali, P. J., and Sim, M. V.,** Critical oxygen requirement of the corneal epithelium as indicated by succinic dehydrogenase reactivity, *Am. J. Optom. Physiol. Optics,* 51, 331, 1974.

31. **Smelser, G. K. and Ozanics, V.,** Structural changes in corneas of guinea pigs after wearing contact lenses, *Arch. Ophthalmol.,* 49, 335, 1953.
32. **Thoft, R. A. and Friend, J.,** Biochemical aspects of contact lens wear, *Am. J. Ophthalmol.,* 80, 139, 1975.
33. **Hamano, H., Hori, M., and Hirayama, K.,** Scanning electron microscopic findings of wearing contact lenses on the rabbit corneal epithelium. II, *J. Jpn. C. L. Soc.,* 14, 69, 1972b.
34. **Burns, P. R., Roberts, H., and Rich, F. L.,** The effect of silicone contact lenses on corneal epithelial metabolism, *Am. J. Ophthalmol.,* 71, 486, 1971.
35. **Ehlers, N.,** Morphology and histochemistry of the corneal epithelium of mammals, *Acta Anat.,* 75, 161, 1970.
36. **Hamano, H., Hori, M., Hamano, T., Kawabe, H., Mikami, M., Mitsunaga, S., and Hamano, T.,** Effects of contact lens wear on mitosis of corneal epithelium and lactate content in aqueous humor of rabbit, *Jpn. J. Ophthalmol.,* 27, 451, 1983.
37. **Sugiura, E., Iwata, S., and Hamano, H.,** Studies on the bioincompatibility of contact lens materials. VIII. Effects of oxygen and collagen substrates on the growth of epithelial cells in culture, *J. Jpn. C. L. Soc.,* 27, 159, 1985.
38. **Boberg-Ans, J.,** Experience in clinical examination of corneal sensitivity, *Br. J. Ophthalmol.,* 39, 705, 1955.
39. **Jalavisto, E., Orma, E., and Tawast, M.,** Aging and relation between stimulus intensity and duration in corneal sensitivity, *Acta Physiol. Scand.,* 23, 224, 1951.
40. **Millidot, M.,** The influence of age on the sensitivity of the cornea, *Invest. Ophthamol.,* 16, 240, 1977.
41. **Byron, H. M. and Wesely, A. C.,** Clinical investigation of corneal contact lenses, *Am. J. Ophthalmol.,* 51, 675, 1961.
42. **Boberg-Ans, J.,** On the corneal sensitivity, *Acta Ophthalmol.,* 35, 149, 1956.
43. **Mandell, R. B.,** *Contact Lens Practice,* Charles C Thomas, Springfield, IL, 1988.
44. **Polse, K. A.,** Etiology of corneal sensitivity accompanying contact lens wear, *Invest. Ophthalmol.,* 17, 1202, 1978.
45. **Polse, K. A.,** Measurement of corneal sensitivity. I, *Optom. Weekly,* 68, 1380, 1977a.
46. **Polse, K. A.,** Measurement of corneal sensitivity. II, *Optom. Weekly,* 68, 1404, 1977b.
47. **Hill, R. M. and Leighton, A. J.,** Temperature changes of human corneas and tears under contact lenses. III. Ocular sensation, *Am. J. Optom.,* 42, 584, 1965.
48. **Hill, R. M.,** How the cornea "takes the heat", *Int. Cont. Lens Clin.,* 5, 65, 1978.

Chapter 15

TEAR ENZYMES AND CONTACT LENSES

Part B
Tear Film Physiology in Contact Lens Wear

The most prominent change induced by contact lens wear is the dramatic increase in tear production rate on lens insertion. Reflex tearing causes an increase in tear volume, as well as a decrease in the tonicity and the osmolarity of the tear film. The effect on tear production is relatively short lived, within a week lacrimal secretion and the levels of anorganic ions in tears are apparently back to normal.[1] These early changes are often followed by alterations in the opposite direction, i.e., slightly decreased tear secretion and raised tonicity and osmolarity in adapted contact lens wearers as compared to normal non-contact lens wearers.[2] Reduced tear flow was suggested to result from "fatigue block" in the efferent nerves to corneal and conjunctival impulses;[3] a study on tear flow rates, however, in soft contact lens wearers showed the absence of this type of reaction.[4]

When a contact lens is placed over the eye it acts as a foreign body and changes the properties of the precorneal tear film to a great extent. It has an immediate thinning effect on the surrounding tear film. Any fluid film thinner than 100 μm tends to develop a depression in the surface around the boundary of a solid object placed in the film.[5] Such elongated depressions can be observed in the form of dark lines in the fluorescein-stained tear film by the lid margins. The same depression of the surface and thinning of the tear film can be observed around contact lenses. In such thinned regions the tear film ruptures easily. The meniscus and the surrounding thinned region is thought to form as a result of a fluid-drawing effect due to unsatiated negative pressure by the margin of the contact lens.[6]

The contact lens divides the tear film into three parts: pre-contact lens, post-contact lens, and peri-contact lens tear film. Both the pre- and post-contact lens part of the tear film are thinner than the surrounding parts of the film layer. When the lens is put over the eye its surface is generally not suitable for forming a continuous fluid layer on its anterior surface. On repeated blinking the surface of the lens will be coated by a thin layer of mucous glycoproteins deriving from the goblet cells of the conjunctiva. After this the surface of the lens, as it happens with that of the cornea, becomes hydrophilic and a relatively stable pre-lens tear film can be formed. The maintainance of a continuous fluid layer over the contact lens, as that over the cornea, is essential to obtain good visual acuity. The pre-lens tearfilm is rather stable, is covered by a superficial lipid layer, and is reformed over the lens during each blink. An important difference however is that the aqueous layer covering the lens is thinner than it would be without a contact lens. A thin aqueous layer makes the diffusion of lipids to the mucus layer easier. This is one reason the pre-lens tear film ruptures earlier than the precorneal tear film. Another difference between the ruptures of the two films is that the break-up of the pre-lens tear film usually occurs at once over the whole lens surface. This sudden drying effect is supposed to be the result of a draining-type action of a nonequilibrium meniscus that surrounds the contact lens.[7] Scratches on the lens surface may also trap lipid contamination and cause the break-up of the film almost instantaneously.[8]

The affinity of the lipid molecules for contact lenses largely depend on the surface-chemical characteristics of the lens surfaces. Lipids that are always present in the tear film and on the lid margins will readily adsorb onto any surface if by doing so they can lower the free energy of the surface. In the case of hydrophobic (PMMA or silicon) contact lenses this change is not very pronounced as their critical surface tension is already low, but even this change, if the critical surface tension becomes lower than the film surface tension of the fluid, will result in an unstable pre-lens tear film. Lipid adsorption on hydrogel lenses causes much greater change in free energy, making the otherwise hydrophilic surfaces hydrophobic, resulting in break-up of the prelens film and blurred vision. Lipid contamination of contact lenses can be present in the form of lipid-protein complexes. Lipid-protein interaction may lead to the denaturation of proteins and the formation of insoluble deposits on the surface of the contact lenses.[9] These lipid-denatured protein deposits, besides causing premature rupture of the tear film, may lead to allergic reactions of various degree.[10]

The surface of hydrophobic contact lenses can be covered with thin hydrophilic coating or chemically modified. These artificial surfaces, however, adsorb lipids tenaciously in the eye and regain hydrophobic characteristics in a short time. Conjunctival mucin (mucous glycoproteins) seem to be the best to coat a hydrophobic contact lens. After a few blinks a continuous mucin layer forms on the anterior lens surface, which makes possible the formation and maintenance of a relatively stable pre-lens tear film. This is a physiologic mechanism that restores the hydrophilic properties of both the cornea and the contact lens on each blink. In cases with poor congruity between lids and lens mucin cannot be distributed evenly and the hydrophility is not completely restored. The lack of reformation of a continuous mucous layer by the edge of the lens results from the fact that the moving lid does not come in contact with this part of the corneal surface. The 3 to 9 o'clock staining observed on contact lens wearers is the result of localized dessication. With very similar mechanism punctate epithelial keratopathy may be formed on the lower part of the cornea due to incomplete wetting.[11]

The break-up of the pre-lens tear film is not directly sensed by the wearer. It can be noticed only through the instant diminution of visual acuity which occurs when the tear film over the lens ruptures. Moreover, corneal sensitivity is usually decreased by contact lens wear. Contact lens wearers have to learn to blink at a necessary rate to have a good vision through maintaining a fairly stable fluid film over the lens surface.

If the contact lens design or the fitting is not optimal for the wearer the lens may cause further problems. When the contact lens has a thick edge, as in the case of highly concave lenses, the lid globe congruity may become poor on the nasal and temporal side of the lens. The sheer action of the lids during blinking becomes insufficient in these areas and as a result the stability of the tear film decreases due to locally increased interfacial tension. This effect has been suggested to be responsible for the development of epithelial defects shown by fluorescent staining at 3 and 9 o'clock by the edge of the corneal contact lenses. Poorly fitted lenses may be either too tight, preventing the necessary movements of the lens over the surface of the cornea, or may be loose resulting in drifts of the lenses to undesired positions where they can hurt the surface of the eye. Both of these conditions will lead to injuries of the otherwise smooth epithelial surface creating spots of tear film instability. The constant rupture of the tear film over these areas will further worsen the condition by creating a vicious cycle.

The oxygen permeability of contact lenses has been shown to be important for the well being of the underlying epithelium, and the wettability of the lens surface is known to determine the stability of the fluid layer over the lens. Different factors have also been suggested to play a role in lens tolerability, which depends basically on the maintainance of the integrity of the ocular surface and that of a stable tear film over the cornea and the contact lens.[12] One of these factors is the exchange of fluid and soluble substances (nutrients, waste materials, oxygen, carbon dioxide, etc.) between the post-lens part and the rest of the precorneal tear film. Physiologic movements of the lenses which occur on each blink play an important role in mixing different parts of the tear film in contact lens wearers. This mixing effect together with a pump mechanism, also thought to be brought about by the blink movements of the upper eyelid, are probably more important for the health of the underlying epithelium than the oxygen permeability of the lens itself.[10,13,13a] The wettability of a lens surface is a necessary, but not satisfactory, condition to provide a stable pre-lens tear film. A hydrophilic lens may be wettable by water because of its high surface free energy, and can still be bioincompatible if resulting from high interfacial tension its high energy surface binds lipids and denatures proteins leading to the formation of deposits with all of their consequences of decreased vision and poor tear film stability.[12] Such deposits would also prevent the constant reformation by the sheer action of the lids of a continuous layer of hydrophilic macromolecules over the lens surface, which is also essential in maintaining tear film stability. The use of biocompatible materials, well-fitted lenses, and optimal wetting solutions can minimize these unfavorable effects and result in good contact lens tolerance.

A contact lens placed on the eye causes changes both in the structure and function of the precorneal tear film. Changes in the lipid phase caused by contact lens wear seem to be of utmost importance. It has been suggested that some of the changes in tear osmolarity and tonicity are the results of increased evaporation which may develop due to contact lens-induced disruption of the superficial lipid layer of the tear film. When a light is thrown on the tear film covering the cornea at the margin of reflection several colors appear. Different colors can be seen depending on the angle of incidence. This color dispersion phenomenon was described by Vogt in 1921,[14] by Koby in 1924,[15] and by Meesmann in 1927.[16] The thickness of the film can be determined using the dispersion colors of the reflected light. Since the thickness of the lipid phase is in the order of magnitude of the wavelengths of visible light, characteristic colors are produced by interference within the tear film. Normally the lipid layer is about 100 nm thick and has a grayish appearance. With increasing thickness fringes of red, orange, yellow, green, and eventually blue may appear.[17] Changes in the lipid layer can be recorded by photographic technique, used by Guillon,[18] utilizing crossed Polaroids and the reflected image of the slit lamp. In addition to the possibility for the evaluation of the thickness of the lipid layer the assessment of lipid layer continuity and lipid layer stability can also be performed.[19]

The appearance and the behavior of the lipid phase with a new and clean soft contact lens was found to be similar to those in the non-lens wearing eye. However, when deposits or lens defects were present structural alterations and unstability of the tear film was observed. When a hard lens is on the eye the lipid phase is absent over the anterior lens surface. Interface colors, however, may be observed due to the presence of an uneven and quickly breaking aqueous layer on the surface of the lens. The colored fringes appear in a narrow zone around the drying spots. A thickening of the lipid layer can be seen around the lens. It appears as if the lens edge forms a barrier that prevents the spreading of the lipid over the anterior surface of the hard lens.[18,20]

The lack of continuity of the lipid layer, besides affecting the stability of the tear film, is thought to be the cause of increased tear evaporation rate observed in contact lens wearers. Increased evaporation rates were measured both in soft and in hard contact lens wear.[21,22] The extent of the increased evaporation rates was found to correlate with osmolarity changes in the tears collected from the same contact lens wearers. It has been suggested that the increased osmolarity may be beneficial in reducing corneal swelling seen as a complication of long contact lens wear.[23] If there is a continuous tear film over the contact lens, water from the film evaporates between blinks. Most contact lenses, however, are not covered by a stable tear film. Hydrogel lenses after the pre-lens tear film ruptures also lose some of their water content through evaporation. Some water from the tear film that separates the lens from the cornea can pervaporate through the lens, too.[24]

Contact lenses can be classified as rigid lenses, elastomeric lenses, and hydrogel lenses. Rigid lenses can be subdivided into non-oxygen permeable such as the poly-methyl methacrylate (PMMA) lenses and oxygen-permeable contact lenses such as the silicon resin lenses, siloxane-methacrylate lenses, cellulose-acetate-butyrate (CAB) lenses, and alkyl-styrene lenses. Essentially two types of elastomeric contact lenses are known: silicone rubber lenses and acrylic rubber lenses. Hydrogel lenses can be classified according to their water content and ionic or nonionic nature. Hydrogel lenses are available from 30 to 80% hydration. Low water content lenses contain less than 45% water when fully hydrated; lenses with 55% water content or more are generally referred to as high water content hydrogel lenses. Some of the above-listed lenses are still used in contact lens practice, others only represent certain stages of the development in the past, still others will be introduced in the future. Other classifications also exist that are based on the possible length of wear and other practical characteristics (extended wear contact lenses, disposable lenses, etc.).[4,25,26]

When a hydrogel lens is placed in the eye it starts to lose some of its water content. Lens dehydration is an unavoidable characteristic of the lens surrounded by the tear film. The decrease in the hydration of the lens results in a decrease in the oxygen permeability of the lens; in an increase in the refractive index, the lens becomes thinner, the radius of curvature will be shorter, so the lens becomes steeper. The refractive power increases in the plus lens and decreases in minus lenses. The dehydration continues until a steady state is reached; once this point is reached the lens retains its physical and optical characteristics for a long time. The extent and the rate of dehydration is determined primarily by the physical properties of the lens, but is also greatly affected by the osmotic characteristics of the surrounding tear film.[27]

A large proportion of the water in a hydrogel lens is not bound to the polymer network and therefore can be lost by evaporation. The rate of evaporation is to a large extent determined by the pore size, chemical nature, and water content, the extent of the hydration of the lens.[28] The rate of evaporation slows down faster in lenses with low water content than in lenses with high water content. A high water content lens reaches the steady state more rapidly than a low water content lens. In addition more water evaporates from a more-hydrated lens than from one with lower hydration.[24] Water evaporates only from the lens surface, but there is a continuous diffusion of water from inside the lens to the surface as the surface dries, and from the tear film behind the lens. Water diffusion through the lens is easier in the high-hydration lens, because its polymer network is looser.[25]

The dehydration process is also affected by the thickness of the lens. More water is lost by the thicker lens in equal time than by the thinner lens. Thinner lenses, on the other hand, due to their lower absolute water content, can reach the steady state faster and may dry out easier than the thicker lenses. In certain cases, when the evaporation at the surface of the lens is faster than the bulk of moisture that diffuses to the surface and the unequal distribution of moisture in the lens develops, the lens will roll up due to forces acting in the network due to the nonuniform swollen state. The partly dried and distorted lens will be soon removed from the surface of the eye by lid motions. Such a situation develops easier with thinner and low water content than with thicker and more-hydrated lenses. This "fast-drying phenomenon" is often seen when bandage lenses are used in dry eye patients, in patients with extended-wear contact lenses, especially if their eyes are not completely closed during sleep.[25] As hydrogel lenses dehydrate, they develop a "water imbition pressure". Such a lens can behave as a hyperosmotic agent on the cornea which may have a therapeutic value. In patients with corneal edema a "hyperosmotic lens" may draw water from the epithelium, reducing edema and improving vision.[29]

The chemical characteristics and the extent of hydration also effect the formation and removal of lens deposits. Rigid lenses, when completely intact, have deposits only on their surface. Such deposits can be removed with detergents or enzymes, or by repolishing the surface of a PMMA lens. Silicon resin lenses cannot be polished, because this would remove the thin hydrophilic coating. Abrasives cannot be used on lenses with hydrophilic coating because such chemicals would remove the superficial layer of the lenses that is needed for wettability and good tolerance. Silicon rubber lenses besides the surface deposits may also adsorb lipids and proteins. Surface deposits can be removed by detergents or enzymes; contamination inside the lens usually cannot be removed. Hydrogel lenses may accumulate contaminants both on the surface and inside the lens. Surface deposits can be removed from hydrogel lenses by detergents, enzymes, or mild abrasives, but contaminants that penetrate these lenses are practically unremovable. The removal of deposits from the surface is easier with low water content lenses than with high water content hydrogels. Highly hydrated lens materials, due to their larger pore sizes, tend to absorb more contaminants (proteins, calcium salts, microorganisms, etc.) than do lenses with lower water content.[25,28]

REFERENCES

1. **Schmidt, P. P., Schoessler, J. P., and Hill, R. M.,** Effect of hard contact lenses on the chloride ion of the tears, *Am. J. Optom. Physiol. Opt.,* 51, 84, 1974.
2. **Farris, R. L., Stuchell, R. N., and Mandel, I. D.,** Basal and reflex human tear analysis. I. Physical measurements. Osmolarity, basal volumes, and reflex flow rate, *Ophthalmology,* 88, 852, 1981.
3. **Jones, L. T.,** The lacrimal secretory system and its treatment, *Am. J. Ophthalmol.,* 62, 47, 1966.
4. **Sorenson, T., Taagehoj, F., and Christensen, U.,** Tera flow and soft contact lenses, *Acta Ophthalmol.,* 58, 182, 1980.
5. **McDonald, J. E. and Brubaker, S.,** Meniscus-induced thinning of tear films, *Am J. Ophthalmol.,* 72, 139, 1971.
6. **Holly, F. J.,** Surface chemistry of contact lens wear, in *The Preocular Film and Dry Eye Syndromes,* Holly, F. J. and Lemp, M. A., Eds., Little, Brown and Company, Boston, 1973, 279.
7. **Lemp, M. A.,** Surfacing abnormalities, in *The Preocular Film and Dry Eye Syndromes,* Holly, F. J. and Lemp, M. A., Eds., Little, Brown and Company, Boston, 1973.
8. **Holly, F. J. and Lemp, M. A.,** Tear physiology and dry eyes, *Surv. Ophthalmol.,* 22, 69, 1977.

9. **Holly, F. J.,** Protein and lipid adsorption by acrylic hydrogels and their relation to water wettability, *J. Polymer Sci.,* 66, 409, 1979.

10. **Refojo, M. F. and Holly, F. J.,** Tear protein adsorption on hydrogels: a possible cause of contact lens allergy, *Contact Intraoc. Lens Med. J.,* 3, 23, 1977.

11. **Abelson, M. B. and Holly, F. J.,** A tentative mechanism for inferior punctate keratopathy, *Am. J. Ophthalmol.,* 83, 866, 1977.

12. **Holly, F. J.,** Basic aspects of contact lens biocompatibility, *Colloid. Surf.,* 10, 343, 1984.

13. **Holly, F. J. and Refojo, M. J.,** Oxygen permeability of hydrogel contact lenses, *J. Am. Optom. Assoc.,* 43, 11, 1972.

13a. **Refojo, M. F., Holly, F. J., and Leong, F.-L.,** Permeability of dissolved oxygen through contact lenses. I. Cellulose acetate butyrate, *Contact Intraoc. Lens Med. J.,* 3, 27, 1977.

14. **Vogt, A.,** *Atlas der Spaltenlampenmikroskopie des lebenden Auges,* Springer-Verlag, Berlin, 1921.

15. **Koby, F. E.,** *Microscopie de l'Oeil Vivant,* Masson et Cie, Paris, 1924.

16. **Meesmann, A.,** *Die Mikroskopie des lebenden Auges an der Guldstrandschen Spaltlampe mit Atlas typischer befunde,* Urban & Schwarzenberg, Berlin, 1927.

17. **Ehlers, N.,** The precorneal film. Biomicroscopical, histological, and chemical investigations, *Acta Ophthalmol.,* 81(suppl.), 5, 1965.

18. **Guillon, J.-P.,** Tear film photography and contact lens wear, *J. Br. Contact Lens Assoc.,* 5, 84, 1982.

19. **Guillon, J.-P.,** Tear film structure and contact lenses, in *The Preocular Tear Film in Health, Disease, and Contact Lens Wear,* Holly, F. J., Ed., Dry Eye Institute, Lubbock, TX, 1986, 914.

20. **Guillon, M. and Lydon, D. P.,** Tear layer thickness characteristics of rigid gas permeable lenses, *Am. J. Optom. Physiol. Opt.,* 63, 527, 1986.

21. **Hamano, H., Hori, M., and Mitsunaga, S.,** Application of an evaporimeter to the field of ophthalmology, *J. Jpn. Contact Lens Soc.,* 22, 101, 1980.

22. **Tomlinson, A. and Cederstaff, T. H.,** *Am. J. Optometry,* 60, 167, 1983.

23. **Larke, J. R.,** *The Eye in Contact Lens Wear,* Butterworth, London, 1985, 22.

24. **Andrasko, G.,** Hydrogel dehydration in various environments, *Int. Contact Lens Clin.,* 10, 22, 1983.

25. **Refojo, M. F.,** The tear film and contact lenses: the effect of water evaporation from the ocular surface, in *Fisiopatologia del Film Lacrimale,* Calabria, G. and Rolando, M., Eds., Societa Oftalmologica Italiana, Genova, 1984.

26. **Mandell, R. B.,** *Contact Lens Practice,* Charles C Thomas, Springfield, IL, 1988.

27. **Hosaka, S., Tanzawa, H., Kenjo, H., and Ovado, S.,** Properties of soft contact lens materials with high water content, *J. Jpn. Contact Lens Soc.,* 22, 367, 1980.

28. **Refojo, M. F. and Leong, F. L.,** Microscopic determination of the penetration of proteins and polysaccharides into poly (hydroxy methacrylate) and similar hydrogels, *J. Polymer Sci.,* 66, 227, 1979.

29. **Baldone, J. A. and Kaufman, H. E.,** Soft contact lenses and clinical disease, *Am. J. Ophthalmol.,* 195, 851, 1983.

TEAR ENZYMES AND CONTACT LENSES

Part C
Changes in Tear Composition in Contact Lens Wear

A contact lens placed in the eye may alter the composition of tears in at least three ways. (1) The contact lens may induce reflex tearing; as a result a higher proportion of the tear samples will be the secretion of the main lacrimal gland, and the levels of locally produced or released substances, like intracellular enzymes deriving from the epithelial cells of the cornea and the conjunctiva, will be lower due to the "diluting effect". (2) There may be an increase in the production and release of compounds of corneal and conjunctival origin due to the mechanical irritation and/or hypoxia (the lack of sufficient oxygen supply) caused by certain types of contact lenses. (3) The irritation caused by the contact lens may induce an increase in the permeability of conjunctival blood vessels resulting in the appearance of plasma proteins in the tear samples.

The analysis of the composition of tear samples collected from the eyes of contact lens wearers both before and after fitting may reveal much about the impact of the lens on the eye. Such changes depend to a large extent on the type of the lens, the length of the wear, and the responses of the patient. The changes in the tear sodium content during the adaptation to a hard lens may reflect the increased reflex tearing caused by the sensation of a "foreign body" (the contact lens) and may return to normal as adaptation occurs.[1] This effect has also been demonstrated for chloride ion levels of hard lens wearers during the period of adaptation accompanied by a slight corneal edema. After the 1st week, the chloride levels in tears gradually normalized and even slightly above normal chloride levels were observed. By this time the corneal edema disappeared. In contrast no such changes were detected in soft contact lens wear.[2] Total tear protein levels as well as glucose and cholesterol concentrations have been shown to undergo the same drop and slow recovery to normal in the period of adaptation to hard contact lens wear. It has been suggested that the decrease in the concentrations of inorganic ions and small molecular weight organic solutes in the tears of contact lens wearers may contribute to the development of corneal edema by lowering the osmotic pressure of the tear fluid in the initial period of adaptation.[3]

Alterations in the lipid composition and raised cholesterol levels in the tears have been suggested as a cause of intolerance in contact lens wear.[4] High levels of cholesterol in the tear fluid were observed in obesity, during pregnancy, and in association with the use of diuretics.[5] Lenses worn by patients suffering from these conditions were shown to have recurrent lipid deposits causing discomfort and disturbances of vision.[6] Tear cholesterol levels in contact lens wearers with oily lens deposits were found to exceed equivalent plasma cholesterol levels, suggesting special mechanisms of accumulation. Under normal conditions tear levels of cholesterol were found to be much lower than plasma levels,[7] in accordance with differences in the protein content and composition of the two fluids, and the proposed function of lipoproteins in the transport of cholesterol in other body fluids. The marked differences in tear cholesterol levels found in various studies indicate the importance of the methods used for the measurements, and suggests that the role of tear cholesterol levels in contact lens intolerance needs further investigation.[6]

Among tear components influencing contact lens wear and contact lens adaptation, tear proteins are thought to be the most important.[8] Tear proteins play a double role in contact lens wear. On the one hand a contact lens on the eye may alter the protein composition of the tear fluid, and on the other hand tear proteins may adhere to the surface of contact lenses and form protein deposits. Both of these events have effects on contact lens tolerance, adaptation, the vision of the patient, and the discomfort caused by the lens. Besides, tear protein changes may alter the surface defense mechanisms of the eye and can serve as diagnostic tools to be used in evaluating the responsiveness of the patient to the lens, in choosing the optimal lens for the patient, in the diagnostics of contact lens-associated eye diseases, as well as in the development of new types of contact lenses.

Gachon and co-workers studied the adherence of lysozyme, tear-specific prealbumin, and albumin to soft contact lenses. They demonstrated that lysozyme has the strongest adherence capacity of the three proteins.[9] Gudmundsson et al.[10] investigated the protein composition of contact lens deposits by means of immunofluorescent microscopy. Lysozyme was found to be the dominant protein in lens deposits. This is in good agreement with the observation that lysozyme is the tear protein that is most difficult to be removed from the surface of the lenses. It has been shown that IgG and IgA also adhere to the lenses. Lactoferrin does as well but only with a small adhering capacity.[10] Several types of interaction by which tear proteins may adhere to the surface of contact lenses have been proposed including hydrogen bonds and hyrophobic and ionic interactions.[9] The binding is basically determined by the surface characteristics of the lens, the reacting groups on the surface of the adhering proteins, and the ion composition, ionic strength, and pH of the surrounding tears. The interaction in the case of hydrogel lenses was shown to depend primarily on the ion-binding capacity of the lens surface.[11]

The formation of deposits, composed of proteins and nonprotein substances, on soft contact lenses and to a lesser extent on gas-permeable hard contact lenses has been found to be one of the major factors that cause problems in contact lens tolerance. Soon after the lens is inserted in the eye a thin mucous layer is formed on the anterior surface which is rich in mucus glycoproteins and adhered tear proteins. These proteins may become denatured due to interactions with the lens surface or due to changes in the inonic strength or pH of the tear film. A drying of the lens surface, changes in the tear composition due to inflammatory or allergic complications, as well the used chemicals and the heating of the hard lens without previously removing the contamination may enhance the precipitation of proteins on the lens surface. It has been suggested that denatured proteins, besides causing irritation and leading to a progressive loss in the wettability of the lens surface, may also cause allergic reactions in some of the contact lens wearers.[9,12,13] Joutsimo and colleagues reported that contact lens deposits, containing proteolytic enzymes (plasmin and plasminogen activator) and viable bacteria, showed proteolytic activity when cut pieces of the lens were cultured on the surface of casein-containing agarose plates. The authors suggested that proteolytic enzymes and bacteria adhering to soft contact lenses may play a role in the development of the rare but very serious complication of contact lens wear: corneal ulceration.[14]

Contact lens wear may as well cause changes in the tear levels of various proteins. Hemmingsen et al.[15] found significantly lower levels of secretory IgA in the tears of contact lens wearers during the first month after fitting, while tear levels of lysozyme and lactoferrin remained unchanged. Vinding et al.[16] made similar observations and suggested that the drop in the level of secretory IgA cannot be the result of a diluting effect as the concentrations of other tear proteins did not have a parallel decline. Manucci and co-workers on the other hand found significantly higher IgA concentrations in the tears of contact lens wearers as compared to normal non-contact lens wearers. They suggested that the difference might result from constant mechanical stimulation of the conjunctiva by permanent contact lens wear.[1] The observation that the levels of IgA and lysozyme did not have a parallel change does not rule out a possible diluting effect. In the case of reflex tearing the conjunctival fluid is diluted by the secretion of the main lacrimal gland. The concentration of lysozyme in this secretion is constant in a wide range of secretion rates.[17] Secretory IgA is secreted by the lacrimal gland,[18] but may get into the tear fluid also via active transport from the subconjunctiva.[19-21] The local production of secretory IgA, as in the case of other mucous membranes, may be increased in various inflammatory diseases of the conjunctiva.[22,23] Secretory IgA levels were found to vary at different flow rates even in the secrete of the lacrimal gland.[24-27] These factors may account for changes in the IgA levels in the tears of contact lens wearers and may also explain the differences observed by various authors.

Farris and co-workers compared basal tear volumes, tear osmolarity values, and tear flow rates of contact lens wearers with those of healthy non-contact lens wearers. Tear samples were collected with filter paper strips. Basal tear volumes (tear volumes collected from the lower cul-de-sac with Periopaper® for 5 min and measured by means of an electronic device Periotron®) were somwhat higher in contact lens wearers than in normal controls but the difference was not statistically significant. Tear osmolarity values in contact lens wearers was significantly higher than in the control persons, both in males and females in all age groups studied. Reflex tear flow rates determined by Schirmer paper strips 6.96 ± 6.07 µl/min in contact lens wearers as compared to 5.71 ± 5.86 µl/min in the control group; the difference was not statistically significant.[28] In further experiments Stuchell et al.[28a] studied the effect of contact lens wear on the tear concentrations of lysozyme and lactoferrin in "unstimulated basal tears" collected with Periopaper® strips causing minimal stimulation and in stimulated tears collected with Schirmer paper strips. No significant difference in lysozyme or in lactoferrin levels was observed between the two groups either in the minimally stimulated "basal" tears or in the stimulated tear samples.[29]

Some of the corneal changes seen after prolonged wearing of contact lenses have been suggested to be secondary to anoxia. Epithelial edema caused by extensive wear or poorly fitted lenses is a recognized indicator of cellular dysfunction.[29] Various enzymes of the energy-producing metabolism have been studied in the tears of contact lens wearers. Lactate dehydrogenase (LDH) and malate dehydrogenase (MDH) levels were found to be changed in contact lens wear, reflecting the metabolic status the corneal epithelial cells under the contact lens.[30,31]

LDH isoenzymes were detected by electrophoretic methods in normal human and rabbit tears, lacrimal gland and corneal epithelium extracts, and in the tears of contact lens wearers. The LDH isoenzyme patterns of normal tears, lacrimal gland, and corneal epithelium were found to be very similar, consisting mainly of isoenzymes built up of subunits M, while the isoenzyme patterns of the same rabbit samples showed a predominance of isoenzymes consisting of subunits H.[32,33]

The enzyme patterns of tear samples were shown to depend on the method of sampling. In samples collected by glass capillaries, tear enzymes of lacrimal gland origin were detected both in case of stimulation and without stimulation. In tear samples collected by Schirmer paper strips the enzymes of the energy-producing mechanisms were also detected in relatively large amounts. These enzymes were suggested to originate from conjunctival epithelial cells disrupted or desquamated by the insertion of Schirmer test paper. It has been suggested that the appearance of intracellular enzymes in the tear fluid may indicate the presence or degree of epithelial cell damage in patients with contact lens problems.[34]

Fullard and Carney[30] used a fluorometric technique to measure the activities of LDH and MDH in tears to assess corneal response to contact lens wear. The absolute values of LDH and MDH proved to be very variable depending on the rate of stimulated lacrimation, while the LDH/MDH ratio proved to indicate changes in the levels of the enzymes in tears of epithelial cell origin. In this way the effects of environmental stresses and the severity of damage caused by them to the corneal epithelial cells could be estimated. Following short-term wear of contact lenses, the LDH and MDH activities were shown to alter so that the LDH/MDH ratio was elevated. The magnitude and time course of this elevation were found to be influenced by contact lens type, fit, and duration of wear. The ratio of these enzymes has been suggested to be used as an indicator of the corneal response to contact lens wear.[30]

Tervo et al. demonstrated the appearance of plasmin in the tears of nonsymptomatic contact lens wearers. Tear samples collected from 50 eyes of healthy non-contact lens wearers, 49 eyes of subjectively symptomless contact lens wearers, and from 41 eyes after the cessation of contact lens wear. Proteolytic activity of the tear samples was analyzed by the radial caseinolytic method. The measured proteolytic activities were shown to be related to plasmin indicated by zymographic analysis and the inhibition by monoclonal antibodies. Plasmin activity was detected in the tears of 75.5% of the lens wearers and in 20% of the healthy controls. Three to 21 d after the cessation of contact lens wear plasmin activity was still detectable in the tears of the persons who previously had worn contact lenses. The authors suggested that the proteolytic activity detected in the tears of contact lens wearers may contribute to the development of corneal epithelial pathology associated with contact lens wear.[35] In further experiments authors of the same group demonstrated that soft contact lenses worn by patients exhibit proteolytic activity due to plasmin, plasminogen activators (PAs), and bacteria adhering to the lens surface. They proposed that bacterial proteases may inactivate the physiological plasmin inhibitors and by doing so contribute to the high and prolonged plasmin activity in the tears of contact lens wearers.[14]

Though no controlled studies relating high plasmin levels to contact lens-induced corneal pathology has been reported so far, as in the cases of other destructive corneal diseases, the PA-plasmin system probably also plays an important role in the development of the serious corneal complications of contact lens wear. PAs released from the affected epithelial cells seem to play a decisive role in initiating the proteolytic events leading to the activation of several proteolytic enzymes including plasmin.[36,37] The action of PAs is controlled by intrinsic PA inhibitors,[38] and when this first line of defense is used up the activation of plasminogen into plasmin occurs. The activity of plasmin can be further regulated by plasmin inhibitors, which constitute the second line of defense. The determination of PAs and their inhibitors in the tears of patients with ocular surface diseases was found to be useful in following up the biochemical changes that precede the appearance of aggressive proteolytic enzymes.[39] The detection of PAs in the tear fluid, besides providing information concerning their pathogenetical role, may also be used for diagnostic purposes. Urokinase-type PA was found to be a good indicator of epithelial cell injury in various external eye diseases.[40,37] Further experiments will reveal whether increased levels of urokinase (in its active form or in the form of complexes with its specific inhibitors) in the tears of contact lens wearers can be used as a sign of lens overwear or lens intolerance.

REFERENCES

1. **Manucci, L. L., Pozzan, M., Fregona, I., and Secchi, A. G.,** The effect of extended wear contact lenses on tear immunoglobulins, *CLAO,* 10, 163, 1984.
2. **Schmidt, P. P., Schoessler, J. P., and Hill, R. M.,** Effect of hard contact lenses on the chloride ion of the tears, *Am. J. Optom. Physiol. Opt.,* 51, 84, 1974.
3. **Callerder, M. and Morrison, P. E.,** A quantitative study of human protein before and after adaptation to non-flexible contact lenses, *Am. J. Optom. Physiol. Opt.,* 51, 939, 1974.

4. **Young, W. H. and Hill, R. M.,** Tear cholesterol levels and contact lens adaptation, *Am. J. Optom. Arch. Am. Acad. Optom.,* 50, 12, 1973.

5. **Young, W. H. and Hill, R. M.,** Cholesterol levels of human tears: case reports, *J. Am. Optom. Assoc.,* 45, 424, 1978.

6. **Larke, J. R.,** *The Eye in Contact Lens Wear,* Butterworth, London, 1985, 22.

7. **van Haeringen, N. J. and Glasius, E.,** Cholesterol in human tear fluid, *Exp. Eye Res.,* 20, 271, 1975.

8. **Boonstra, A.,** *Tear Proteins,* Academisch Proefschrift, University of Amsterdam, Amsterdam, 1988.

9. **Gachon, A. M., Bilbaut, T., and Dastuge, B.,** Adsorption of tear proteins on soft contact lenses, *Exp. Eye Res.,* 40, 105, 1985.

10. **Gudmundsson, O. G., Woodward, D. F., Fowler, S. A., and Allansmith, M. R.,** Identification of proteins in contact lens surface deposits by immunofluorescence micros-copy, *Arch. Ophthalmol.,* 103, 196, 1985.

11. **Sack, R. A., Jones, B., Antignani, A., Libow, R., and Harvey, H.,** Specificity and biological activity of the protein deposits on the hydrogel surface. Relationship of polymer structure to biofilm formation, *Invest. Ophthalmol. Vis. Sci.,* 28, 842, 1987.

12. **Mondino, B. J., Samuel, M. S., and Zaidman, R. W.,** Allergic and toxic reactions in soft contact lens wearers, *Surv. Ophthalmol.,* 26, 337, 1982.

13. **Refojo, M. F. and Holly, F. J.,** Tear protein adsorption on hydrogels: a possible cause of contact lens allergy, *Contact Intraoc. Lens. Med. J.,* 3, 23, 1977.

14. **Joutsimo, L., van Setten, G. B., Renkonen, O. V., Tarkkanen, A., Paivarinta, H., and Tervo, T.,** On the proteolytic activity of contact lenses and bacteria, *Acta Ophthalmol.,* 68, 390, 1990.

15. **Hemmingsen, L., Holm, J., Ericksen, J., and Persson, B. E.,** Secretory immunoglobulin A, immunoglobulin G and lysozyme levels in tear fluid from subjects wearing contact lenses, *IRCS Med. Sci.,* 14, 950, 1986.

16. **Vinding, J. T., Ericksen, S. J., and Nielsen, V. N.,** The concentration of lysozyme and secretory IgA in tears from healthy persons with and without contact lens use, *Acta Ophthalmol.,* 65, 23, 1987.

17. **Berta, A.,** Standardization of tear protein determinations. The effects of sampling, flow rate, and vascular permeability, in *The Preocular Tear Film in Health, Disease, and Contact Lens Wear,* Holly, F. J., Ed., Dry Eye Institute, Lubbock, TX, 1986, 418.

18. **Franklin, R. M., Kenyon, K. R., and Tomasi, T. B.,** Immunohistologic studies of human lacrimal gland: localization of immunoglobulins, secretory component and lactoferrin, *J. Immunol.,* 110, 984, 1973.

19. **Allansmith, M. R., Kajiyama, G. A., Abelson, M. B., and Simon, M. A.,** Plasma cell content of main and accessory lacrimal glands and conjunctiva, *Am. J. Ophthalmol.,* 82, 819, 1976.

20. **Berta, A. and Kálmán, K.,** Tear protein tests in Graves' ophthalmopathy, *Rad. Diagn. Radiobiol. Radiother.,* 28, 546, 1987.

21. **Bracciolini, M.,** Le immunoglobuline della lacrime, *Ann. Otal. Clin. Oculist,* 94, 490, 1968.

22. **Sen, D. K. and Sarin, G. S.,** Immunoglobulin concentrations in human tears in ocular diseases, *Br. J. Ophthalmol.,* 63, 297, 1979.

23. **Tragakis, M., Economidis, I., Pollalis, S., and Gartaganis, S.,** Tear immunoglobulin levels in normal persons and patients with staphylococcal, trachomatous and allergic diseases, in *The Cornea in Health and Disease,* Trevor-Roper, P. D., Ed., Grune and Stratton, New York, 1981.

24. **Chandler, J. W., Leder, R., Kaufman, H. E., and Caldwell, J. R.,** Quantitative deter-minations of complement components and immunoglobulins in tears and aqueous humor, *Invest. Ophthalmol.,* 13, 151, 1974.

25. **Horowitz, B. L., Christensen, G. R., and Rotzmann, S. R.,** Diurnal profiles of tear lysozyme and gamma A globulin, *Am. J. Ophthalmol.,* 10, 75, 1978.

26. **Liotet, S. and Rouchy, J.-P.,** Etude des immunoglobulines des larmes humaines, *Arch. Ophthalmol.,* 30, 799, 1970.

27. **McClellan, B. H., Whitney, C. R., Newman, L. P., and Allansmith, M. R.,** Immunoglobulins in tears, *Am. J. Ophthalmol.,* 76, 89, 1973.

28. **Farris, R. L., Stuchell, R. N., and Mandel, I. D.,** Basal and reflex human tear analysis. I. Physical measurements. Osmolarity, basal volumes, and reflex flow rate, *Ophthalmology,* 88, 852, 1981.

28a. **Stuchell, R. N., Farris, R. L., and Mandel, I. D.,** Basal and reflex tear analysis. II. Chemical analysis: lactoferrin and lysozyme, *Ophthalmology,* 88, 858, 1981.

29. **Thoft, R. A. and Friend, J.,** Biochemical aspects of contact lens wear, *Am. J. Ophthalmol.,* 80, 139, 1975.

30. **Fullard, R. J. and Carney, L. G.,** Use of tear enzyme activities to assess the corneal response to contact lens wear, *Acta Ophthalmol.,* 64, 216, 1986.

31. **Kahán, I. L., Tóth, M., and Nádrai, Á.,** A szaruhártya anyagcseréjének változása contactlencse viselés során. (Alteration of corneal metabolism during contact lens wear), *Szemeszet,* 120, 23, 1983.

32. **Kahán, I. L.,** Prognostic value of tear lactate dehydrogenase isoenzyme determinations, in *The Preocular Tear Film in Health, Disease, and Contact Lens Wear,* Holly, F. J., Ed., Dry Eye Institute, Lubbock, TX, 1986, 182.

33. **Kahán, I. L. and Ottoway, É.,** Lactate dehydrogenase of tears and corneal epithelium, *Exp. Eye Res.,* 20, 129, 1975.

34. **van Haeringen, N. J. and Glasius, E.,** The origin of some enzymes in tear fluid, determined by comparative investigation with two collection methods, *Exp. Eye Res.,* 22, 267, 1976.

35. **Tervo, T., van Setten, G.-B., Andersson, R., Salonen, E.-M., Vaheri, A., Immonen, I., and Tarkkanen, A.,** Contact lens wear is associated with the appearance of plasmin in the tear fluid — preliminary results, *Graefe's Arch. Clin. Exp. Ophthalmol.,* 227, 42, 1989.

36. **Punyiczki, M., Berta, A., and Tözsér, J.,** Study of neutral proteinases of human tear samples using gels containing substrates, *Clin. Chem. Enzym. Comns.,* 1, 115, 1988.

37. **Tözsér, J., Berta, A., and Punyiczki, M.,** Plasminogen activator activity and plasminogen independent amidolytic activity in tear fluid from healthy persons and patients with anterior segment inflammation, *Clin. Chim. Acta,* 183, 323, 1989.

38. **Tözsér, J. and Berta, A.,** Plasminogen activator inhibitors in human tears, *Acta Ophthalmol.,* 69, 426, 1991.

39. **Berta, A., Tözsér, J., and Holly, F. J.,** Determination of plasminogen activator activities in normal and pathological human tears. The significance of tear plasminogen activators in the inflammatory and traumatic lesions of the cornea and the conjunctiva, *Acta Ophthalmol.,* 68, 508, 1990.

40. **Berta, A., Holly, F. J., Tözsér, J., and Holly, T. F.,** Tear plasminogen activators — indicators of epithelial cell destruction. The effects of scraping, n-heptanol treatment, and alkali burn of the cornea on the plasminogen activator activity of rabbit tears, *Int. Ophthalmol.,* 15, 363, 1991.

Chapter 16

ENZYMATIC CHANGES IN TEARS IN GENETICALLY DETERMINED METABOLIC DISEASES

I. INBORN ERRORS OF METABOLISM

In 1908 Garrod described alkaptonuria, cystinuria, albinism, and pentosuria in his paper entitled "Inborn errors of metabolism". His hypothesis that alkaptonuria resulted from a congenital deficiency of the specific enzyme normally responsible for the degradation of homogentisic acid in the metabolism of phenylalanine and tyrosine was a novel idea, and introduced a completely new concept and a new class of diseases into medicine. He described some important characteristics of these diseases such as the familial distribution, the retarded manifestation in seemingly healthy newborns, and realized that these diseases followed the pattern of recessive Mendelian inheritance. Therefore, he not only introduced the concept of diseases that develop due to the missing activity of specific enzymes, but also pointed out that they are brought about by genetic defects that follow a recessive route of inheritance, and that such genetic defects may be the result of harmful environmental influences.[1]

Since Garrod's first paper, more than 200 metabolic diseases have been described that are casued by enzyme deficiency brought about primarily by gene defects. The first groups of these diseases were the aminoacidopathies, mucopolysaccharidoses, sphingolipidoses, and the inborn errors of carbohydrate metabolism (galactosemia, carbohydrate intolerancies, and glycogenoses).[2] In the past 25 years lysosomal enzymopathies, organic acidurias, glycoproteinoses, mucolipidoses, and genetic disorders of the urea, purine, and pirimidine metabolism were added to the list of genetically determined metabolic diseases.[3] This was followed by the recent discovery of diseases caused by mitochondrial and peroxisomal disturbances, and the defects of the membrane transport mechanisms.[4-6]

Genetically determined enzymopathies are caused by structural changes (mutations) of specific genes coding the amino acid sequence of one of the metabolic enzymes. As a result the enzyme is either not produced (is completely missing) or is produced, but in an inactive form (being unable to perform the specific catalytic function), or the activity and/or stability of the produced enzyme is greatly reduced. All these possibilities lead to the same result: the specific function of the disturbed enzyme (the conversion of one or more substrates into an end product) is either missing or greatly retarded. Such a condition may be detected by methods determining the level or activity of the enzyme, an increase in the level of its substrate(s), or a decrease in the level of the product of enzymatic action. Either of the three disorders (missing enzyme action, increased level of substrate, or decreased level of product) or the block of an enzyme cascade (if the enzyme is a part of a cascade mechanism) may form the pathological basis for the development of the symptoms of genetically determined metabolical diseases (enzymopathies).[2,7]

Recently another pathomechanism was described, which is characterized by the disturbance of the transport of enzymes through the membranes of lysosomes leading to the same result (missing enzyme activity), as caused by the defects of the enzyme structure.[4] Lysosomal storage diseases are brought about by specific mutations either causing the lack of enzyme production or affecting the transport of lysosomal enzymes. More than 30 different lysosomal enzymopathies have been described. Some, such as inclusion cell disease and metachromatic leukodystrophy, are caused by disturbances of the enzyme transport or enzyme stability without affecting the catalytical function of the lysosomal enzyme.[8] Mutations affecting the transport mechanism may be mannose-6-phosphate (M-P-6)-receptor dependent (mucolipidosis II) or M-P-6-receptor independent (mucolipidosis III). In the case of mutations causing the decreased stability of the enzyme, the disturbance is brought about by the increased activity of lysosomal cysteine proteinases leading to the fast metabolism of the otherwise-active enzymes.[3]

II. METABOLIC DISEASES WITH EYE MANIFESTATIONS

From the disorders of protein and amino acid metabolism cystinosis, tyrosinemia, alkaptonuria, Wilson's disease, and amyloidosis are the most important ones with ocular manifestations.

Cystinosis, an autosomal recessive disorder of cystine storage, is probably caused by an abnormality of the lysosomal transport of cysteine or cystine, and occurs in three forms: infantile or nephropathic, adolescent or intermediate, and adult.[9] Cystine is accumulated in the cells in intracytoplasmic, membrane-bound granules. The conjunctiva and the cornea are involved in all forms of the disease. Cystine deposits form characteristic needle-shaped, refractile, polychromatic crystals in the anterior corneal stroma, the conjunctiva, and sometimes the retina.[10,11]

Tyrosinemia, an autosomal recessive disorder of tyrosine metabolism, is caused by the deficiency of the enzyme tyrosine aminotransferase, and is characterized by high levels of tyrosine in the blood and urine, associated with severe lesions in the liver and the kidneys.[12] Ocular symptoms include characteristic lesions of the corneal and conjunctival epithelium, chronic irritation, and photophobia. Epithelial lesions in the cornea appear as superficial drendritic or snowflake-like opacities that do not stain with fluorescein.[13] Histology of the conjunctiva shows cytoplasmic, membrane-bound inclusions in the epithelial cells, thickening of the subconjunctival connective tissue, and infiltration with plasma cells.[14]

Alkaptonuria, another recessive disorder of tyrosine metabolism, is caused by the deficiency of homogentisic acid oxidase, resulting in the accumulation of homogentisic acid in the connective tissues in the form of characteristic pigmented extracellular granules.[15] General symptoms are pain in the joints due to arthropathy, dark pigmentation in the connective tissues (ochronosis), and the appearance of dark pigment in the urine when alkali is added to the urine sample. Ocular signs include triangular patches of intrascleral pigmentation near the insertion of horizontal rectus muscles, oil-droplet opacities in the peripherial corneal epithelium and Bowman's membrane, pigmented pinguecula, and irregularly pigmented granules of the sclera. The eye is free of irritative signs.[16]

Wilson's disease is also a recessively inherited metabolic disorder, characterized by reduced ceruloplasmin plasma levels, copper deposits, hepatolenticular degeneration, hemolytic changes, and progressive lesions of the central nervous system. Ocular findings often help the diagnosis of the disease. The Kayser-Fleischer ring is a crescent-shaped or circumferential yellowish-green copper deposit in the Descemet membrane near the limbus.[17] Histologically the granules (copper deposits) located in the Descemet's membrane are smallest near the endothelium, suggesting a possible role of the endothelial cells in copper deposition. Copper is often deposited also in the lens, causing green or brown "sunflower" cataract.[18]

Congenital amyloidosis is an autosomal dominant genetical disorder of protein metabolism. No deficient enzyme is known characteristic of the disease. Amyloid is a group of protein-like substances deposited in a number of tissues in various diseases. Inherited forms of amyloidosis cause lattice or gelatinous dystrophy of the corneal stroma.[19]

Mucopolysaccharidoses are the most important from the inherited disorders of carbohydrate metabolism that cause ophthalmic lesions. Mucopolysaccharidoses, among others, include Hurler's syndrome, Scheie's syndrome, Hunter's syndrome, Sanfilippo's syndrome, and Maroteaux-Lamy syndrome. These diseases are caused by deficiencies of lysosomal acid hydrolases normally responsible for the catabolism of mucopolysaccharides (glycosaminoglycans). The defective enzymes are α-L-iduronidase, N-acetyl-glucosaminidase, N-acetyl-galactosaminidase, β-galactosidase, β-glucuronidase, iduronate sulfatase, and other sulfatases. The missing or reduced enzyme activity results in the accumulation in intracellular lysosomes of heparan sulfate, and/or dermatane sulfate or keratane sulfate. Most mucopolysaccharidoses follow autosomal recessive inheritance with the exception of Hunter's syndrome which is an X-linked recessive disease. The most characteristic ocular findings in these diseases are corneal stromal opacities. The opacities are usually present in the form of diffuse clouding of the stroma, with the exception of Sanfilippo's syndrome in which the stroma is either clear or small corneal opacities are present. Diffuse clouding is caused by the presence of large quantities of glycosaminoglycans in the extracellular matrix, while small corneal opacities are characterized by the accumulation of glycosaminoglycans in intracellular cytoplasmic vacuoles in keratocytes, endothelial, and epithelial cells. Other occasional ocular findings in mucopolysaccharidoses are pigmentary retinopathy, optic atrophy and glaucoma.[20-22]

Sphingolipidoses are disorders of lipid metabolism and storage. Most of them, except for Fabry's disease which is X-linked recessive follow autosomal recessive route of inheritance. They are caused by the deficiency of one of the specific enzymes involved in the metabolism of various sphingolipids. The lack of enzyme activity leads to the intracellular accumulation of a specific sphingolipid molecule, in those tissues in which the involved sphingolipid is an important constituent. Neural sphingolipids are involved in G_{M1} gangliodosis and metachromatic leukodystrophy, and lead to central nervous system disease. Visceral sphingolipids are involved in Fabry's disease and lead to renal failure without central nervous system involvement. Diseases caused by the accumulation of several sphingolipids have been described. Among these the following cause ocular lesions, beginning with corneal opacification: G_{M2} gangliosidosis II, metachromatic leukodystrophy, Fabry's disease, lecithin cholesterol acyltransferase deficiency, Tangier disease, fish eye disease, and hyperlipoproteinemias.[19,20,23-24a]

Mucolipidoses are combined disorders of the carbohydrate and lipid metabolism. Pathological changes seen in mucopolysaccharidoses and visceral involvement characteristic of sphingolipidoses may be present at the same time in these patients. The deficiency of one or more enzymes (acidic lysosomal hydrolases) results in the abnormal catabolism of glycoproteins and glycolipids. Histologically the diseases are characterized by the accumulation of oligosaccharides in the tissues; therefore, they are also called oligosaccharidoses. The most prominent eye symptom they share is corneal clouding. The mode of inheritance is autosomal recessive in the case of all mucolipidoses. Corneal involvement is present in Spangler's syndrome (MLS I), I-cell disease (MLS II), pseudo-Hurler's dystrophy (MLS III), Berman syndrome (MLS IV), and Goldberg syndrome.[25-27]

Several other inherited metabolic diseases exist that cause ocular lesions. Oculocerebrorenal syndrome (Lowe syndrome)[28] and familial dysautonomia (Riley-Day syndrome)[29] are the best known among them. The specific biochemical mechanism underlying these diseases is not known. They are inherited in a recessive way.[19]

III. TEAR ENZYMES IN THE DIAGNOSTICS OF LYSOSOMAL STORAGE DISEASES AND OTHER GENETICALLY DETERMINED METABOLIC DISEASES

The diagnosis of lysosomal storage diseases is based on the demonstration of an enzymatic deficiency (the lack of sufficient enzyme activity in one of the secretions or body fluids) and on the demonstration of lysosomal overloading (the ultrastuctural analysis of lysosomes detected in biopsy materials).[30] As lysosomal diseases affect practically all cell types in the human body, both enzymatic and ultrastructural analysis can be performed on a number of different materials including those obtained from the eye.

Tear enzyme activity determinations were first used for the diagnosis of Tay-Sachs' disease[30,31] and for the detection of Fabry's disease.[32] Conjunctival biopsies also have been successfully used in the diagnostics of lysosomal diseases such as Fabry's disease,[33] mucolipidosis II,[34] systemic mucopoly-saccharidoses,[35] and mucolipidosis IV.[36]

van Hoof et al.[37] used tear enzyme activity determinations and the ultra-structural analysis of conjunctival biopsy materials for the diagnosis of various lysosomal storage diseases. They reported 44 patients suffering from different lysosomal diseases all with decreased or missing tear activities of a specific hydrolytic enzyme. They collected 100 to 150 µl of tears from each patient with filter paper strips in 15 to 25 min. Tear proteins were washed out of the filter paper, and different hydrolyses were determined in the eluant using synthetic substrates. With this technique the same group demonstrated specific enzyme deficiency in the tears of 3 patients with fucosidosis (α-L-fucosidase), of 13 patients with Fabry's disease (thermolabile α-D-galactosidase), of 6 patients with type II glycogenosis (acidic α-D-glucosidase), of 5 patients with Hurler's disease (α-L-iduronidase), of 5 patients with mannosidosis (α-D-mannosidase), of 3 patients with metachromatic leukodystrophy (sulfatase A), of 3 cases with mucosulfatidosis (sulfatase a and sulfatase B), of 2 cases with Sandhoff's disease (*N*-acetyl-β-glycosaminidase), and in the tears of 1 patient with Tay-Sachs' disease.[27,38,39]

In some of the lysosomal storage diseases, besides the missing activity of one hydrolytic enzyme, the hyperactivity of other hydrolases also can be demonstrated. Libert et al.[40] measured elevated (four to five times more than normal) levels of α-fucosidase, α-mannosidase, and β-hexosaminidase in the tears of patients with type II mucolipidosis. The same authors tested whether tear enzyme activity determinations can be used in the detection of heterozygoticity. They performed specific enzyme activity determinations in the tears of 25 heterozygotes (15 patients were heterozygotes for Fabry's disease, 4 patients for fucosidosis, 3 patients for mannosidosis, 2 patients for metachromatic leukodystrophy, and 1 patient for Tay-Sachs' disease). In 18 out of the 25 heterozygotes the tear levels of the specific enzymes were intermediate between normal levels and those found in homozygotes; the rest of the heterozygotes could not be differentiated by tear enzyme determinations from normal controls.[27,38,39]

Lysosomal enzymes are very active in tears. The activity of β-hexosamini-dase, for example, is five times higher in normal tears than in normal serum. The relatively high level of lysosomal enzymes in tears and the observation that their concentration and activity in tears does not depend on tear secretion rate suggest that they are secreted by the lacrimal gland and are not released from injured epithelial cells of the conjunctiva or the cornea. The role of lysosomal enzymes in tears is rather obscure. It has been suggested that besides lysozyme other hydrolytic enzymes also contribute to the antibacterial properties of human tears.

Tear enzyme determinations were successfully used in the diagnosis of mucopolysaccharidoses, sphingolipidoses, glycogenoses, and mucolipidoses (oligosaccharidoses). The diagnosis was based on the demonstration of decreased or missing enzyme activity of the enzyme in the tears whose defect is known to be responsible for the development of the disease. In certain diseases (I-cell disease, type II mucolipidosis) also the hyperactivity of other enzymes may be of diagnostic value. In a series of studies reviewed by van Hoof et al.,[37] tear enzyme activity values were in perfect accordance with the detection of the enzyme deficiency in other biological materials, and by the morphological findings gained from conjunctival biopsies. Therefore, they suggested that tear enzyme activity determinations should be used in the screening of genetically determined metabolic diseases. Such assays may be used also in the detection of heterozygoticity in the same diseases, though the reliability of tear tests in heterozygote screening is lower than in the detection of manifest enzymopathies. A very promising new possibility for the use of tear enzyme determinations is the demonstration of the beneficial effects of newly developed forms of therapy, already used in the treatment of some of these genetically determined metabolitic diseases.[41]

REFERENCES

1. **Garrod, A. E.,** Inborn errors of metabolism, *Lancet,* II, 73, 142, 1908.
2. **Bearn, A. G.,** Inborn errors of metabolism and molecular disease, in *Textbook of Medicine,* Beeson, P. B. and McDermott, W., Eds., W.B. Saunders, Philadelphia, 1975, 22.
3. **László, A.,** *Enzimopathiák,* Medicina, Budapest, 1990.
4. **Gahl, W. A.,** Disorders of lysosomal membrane transport. Cystinosis and Salla disease, *Enzyme,* 38, 154, 1987.
5. **Menkes, J. H.,** Genetic disorders of mitochondrial function, *J. Pediatr.,* 110, 255, 1987.
6. **Moser, H. W.,** Peroxisomal disorders, *J. Ped.,* 108, 89, 1986.
7. **Harris, H.,** The principles of human biochemical genetics, in *Frontiers in Biology,* Vol. 19, Neuberger, A. and Tatum, E. L., Eds., American Elsevier, New York, 1970.
8. **von Figura, K., Hasilik, A., Pohlman, R., et al.,** Mutations affecting transport and stability of lysosomal enzymes, *Enzyme,* 38, 144, 1987.
9. **Kroll, W. and Lichte, K.,** Cystinosis: a review of the different forms and of recent advances, *Humangenetik,* 20, 75, 1973.
10. **Wong, V.,** Ocular manifestations in cystinosis, *Birth Defects,* 12, 181, 1976.
11. **Yamamoto, G. K., Schulman, J. D., Schneider, J. A., and Wong, V. G.,** Long-term ocular changes in cystinosis: observation in renal transplant recipients, *J. Pediatr. Ophthalmol. Strab.,* 16, 21, 1979.
12. **Goldsmith, L. A., Kang, E., Bienfang, D. C., et al.,** Tyrosinemia with plantar and palmar keratosis and keratitis, *J. Pediatr.,* 83, 798, 1973.
13. **Burns, R. P., Gipson, I. K., and Murray, M. J.,** Keratopathy in tyrosinemia, *Birth Defects,* 12, 169, 1976.
14. **Bienfang, D. C., Kuwabara, T., and Pueschel, S. M.,** The Richner-Hanhart syndrome, *Arch. Ophthalmol.,* 94, 1133, 1976.
15. **Kampik, A., Sani, J. N., and Green, W. R.,** Ocular ochronosis. Clinicopathological, histochemical, and ultrastructural studies, *Arch. Ophthalmol.,* 98, 1441, 1980.

16. **Wirtschafter, J. D.,** The eye in alkaptonuria, *Birth Defects,* 12, 279, 1976.

17. **Welsche, J. M.,** The eye in Wilson disease, *Birth Defects,* 12, 187, 1976.

18. **Tso, M. O. M., Fine, B. S., and Thorpe, H. E.,** Kayser-Fleischer ring and associated cataract in Wilson's disease, *Am. J. Ophthalmol.,* 79, 479, 1975.

19. **Sugar, J.,** Metabolic disorders of the cornea, in *The Cornea,* Kaufman, H. E., McDonald, M. B., Barron, B. A., and Waltman, S. R., Eds., Churchill Livingstone, New York, 1988, 361.

20. **Francois, J.,** Metabolic disorders and corneal changes, *Dev. Ophthalmol.,* 4, 1, 1981.

21. **Kenyon, K. R.,** Ocular manifestations and pathology of systemic mucopolysaccharidoses, *Birth Defects,* 12, 133, 1976.

22. **Spranger, J.,** The mucopolysaccharidoses, in *Principles and Practice in Medical Genetics,* Emery, A. E. H. and Rimoin, D. L., Eds., Churchill Livingstone, New York, 1983, 1339.

23. **Francheschetti, A. T.,** Fabry's disease: ocular manifestations, *Birth Defects,* 12, 195, 1976.

24. **Weingeist, T. A. and Blodi, F. C.,** Fabry's disease: ocular findings in a female carrier. A light and electron microscopic study, *Arch. Ophthalmol.,* 85, 169, 1971.

24a. **Francois, J., Hanssens, M., and Teuchy, H.,** Corneal ultrastructural changes in Fabry's disease, *Ophthalmologica,* 176, 313, 1978.

25. **Cibis, G. W., Harris, D. J., Chapman, A. L., and Tripathi, R. C.,** Mucolipidosis I, *Arch. Ophthalmol.,* 101, 933, 1983.

26. **Leroy, J. G.,** The oligosaccharidoses (formerly mucolipidoses), in *Principles and Practice in Medical Genetics,* Emery, A. E. H. and Rimoin, D. L., Eds., Churchill Livingstone, New York, 1983, 1348.

27. **Libert, J., van Hoof, F., and Tondeur, M.,** Fucosidosis: ultrastructural study of conjunctiva and skin, and enzyme analysis of tears, *Invest. Ophthalmol.,* 15, 626, 1976.

28. **Ginsberg, J., Bore, K. E., and Fogelson, M. H.,** Pathological features of the eye in oculocerebrorenal (Lowe) syndrome, *J. Pediatr. Ophthalmol. Strab.,* 18, 16, 1981.

29. **Goldberg, M. F., Payne, J. W., and Brunt, P. W.,** Ophthalmological studies of familial dysautonomia. The Riley-Day syndrome, *Arch. Ophthalmol.,* 80, 732, 1968.

30. **Singer, D., Cotlier, E., and Krimmer, R.,** Hexosaminidase A in tears and in saliva for rapid identification of Tay Sachs' disease and its carriers, *Lancet,* II, 1116, 1973.

31. **Carmody, P., Rattazzi, M., and Davidson, R.,** Tay-Sachs disease — the use of tears for the detection of heterozygotes, *N. Engl. J. Med.,* 289, 1072, 1973.

32. **Del Monte, M. S., Johnson, D. L., Cotlier, E., Krivit, W., and Desnick, R. J.,** Diagnosis of Fabry's disease by tear galactosidase A, *N. Engl. J. Med.,* 290, 57, 1974.

33. **Spaeth, G. L. and Frost, P.,** Fabry's disease. Its ocular manifestations, *Arch. Ophthalmol.,* 74, 760, 1965.

34. **Kenyon, K. R. and Sensenbrenner, J. A.,** Mucolipidosis II (I cell disease). Ultrastructural observation of conjunctiva and skin, *Invest. Ophthalmol.,* 10, 555, 1971.

35. **Kenyon, K. R., Quigley, H. A., Hussels, I. E., Wyllie, R. G., and Goldberg, M. F.,** The systemic mucopolysaccharidoses, *Am. J. Ophthalmol.,* 73, 811, 1972.

36. **Merin, S., Livni, N., Berman, E. R., and Yatziv, S.,** Mucolipidosis IV: ocular, systemic and ultrastructural findings, *Invest. Ophthalmol.,* 14, 437, 1975.

37. **van Hoof, F., Libert, J., Aubert-Tulkens, G., and Sera, M. V.,** The assay of lacrymal enzymes and the ultrastructural analysis of conjunctival biopsies: new techniques for the study of inborn lysosomal diseases, in *Metabolic Ophthalmology,* Vol. 1, Pergamon Press, Elmsford, NY, 1977, 165.

38. **Libert, J., van Hoof, F., Farriaux, J. P., Tondeur, M., Vamos-Hurvitz, E., Mandelbaum, I., and Danis P.,** La mannosidose: diagnostic par biopsie conjunctivale et analyse enzymatique des larmes, *Bull. Mem. Soc. Franc. Ophthal.,* 1977.

39. **Libert, J., Tondeur, M., and van Hoof, F.,** The use of conjunctival biopsy and enzyme analysis in tears for the diagnosis of homozygotes and heterozygotes with Fabri disease, *Birth Defects* Original Article Series, 12(3), 221, 1976.

40. **Libert, J., van Hoof, F., Farriaux, J. P., and Toussaint, D.,** Ocular findings in I-cell disease (Mucolipidosis type II), *Am J. Ophthalmol.,* 83, 617, 1977.
41. **Yamaguchi, K., Hayasaka, S., Hara, S., Kurobane, I., and Tada, K.,** Improvement of tear lysosomal enzyme levels after treatment with bone marrow transplantation in a patient with I-cell disease, *Ophthalmic Res.,* 21, 226, 1989.

ENZYMATIC CHANGES IN TEARS IN CAUSTIC INJURIES

I. EPIDEMIOLOGY AND PATHOGENESIS OF CHEMICAL INJURIES

Chemical burns of the cornea and the conjunctiva constitute a group of ocular injuries that frequently occur and can be devastating leading to permanent visual loss and even to the total destruction of the eye. The immediate and appropriate treament is most important, determining the visual outcome in the majority of the cases. According to epidemiological surveys, chemical injuries represent 7 to 10% of all cases with ocular trauma.[1,2] Alkali burns cause 26% of all chemical injuries.[3] The severity of the injury is determined by the chemical nature and the concentration of the agent causing the injury, by the form and length of the exposure, and by the effectiveness of the rapid "first aid", which means a thorough washing of the eye with water or appropriate solutions, diluting the chemicals, and removing fluid or solid parts of the affecting agent.

Chemicals can be organic or inorganic, acidic, alkaline, or neutral. The various chemicals interact with cellular and extracellular components and cause tissue destruction in specific ways. Alkali and acids, for example, differ in tissue penetration, in the chemical reactions they generate, and therefore in the form of tissue necrosis they cause. Alkali act by raising the concentration of hydroxyl ions, saponificating fatty acids, and dissociating cell membranes; they also denaturate collagen and cause the swelling of collagen fibers and the loss of glycosaminoglycans in the corneal stroma. They lead to colliquating necrosis, deeply penetrate into the cornea and the conjunctiva, and in severe cases also into the anterior chamber, with a potential of causing damage to other structures inside the eye. Acids on the other hand coagulate proteins, providing a barrier in the superficial tissues, and penetrate deep into the eye only in case of severe necrosis. Other chemicals bind to proteins, inactivate enzymes, tie up sulfhydryl groups, break disulfide bonds, or destroy cells and intracellular proteins in a detrimental fashion.

II. CLINICAL SYMPTOMS AND THE SEVERITY OF THE CHEMICAL BURN

While the identification of the causative factor may help in understanding the pathogenesis of the injury, in finding appropriate forms of treatment, and may be informative about the prognosis, the clinical symptoms caused by different chemicals are often uniform and undistinguishable.[4] Several classifications have been developed to describe clinical findings, to determine the severity and the prognosis of the injury. Most of these schemes were proposed for the classification of alkali burns, but they also can be used for the evaluation of other chemical injuries.

Huges[5] classified alkali burns in three stages: (1) mild: erosion of the corneal epithelium, with or without slight haziness of the superficial cornea, and no ischemic necrosis of the conjunctiva and sclera; (2) moderately severe: corneal opacity, the iris structure is blurred, minimal ischemic necrosis of the conjunctiva and the sclera; and (3) severe: blurred pupillary outline, and blanching of the conjunctiva and the sclera. Other classifications include the prognosis in a four-step sheme.[6,7] The assessment of severity involves corneal epithelial damage, initial stromal haziness, conjunctival ischemia, and necrosis. Other classifications also take into consideration damage to the lids and adnexae. The prognosis in groups I and II are "good", in group III "guarded", and in group IV "poor".[7,8]

The clinical course of the alkali-wounded eye can be described in three phases: (1) acute stage, (2) repair stage, and (3) late complication stage.[9,10]

1.   The acute stage is characterized by severe pain, photophobia, and blepharospasm, accompanied by the destruction of conjunctival and corneal epithelial cells, and depending on the severity of the injury by stromal haziness and destruction of the keratocytes and the endothelium, ischemic necrosis of the subconjunctiva and sclera, and the signs of acute iritis.
2.   The reparative stage involves the regeneration of the epithelium, the resurfacing of the denuded part of the cornea from remnants of the corneal epithelium or from the conjunctival epithelium, vascularization, and scarring of the stroma, deminution of iritis.
3.   In the late phase the complications develop: dry eyes, symblepharon, vascularized (pseudo)pannus of the cornea, recurrent epithelial defects, corneal ulceration, in severe cases perforation of the cornea, and secondary effects such as uveitis, cataract, and secondary glaucoma.

III. PATHOLOGICAL AND BIOCHEMICAL CHANGES IN THE CORNEA AND THE CONJUNCTIVA

Tissue destruction caused by alkalies is primarily mediated by hydroxyl ions, which by saponifying the fatty acids disintegrate cell membranes and lead to the death of various cells in the cornea and the conjunctiva. Hydroxyl ions were shown to bind to glycosaminoglycans and cause a marked decrease in the level of these macromolecules in the corneal stroma.[11] The severity of the injury is also determined by the cation of the alkali which determines the penetrability of the alkali molecule into the tissues. Cations bind to collagen and glycosaminoglycans at high pH by reacting with the carboxyl groups.[12] This together with changes in pH causes the swelling of the corneal stroma which is characterized by thickening and shortening of collagen fibrils without visible changes on the collagen molecules detected by electron microscopy.[13]

Changes in the morphology of individual collagen fibers, besides the direct action of alkalies, may be brought about by the loss of glycosaminoglycans in the surrounding matrix. The degeneration of collagen and the decrease in the glycosaminoglycan content of the extracellular matrix result in the increased susceptibility of stromal collagen to proteolytic effects.[11,14] These chemical and pathological changes, together with necrosis due to coagulation of blood in the affected blood vessels, lead to the destruction of cornea, conjunctiva, iris, trabecular meshwork, lens capsule, and lens epithelium.[4]

Increased pH in the tears, the tissues, and the aqueous humor correlates with increasing tissue damage. Irreversible tissue necrosis usually occurs at pH values above pH 11.[15] The pH increase after the exposure peaks a few minutes after the exposure and returns to normal in $^1/_2$ to 3 h, depending on the chemical nature and the concentration of the alkali, the form, and length of the exposure, and on the effectiveness of therapeutic external irrigation.[16] If the effect of the injuring agent reaches the trabecular meshwork and the ciliary body the intraocular pressure may rise shortly after the injury.[17,18] Besides pH changes, both aqueous glucose and ascorbate levels are decreased. The ascorbate deficiency was suggested to be one of the causes of subsequent reduced collagen synthesis.[19,20]

Severe chemical burns often lead to corneal ulceration. The ulceration usually occurs 2 to 3 weeks after the exposure. The ulcer formation is primarily effected by proteolytic enzymes deriving from leukocytes invading the cornea. Enzymes and enzyme activators produced by the epithelial cells and stromal fibroblasts also contribute to the proteolytic effect. The invasion of leukocytes starts a few hours after the injury and usually abates by the end of the 1st week.[21] Leukocytes, first of all polymorphonuclear leukocytes (PMNs), originate from the capillaries of the limbal conjunctiva and reach the injured part of the cornea through the tear film in a very short time. The transport of cells in the tear film is promoted by fluid movements brought about by the blinking mechanism.[22] Additional PMNs reach the wound by migration through the corneal stroma.[23] The ulceration may lead to perforation and to the total destruction of the eye, or may be limited by self-defensive mechanisms and/or by successful therapy. This is followed by the healing of the ulcer, the formation of a vascularized scar, and the re-epithelization of the denuded stromal surface.

Based on the pathological and biochemical changes observed in the eye following chemical injuries, we propose a new classification of chemical injuries in which we attempt to integrate the elements of the presently used classifications and also the results of some of the recent clinical and experimental observations. The events occurring in the cornea following severe chemical burns, according to this classification, can be described in four stages: (1) immediate effect (early necrotic) stage, (2) invasive (inflammatory) stage, (3) ulcerative stage, and (4) healing stage.

1. In the early necrotic stage the immediate effects of the chemical injury cause a "metabolic shock", a complete block of metabolic events, which

is soon followed by the disintegration of cell membranes and the release of intracellular substances. All the early chemical and pathological changes are brought about by the presence and direct effects of the chemical (increased or decreased pH in the tears, in tissues, and in the aqueous humor; toxic effects, etc.). This phase starts at the time of the exposure and lasts from a few minutes to a few hours depending on the severity of the injury. The production of degradative and the release of intracellular substances with vasoactive and chemotactic action initiates the processes of (inflammatory) phase 2.

2. The inflammatory stage is primarily characterized by the formation of edema in the corneal stroma and by the invasion of leukocytes. These events are brought about by secondary inflammatory mediators and not by the direct effects of the chemical, which is mostly eliminated by this time. The inflammatory mediators may derive from the injured cells, be produced by the released proteolytic enzymes in the extracellular space, get into the wound area from the blood due to increased permeability, and/or be produced by the invading leukocytes themselves. By this time the "metabolic shock" is over, which may be shown by the starting regeneration of the epithelium and by the renewal of circulation in the limbal blood vessels. Epithelial defects are formed in stage 1, the epithelium usually starts to regenerate in stage 2, and the defects often get larger or newly appear in the ulcerative stage (3). Stage 2 lasts for a few days until the epithelium completely regenerates (in mild cases without ulceration) or until the ulceration begins in severe cases, usually 2 or 3 weeks after the injury.

3. The ulcerative stage is the period in which the progressive destruction (melting) of the cornea actually occurs and a stromal ulcer is formed. The ulceration is mediated by hydrolytic (firstly by proteolytic) enzymes. These enzymes are produced by and released from the lysosomes of PMNs. Other sources, epithelial cells, fibroblasts, keratocytes, and microorganisms (superinfection), may also contribute to the enzymes deriving from leukocytes. Immunological processes (immunoglobulin- or cell-mediated cytotoxicity, other humoral mediators, lymphocytes, macrophages, and mast cells), specific transport mechanisms, as well as aspecific release from disintegrated cells may play a role in the production of proteolytic enzymes that lead to the enzymatic digestion of collagen, glycosaminoglycans, and cause further destruction of cells in the cornea. Certain enzymes may activate inactive precursors of other enzymes, forming a proteinase cascade and creating positive feedback mechanisms, regulated by a number of activators and inhibitors.

4. The healing stage is characterized by the appearance and action of reparative processes. These processes include the regeneration of the epithelium gradually covering the denuded surface of the stroma, the synthesis of collagen and glycosaminoglycans, scar formation, and the vascularization of the cornea. Leukocytes gradually disappear, stromal

edema diminishes, and excessive proteolytic enzyme action can no longer be detected. In cases with persistent or recurring epithelial defects and nonhealing ulcers the long-standing presence of leukocytes and the prolonged action of proteolytic enzymes is often observed. The histological and biochemical characteristics of the normal healing process may differ in several aspects from those cases in which the reparation is retarded or does not occur.

We found this classification very useful both in evaluating experimental data, and in classifying clinical cases. It proved to be in good accordance with clincal findings and with the results of tear enzyme determinations. The therapeutic regimen used at our department is also based on this classification.

IV. TEAR ENZYMES IN CAUSTIC INJURIES

Various hydrolytic enzymes accounting for ulcer formation were detected in the alkali-burned ulcerating corneas and in the tears of patients with severe chemical burns and with corneal ulcers.[24-28] These enzymes include acid glycosidases, elastase, cathepsin G, plasmin, plasminogen activators (PAs), and collagenolytic enzymes (some of which proved to be true mammalian collagenases).[29-35]

Collagenases are the only enzymes known to cleave intact nondenaturated collagen. Other proteases may also play a role in the degradation of denatured collagen and in the further digestion of collagen fragments produced by collagenases. In severe chemical burns both denatured and fragmented collagen is present in large quantities; therefore, a variety of proteinases are thought to play a role in the cleavage of collagen during the ulceration of the corneal stroma.[36] Epithelial cells of alkali-burned ulcerating corneas, normal epithelial cells, PMNs, and stromal fibroblast were shown to secrete or release collagenolytic enzymes.[37] The epithelium of alkali-burned corneas has been shown to produce true collagenase.[38,39] It has been suggested, however, that the epithelium probably plays a greater role in the modulation of the collagenase production of PMNs than it does in effecting the ulceration of the corneal stroma.[40] The epithelium may also stimulate corneal fibroblasts to produce latent collagenase. Different cells in the alkali-burned ulcerating or regenerating cornea are the sources of collagenolytic and other proteolytic enzymes detected in the tears of these patients.[41]

Fibroblasts and epithelial cells produce collagenase in a latent, inactive form.[42] The latent collagenase (procollagenase) usually remains inactive in the absence of epithelial defects and PMNs. In case of persistent epithelial defects or when PMNs reach the wound area latent collagense may be activated, and the active enzyme by digesting collagen may contribute to the formation of a stromal ulcer. Berman et al.[43] provided evidence showing that the activation process is mediated by PAs deriving from epithelial cells and probably also from PMNs. PAs activate plasminogen into plasmin, which in turn, besides digesting fibronectin and fibrin, and producing chemotactic factors for PMNs, is capable of activating procollagenase into collagenase.[36] A proteolytic enzyme cascade is formed consisting of PAs, plasmin, and collagenase, which, together with other hydrolases digesting glycosaminoglycans, is effecting the tissue destruction in corneal ulceration.[44,45]

In rabbit experiments we studied the early changes of tear PA activities occurring after alkali burns of the cornea, and compared the tear urokinase activity profiles with those observed after mechanical scrape and *n*-heptanol debridement of the corneal epithelium. All three types of injuries resulted in an increase in the tear PA (u-PA) activities. The maximum values of activator activities were reached at 24 h after alkali burn and *n*-heptanol treatment, while mechanical scrape resulted in an earlier increase with the highest activities observed at 5 h. The increase in u-PA activity was highest in alkali burn (four to five times the normal level), was lowest for *n*-heptanol, and the changes in tear u-PA activities caused by mechanical scapes were in between. Tear PA activities returned to the normal range by the time of reepithelization after mechanical scrape, and 1 to 3 d following reepithelization after *n*-heptanol debridement and alkali burns.[46]

We also studied the PA, plasmin, and collagenase activities in the tears of patients with severe alkali burns during the four phases in the course of the disease. u-PA and t-PA were found to be present in active forms in the tears of alkali-burned patients in the phase 1, immediate effect (early necrotic) stage. The quantitative determinations and comparisons, however, were extremely difficult because of the diluting effect of excessive tearing induced by the chemical injury. The variability of urokinase (u-PA) activity depending on the tear secretion rate suggests a local production; the early rise of the tear levels suggests a quick release of this activator from the injured epithelial cells rather than an active secretion. During the phase 2, invasive (inflammatory) stage the protein content of the tear samples was usually very high due to increased vascular permeability and the release of intracellular substances from the injured cells. From the studied enzymes PAs were detected in large quantities in these tear samples, but the activator was present predominantly in complexes formed with plasma inhibitors. The appearance of plasmin activity in some of the samples was also demonstrated showing that, when the inhibitors of the PA-plasmin system are used up, active plasmin may appear already in this stage of the disease. An overwhelming majority of the tear samples collected in the phase 3, ulcerative stage showed high PA and plasmin activities, and most of them were positive for collagenase. Both plasmin and collagenase activities gradually disappeared from the tears of the patients as the disease entered the phase 4, healing stage; u-PA activity, however, still remained above normal. High u-PA and plasmin activities and occasionally collagenase were detected in the tears of patients with nonhealing corneal defects. Based on these data and on the observation that tear enzyme levels correlated with clinical findings, we are of the opinion that tear enzyme determinations can be used in the follow-up of alkali-burned patients, in making therapeutic decisions, and to a certain extent also in determining the prognosis of the injury (Berta et al., unpublished data).

V. ENZYME INHIBITORS AND OTHER FORMS OF THERAPY IN THE TREATMENT OF CAUSTIC INJURIES

The successful therapy of severe alkali burns, in our opinion, should be based on (1) the understanding of various pathological and pathobiochemical events going on in the alkali-burned cornea during the four phases following the injury, (2) on careful examination of the clinical symptoms in order to determine the severity of the injury, and to decide in which of the above-described phases is the treated eye, and (3) on the detection of proteolytic and collagenolytic activities in the tears of patients in order to decide whether proteinase inhibitors should be used in the particular case at a certain time during the course of the disease.

Severe chemical burns are true emergencies. To start the appropriate therapy as soon as possible is of utmost importance. The fate of the eye and the visual outcome is decided, to a large extent, by the swiftness and effectiveness of the first aid given to the patient within a few minutes after the injury. The eye should be washed, irrigated with clean water or if available with solutions containing Na citrate and ascorbate.[47] The irrigation should last for several minutes and must be accompanied by the mechanical removal of solid particles, if any, from the surface of the eye and from the superior and inferior fornices.

In our practice, following the thorough irrigation, we use the "good old" subconjunctival autologous blood or autologous serum injection, which helps by diluting and by neutralizing the chemicals within the tissues, besides temporarily separating the conjunctiva from the sclera (helps to prevent conjunctival necrosis, and its late complications, perilimbal scarring and secondary glaucoma), and probably acts by supplying a variety of plasma proteins including proteinase inhibitors.[48,49] This form of treatment had been widely used in the first part of this century in eye clinics all over Europe but has been forgotten and neglected during past decades. Based on the results of tear enzyme determinations and on the continuous clinical observation of its beneficial effects, we decided to continue the use of subconjunctival autologous serum injections in the early management of severe chemical eye burns.

The use of corticosteroids during the 1st 10 d following severe chemical injuries has been shown to have a beneficial effect.[50] This form of treatment was found to alleviate clinical symptoms and seems to be effective in the prevention of severe complications including stromal ulceration. Based on the understanding the pathological events occurring in stage 2 of the disease (invasive or inflammatory stage) the use of antiinflammatory drugs can be looked upon as "causative" therapy. Corticosteroids and nonsteroid antiinflammatory drugs may arrest the progression of the disease by limiting the invasion of leukocytes and the production, release, and action of various inflammatory mediators. It is important to point out however, that the use of corticosteroids in the ulcerating stage (3) can be harmful (may lead to perforation of the cornea by promoting the release of proteolytic enzymes, potentiating the effects of tissue collagenases, and decreasing the rate of collagen synthesis). Therefore, the use of corticosteroids must be stopped as soon as the first clinical signs of ulceration are observed and/or when proteolytic and collagenolytic activity is detected in the tears, i.e., when the course of the disease reaches the (ulcerative) stage (3).

The use of proteinase inhibitors can be justified only in stage 3, which is characterized by the excessive production of proteolytic enzymes and the progressive melting of the corneal stroma. Besides observing the clinical signs of stromal ulceration, the therapeutic decision should be based on the results of tear enzyme determinations and the detection of proteolytic and collagenolytic activity in the tears of the patients. Such *in vitro* tests, besides determining the optimal time interval during which the inhibitor therapy should be continued, may also help in finding out which of the available inhibitors (Ca-EDTA, aprotinin, L-cysteine, *N*-acetylcysteine, L-penicillamine, amiloride), alone or in combination, are most effective in blocking the proteolytic action of the enzymes present in the tears of the patient.[38,51–57] The appropriate use of proteinase inhibitors may help to arrest the ulcerating process and to prevent the development of severe complications, such as corneal perforation.

Proteinase inhibitors do not have any direct beneficial effect on corneal wound healing. Rather, the opposite is true; by blocking the action of proteinases, some of which play a crucial role in epithelial resurfacing and other cellular events in corneal wound healing, these drugs may hinder the actual healing processes. Therefore, the use of proteinase inhibitors should be discontinued when the disease enters stage 4 (healing stage), when the clinical signs of epithelial regeneration appear, and when excessive proteolytic and collagenolytic action is no longer detected in the tears of the patients. Corticosteroids are also not to be used in the healing stage as they are known to hinder most of the processes involved in wound healing. In this stage of the disease beneficial effects can be expected only from topically applied agents that directly (fibronectin, growth factors, neuropeptides, and vitamins) or indirectly (Healon®) promote epithelial and stromal wound healing.[58]

In some of the cases wound healing is abnormal, retarded, or the signs of healing are completely missing. The treatment of these patients is extremely difficult. In some of the patients chronic (trophic) ulcers or late complications (vascularized scars, symblepharon, lid abnormalities, and severe dry eyes) develop. In other cases the destruction of the cornea is progressing and cannot be stopped by conservative therapy. Surgical interventions (conjunctival flaps, tenoplasty, lamellar or penetrating corneal grafts) may help in some of the cases.[48,59] The eye may be lost, in spite of all the efforts, either due to perforation and infection, or as a result of complications such as uveitis, glaucoma, or phthisis of the eye.

REFERENCES

1. **Pfister, R. R.,** Chemical injuries of the eye, *Ophthalmology,* 90, 1246, 1983.
2. **Pfister, R. R.,** Chemical corneal burns, *Int. Ophthalmol. Clin.,* 24, 157, 1984.
3. **Morgan, S. J.,** Chemical burns of the eye: causes and management, *Br. J. Ophthalmol.,* 71, 854, 1987.
4. **Parrish, C. M. and Chandler, J. W.,** Corneal trauma, in *The Cornea,* Kaufman, H. E., McDonald, M. B., Barron, B. A., and Waltman, S. R., Eds., Churchill Livingstone, New York, 1988, 599.
5. **Huges, W. F., Jr.,** Alkali burns of the eye. II. Clinical and pathologic course, *Arch. Ophthalmol.,* 36, 189, 1946.
6. **Ballen, P. H.,** Treatment of chemical burns of the eye, *Ear Eye Nose Throat Mon.,* 43, 57, 1964.
7. **Roper-Hall, M. J.,** Thermal and chemical burns, *Trans. Ophthalmol. Soc. U.K.,* 85, 631, 1965.
8. **Dohlman, C. H. and Pfister, R. R.,** Management of chemical burns of the eye, (Symposium on the Cornea), *Trans. New Orleans Acad. Ophthalmol.,* 105, 1972.
9. **Huges, W. F., Jr.,** Alkali burns of the eye. I. Review of the literature and summary of the present knowledge, *Arch. Ophthalmol.,* 35, 423, 1946.
10. **Lemp, M. A.,** Cornea and sclera, *Arch. Ophthalmol.,* 92, 158, 1974.
11. **Cejkova, J., Lojda, Z., Obenberger, J., and Havrankova, E.,** Alkali burns of the rabbit cornea. II. A histochemical study of proteoglycans, *Histochemistry,* 45, 71, 1975.
12. **Grant, W. M. and Kern, H. L.,** Action of alkalies on the corneal stroma, *Arch. Ophthalmol.,* 54, 931, 1955.
13. **Henriquez, A. S., Pihlaja, D. J., and Dohlman, C. H.,** Surface ultrastucture in alkali-burned rabbit corneas, *Am. J. Ophthalmol.,* 81, 324, 1976.
14. **Gnadinger, M., Itoi, M., Slansky, H., and Dohlman, C.,** The role of collagenase in the alkali burned cornea, *Am. J. Ophthalmol.,* 68, 478, 1969.
15. **Pfister, R. R., Friend, J., and Dohlman, C. H.,** The anterior segments of rabbits after alkali burns, *Arch. Ophthalmol.,* 86, 189, 1971.
16. **Paterson, C. A., Pfister, R. R., and Levinson, R. A.,** Aqueous humor pH changes after experimental alkali burns, *Am. J. Ophthalmol.,* 79, 414, 1975.
17. **Paterson, C. A. and Pfister, R. R.,** Ocular hypertensive response to alkali burns in the monkey, *Exp. Eye Res.,* 17, 449, 1973.
18. **Paterson, C. A. and Pfister, R. R.,** Intraocular pressure changes after alkali burns, *Arch. Ophthalmol.,* 91, 211, 1974.
19. **Levinson, R. A., Paterson, C. A., and Pfister, R. R.,** Ascorbic acid prevents corneal ulceration and perforation following experimental alkali burns, *Invest. Ophthalmol.,* 15, 986, 1976.
20. **Pfister, R. R. and Paterson, C. A.,** Additional clinical and morphological in the favourable effect of ascorbate in experimental ocular alkali burns, *Invest. Ophthalmol. Vis. Sci.,* 16, 478, 1977.
21. **Thomas, C. I.,** *The Cornea,* Charles C Thomas, Springfield, IL, 1955, 848.
22. **Robb, R. and Kuwabara, T.,** Corneal wound healing. I. The movement of polymorphonuclear leukocytes into corneal wounds, *Arch. Ophthalmol.,* 68, 636, 1962.
23. **Kenyon, K. R.,** Morphology and pathologic responses of the cornea to disease, in *The Cornea. Scientific Foundations and Clinical Practice,* Smolin, G. and Thoft, R. A., Eds., Little, Brown and Company, Boston, 1983, 43.
24. **Berta, A., Punyiczki, M., and Tözsér, J.,** Collagenase in normal and pathological human tears determined by microtiter plate method, in *Ophthalmology Today,* Ferraz de Oliveira, N. L., Ed., Elsevier Science, Amsterdam, 1988, 503.
25. **Brown, S. I. and Weller, C. A.,** Cell origin of collagenase in normal and wounded corneas, *Arch. Ophthalmol.,* 83, 74, 1970.

26. **Brown, S. I., Weller, C. A., and Akiya, S.,** The pathogenesis of ulcers of the alkali burned cornea, *Arch. Ophthalmol.,* 83, 205, 1970.

27. **Itoi, M., Gnadinger, M. C., Slansky, H. H., Freeman, M. I., and Dohlman, C. H.,** Collagenase in the cornea, *Exp. Eye. Res.,* 8, 369, 1969.

28. **Slansky, H. H., Freeman, M. I., and Itoi, M.,** Collagenolytic activity in bovine corneal epithelium, *Arch. Ophthalmol.,* 80, 496, 1968.

29. **Berman, M. B., Dohlman, C., Gnadinger, M., and Davison, P.,** Characterization of collagenolytic activity in the ulcerating cornea, *Exp. Eye Res.,* 11, 255, 1971.

30. **Berman, M., Gordon, J., Garcia, L. A., and Gage, L.,** Corneal ulceration and the serum antiproteases. II. Complexes of corneal collagenases and alpha-macroglobulins, *Exp. Eye Res.,* 20, 231, 1975.

31. **Cejkova, J., Lojda, Z., Obenberger, J., and Havrankova, E.,** Alkali burns of the rabbit cornea. I. A histochemical study of beta-glucuronidase, beta-galactosidase, and N-acetyl-beta-D-glucosaminidase, *Histochemistry,* 45, 65, 1975.

32. **Gordon, J. M., Bauer, E. A., and Eisen, A. Z.,** Collagenase in human cornea: immunologic localization, *Arch. Ophthalmol.,* 93, 341, 1980.

33. **Prause, J. U.,** Cellular and biochemical mechanisms involved in the degradation and healing of the cornea, *Acta Ophthalmol.,* 168(suppl.), 7, 1984.

34. **Punyiczki, M., Berta, A., and Tözsér, J.,** Study of neutral proteinases of human tear samples using gels containing substrates, *Clin. Chem. Enzym. Comns.,* 1, 115, 1988.

35. **Wang, H. M., Berman, M., and Law, M.,** Latent and active plasminogen activator in corneal ulceration, *Invest. Ophthalmol. Vis. Sci.,* 26, 511, 1985.

36. **Berman, M.,** Regulation of collagenase. Therapeutic considerations, *Trans. Ophthalmol. Soc. U.K.,* 98, 397, 1978.

37. **McCulley, J. P.,** Chemical injuries, in *The Cornea. Scientific Foundations and Clinical Practice,* Smolin, G. and Thoft, R. A., Eds., Little, Brown and Company, Boston, 1983, 422.

38. **Brown, S. I. and Weller, C. A.,** Collagenase inhibitors in prevention of ulcers of alkali burned cornea, *Arch. Ophthalmol.,* 83, 352, 1970.

39. **Pfister, R. R., McCulley, J. P., Friend, J., and Dohlman, C. H.,** Collagenase activity of intact corneal epithelium in peripheral alkali burns, *Arch. Ophthalmol.,* 86, 308, 1971.

40. **Kenyon, K., Berman, M., Rose, J., and Gage, J.,** Prevention of stromal ulceration in the alkali-burned rabbit cornea by glued-on contact lens: evidence for the role of polymorphonuclear leukocytes in collagen degradation, *Invest. Ophthalmol. Vis. Sci.,* 18, 570, 1979.

41. **Berman, M. B.,** Collagenase and corneal ulceration, in *Collagenase in Normal and Pathological Connective Tissues,* Woolley, D. E. and Evanson, J. M., Eds., John Wiley & Sons, Chichester, 1980, 141.

42. **Berman, M. B., Leary, R., and Gage, J.,** Latent collagenase in the ulcerating rabbit cornea, *Exp. Eye Res.,* 24, 435, 1977.

43. **Berman, M., Leary, R., and Gage, J.,** Evidence for a role of the plasminogen activator-plasmin system in corneal ulceration, *Invest. Ophthalmol. Vis. Sci.,* 19, 1204, 1980.

44. **Berta, A., Tözsér, J., and Holly, F. J.,** Multilevel inhibition of proteases in tears. Changing concepts in and a practical approach for the protease-inhibitor treatment of corneal ulcers, *Proc. 6th Int. Symp. Lacrimal System,* Kugler, Amsterdam, accepted.

45. **Berta, A., Tözsér, J., and Holly, F. J.,** Determination of plasminogen activator activities in normal and pathological human tears. The significance of tear plasminogen activators in the inflammatory and traumatic lesions of the cornea and the conjunctiva, *Acta Ophthalmol.,* 68, 508, 1990.

46. **Berta, A., Holly, F. J., Tözsér, J., and Holly, T. F.,** Tear plasminogen activators — indicators of epithelial cell destruction. The effects of scraping, n-heptanol treatment, and alkali burn of the cornea on the plasminogen activator activity of rabbit tears, *Int. Ophthalmol.,* 15, 363, 1991.

47. **Reim, M.,** Zur Pathophysiologie und Therapie der Veratzungskrankheit, *Fortschr. Ophthalmol.,* 84, 46, 1987.
48. **Alberth, B.,** *Surgical Treatment of Caustic Injuries of the Eye,* Akadémia, Budapest, 1968.
49. **Sallai, S., Valu, L., Fehér, J., and Podhorányi, G.,** On subconjunctival autochemotherapy of experimental lime burns, *Dtsch. Gesundheitsw.,* 22, 1994, 1967.
50. **Donshik, P., Berman, M., Dohlman, C., Gage, J., and Rose, J.,** Effect of topical steroids on ulceration in alkali-burned corneas, *Arch. Ophthalmol.,* 96, 2117, 1978.
51. **Berman, M.,** Collagenase inhibitors: rationale for their use in treating corneal ulceration, *Int. Ophthalmol. Clin.,* 15(4), 49, 1975.
52. **Brown, S. I., Akiya, S., and Weller, C. A.,** Prevention of ulcers of the alkali-burned cornea: preliminary studies with collagenase inhibitors, *Arch. Ophthalmol.,* 82, 95, 1965.
53. **Brown, S. I., Tragakis, M. P., and Pearce, D. B.,** Treatment of the alkali-burned cornea, *Am. J. Ophthalmol.,* 74, 316, 1972.
54. **Francois, J. et al.,** Collagenase inhibitors (penicillamine), *Ann. Ophthalmol.,* 5, 391, 1973.
55. **Salonen, E.-M., Tervo, T., Törma, E., Tarkkanen, A., and Vaheri, A.,** Plasmin in tear fluid of patients with corneal ulcers: basis for new therapy, *Acta Ophthalmol.,* 65, 3, 1987.
56. **Slansky, H. H., Berman, M., Dohlman, C., and Rose, J.,** Cysteine and acetylcysteine in the prevention of corneal ulcerations, *Ann. Ophthalmol.,* 2, 488, 1970.
57. **Slansky, H. H., Dohlman, C. H., and Berman, M. B.,** Prevention of corneal ulcers, *Trans. Am. Acad. Ophthalmol. Otolarygol.,* 75, 1208, 1971.
58. **Reim, M. and Lenz, V.,** Behandlung von schweren Veratzungen mit hochpolymerer Hyaluronsaure (Healon®), *Fortschr. Ophthalmol.,* 81, 323, 1984.
59. **Reim, M.,** Zur Behandlung schwerster Veratzungen und Verbrennungen der Bindehaut, *Fortschr. Ophthalmol.,* 84, 65, 1987.

Chapter 18

ENZYMATIC CHANGES IN THE TEARS IN VITAMIN A DEFICIENCY

I. VITAMIN A DEFICIENCY: EPIDEMIOLOGY, GENERAL CHARACTERISTICS, AND CLINICAL MANIFESTATIONS

The prevalence of the ocular manifestations of malnutrition is very variable in different countries of the Third World.[1] Vitamin A deficiency often occurs as a part of general protein-energy deficiency (marasmus) or associated with selective protein deficiency (kwasiorkor). Besides these pronounced deficiency states, the number of less severe cases is much higher. In these cases, besides slight symptoms of general malnutrition, vitamin A deficiency may be the leading cause for the development of nutritive disease. Disorders of the protein metabolism (inadequate protein synthesis) may also affect the vitamin A uptake and storage of the body by reducing the level of proteins involved in the transport, storage, and utilization of this important vitamin. The major sources of vitamin A for the body are β-carotene from plants and retinol obtained from food of animal origin. β-Carotene is hydrolyzed immediately after absorption into two molecules of retinol aldehyde and subsequently reduced into vitamin A alcohol (retinol). Retinol circulates in the blood and is stored in the liver in esterified form (retinyl palmitate ester). Vitamin A is transported from the liver to different tissues bound to a specific transport protein (retinol-binding protein, RBP). The normal level of RBP in the plasma is 40 to 50 μg/ml. The normal level of circulating retinol in the free or RBP-bound form is more than 20 μg/100 ml. Lower levels of RBP or retinol are the signs of vitamin A deficiency.[2] Vitamin A seems to reach epithelial cells of the corneal and conjunctiva via the preocular tear film. Binding sites and specific receptors for complexes of retinol and RBP have been found on corneal epithelial cells. Retinoic acid binding sites were not found on corneal epithelial cells, but they may be present in the conjunctiva.[3-5a]

Besides its inevitable role in the visual circle in the retina, vitamin A also plays a role in the maintenance of corneal and conjunctival integrity, metabolical processes, the differentation of epithelial cells, and probably in corneal wound healing, too.[6] Morphological studies of the conjunctiva and cornea of vitamin A-deficient animals demonstrated thickening and keratinization of the epithelium, decreased number of goblet cells, loss of microplicae of the cell surface, increased density of keratofilaments, and a large number of exfoliating cells.[7-10] Vitamin A plays a role in the glycoprotein synthesis of corneal cells as it does in other tissues. It stimulates the incorporation of glucose and glycosaminoglycans into some of the glycoproteins of the corneal epithelium.[11,12]

Vitamin A was shown both *in vitro* and *in vivo* to play a key role in the differentiation of mucus-secreting and keratinizing tissues including those covering the surface of the eye.[13-15d] The control is achieved inducing basal epithelial cells of the conjunctival epithelium to differentiate into goblet cells and produce mucus glycoproteins and inhibiting the process of keratinization both in the cornea and the conjunctiva. It has been proposed by Liau et al.[16] that retinol performs its action by binding to a cellular carrier protein which then enters the cell nucleus, dissociates from the carrier, and binds onto the chromatin to modulate gene expression. Receptors for retinol and for retinoic acid are found also in the cytosol of corneal cells, indicating the presence of a similar mode of action in the eye.[4] Retinol appears to have two effects in the corneal and conjunctival epithelium. First, it regulates the synthesis of keratin, by preventing the crosslinking of prokeratin and the formation of a complex keratin matrix.[17] Second, it facilitates the synthesis of mannose- and galactose-containing glycoproteins, by forming a glycosylated derivative (mannosyl retinyl phosphate) that in turn mediates the transport and incorporation of mannose into specific cell surface glycoproteins.[18]

Experimental studies in nondeficient animals have shown the beneficial effect of various retinoids on wound healing in rabbits,[19] though the efficacy of such therapy in humans has not yet been established.[1,20,21] Smolin and Okumoto[19] suggested that this effect may, in part, be a result of irritative action of the retinoid acid, which would stimulate cellular turnover. Ubels et al.[22] demonstrated that *all-trans* retinoic acid enhanced wound healing in rabbits in a dose-related fashion while other derivatives of vitamin A had no therapeutic effect. In another study Ubels et al.[23] found that 0.1% *all-trans* retinoic acid ointment significantly enhanced the healing rate of corneal epithelial wounds in Cynomologus monkeys. The observation that retinoic acid promoted epithelial resurfacing only when the epithelial cells were removed without damaging the basement membrane and that no beneficial effect was observed in diabetic rabbits (the epithelial basement membranes are known to be affected in diabetes) lead to the assumption that the beneficial effect of vitamin A may be related to basement membranes.[14] Besides irritation and enhancement of epithelial growth on basement membranes there are other possibilities how vitamin A may promote corneal wound healing. Studies of epithelial cells *in vitro,* for example, have shown that vitamin A stimulates the synthesis of DNA.[24,25] Vitamin A was found to enhance the synthesis of collagen in cultured corneal endothelial cells.[26] Perhaps the most significant effect of vitamin A is the ability to stimulate the incorporation of glucose and glycosamines into glycoproteins, which are involved in various processes leading to the formation of cellular adhesions and take part also in the process of the reepithelization of the cornea.[11] Retinoic acid also stimulates the attachment and proliferation of epidermal keratinocytes *in vitro*[15] and in the eye is taken up[27] and concentrated specifically in the sliding corneal epithelial cells.[28]

The inevitable role of vitamin A in the eye can be clearly demonstrated in vitamin A-deficiency diseases. As vitamin A deficiency usually occurs as a part of a general protein or energy-protein deficiency, it is difficult to distinguish which part of the symptoms is caused specifically by the missing vitamin

A. The most specific clinical symptoms seen in patients with vitamin A deficiency are xerophthalmia (conjunctival xerosis, Bitot's spots, corneal xerosis), keratomalatia (corneal ulceration and progressive melting of the cornea), and night blindness. The widely used classification of the various eye symptoms of vitamin A deficiency was accepted at a joint meeting of the World Health Organization and the U.S. Agency for International Development held in Jakarta in 1974.[29]

II. PATHOLOGICAL AND BIOCHEMICAL CHANGES IN THE CORNEA IN VITAMIN A DEFICIENCY

Xerophthalmia and keratomalatia due to severe vitamin A deficiency is the leading cause of blindness in childhood in the developing world. Although the role of vitamin A deficiency in the development of xerophthalmia is well established, there are a number of questions as to whether the deficiency alone is a sufficient cause for the development of keratomalatia. It has been suggested that additional factors, minor trauma causing epithelial defect, chemical or thermal burns, and the presence of microorganisms (not necessarily highly pathogenic) are also needed for the development of the severe corneal complications often seen in vitamin A deficiency.[30-33a]

Clinical, histologic, and microbiologic changes occurring in eyes of vitamin A-deficient rats were described by Sendele et al.[34] when keratomalacia-like stromal ulceration was induced by epithelial injury. The corneal epithelia of severely vitamin A-deficient rats and pair-fed controls were totally removed by scraping or *n*-heptanol; 96 h after epithelial removal 93% of the deficient animals showed stromal ulceration. Histological signs of intense acute inflammatory reaction, leukocyte infiltration, and numerous bacteria were found in the sections of these corneas. *Staphylococcus aureus* and *Streptococcus fecalis* were the most frequent pathogens cultured from these ulcerating eyes. In contrast, the control corneas showed complete reepithelization, with no ulceration, minimal inflammatory reaction, and the absence of morphologically demonstrable bacteria. Bacterial cultures made from control eyes showed the presence of both pathogenic and apathogenic bacteria. On the basis of these findings it has been suggested that abnormal epithelial recovery following epithelial trauma may be an important factor which together with the inflammatory reaction and the frequent bacterial infections leads to corneal ulceration in vitamin A deficiency.[34]

Relatively little is known about the hydrolytic enzymes in the cornea and in the tears of patients effecting ulceration and the generalized melting of the cornea (keratomalatia). Signs of autolysis of superficial epithelial cells were observed already in early stages of xerophthalmia. In ulcerating corneas the stroma was found to be infiltrated heavily with inflammatory cells and extensive stromal degradation was observed in the central necrotic region of the lesions. Corneal extracts from rabbits with xerophthalmia and ulcerating xerophthalmia corneas showed proteolytic activity toward hemoglobin at pH

3.3. This activity proved to be due to the presence of a cathepsin D- and cathepsin B-like enzymes. The activity of cathepsin B-like activity was twice and ten times higher than that in normal corneas in the extracts of the xerophthalmic and ulcerating corneas, respectively.[33]

A mild trauma in the form of a thermal burn was applied to corneas of vitamin A-deficient rats and their pair-fed controls. The control corneas routinely showed rapid reepithelialization without stromal changes. The corneas of deficient rats recovered more slowly, frequently exhibiting stromal edema, leukoma, and sometimes ulceration. Because collagenase is thought to initiate collagen destruction in corneal ulceration, the relationships among vitamin A status, severity of trauma, and collagenase levels were determined. Mild thermal burns were found to cause corneas from less severely deficient rats to ulcerate rarely, accompanied by slightly increased levels of collagenase, mainly on the 1st day of culture, as in the case of nonburned, severely deficient rats. Comparable burns of corneas of pair-fed control rats resulted in no ulceration and in very little collagenase release. Severe burns of either pair-fed control or normal rat corneas caused ulceration and collagenase release, but collagenase activity was maximal on the 2nd and 3rd days of culture. Differences in vitamin A status at time of burning gave rise to different patterns of collagenase release. By following the development of the vitamin deficiency, it was determined that little active collagenase is released after mild burns of corneas in animals in the preweight plateau stage but that much more active enzyme is released when animals are in weight plateau or 5% weight loss stages. Studies of the effect of recovery from vitamin A deficiency on the response to mild thermal burn indicated that, the longer the interval between feeding vitamin A and the burn, the lower the postburn level of collagenase in the day-1 medium. Thus it would appear that restitution of vitamin A status decreased the level of active collagenase after the mild thermal burn. The system developed here can be used to study the biochemical basis for ulceration in vitamin A deficiency, and the possibility exists that the ulceration characteristic of keratomalacia in people can be initiated by an environmental trauma.[35]

Plasminogen activator (PA) activity was studied by Hayashi et al. in the corneas of vitamin A-deficient rats during epithelial wound healing. Central 3-mm diameter epithelial defects were made by scraping on the corneas of vitamin A-deficient and pair-fed control rats. Cryostat sections of these corneas were studied, at various times postscrape, by overlaying with fibrin films containing plasminogen to examine the distribution of PA activity. Antibodies to tissue PA (t-PA) or to urokinase-type PA (u-PA) were also incorporated in the films, to determine the type of PAs in the corneal sections by blocking their activity with specific antibodies. Control corneas showed t-PA-dependent lysis around the edge of the regenerating epithelium as well as in the defect zone of the epithelial wound. Corneas from vitamin A-deficient rats also demonstrated t-PA activity in association with the corneal epithelium neighboring the defect zone but showed no detectable t-PA activity on the surface of the wound. Histological examination of wounded corneas of vitamin A-deficient rats proved the formation of a pseudomembrane composed of polymorphonuclear leukocytes, cell debris, and fibrinous exudate on the scraped surface. The leading edge of the regenerating epithelium was shown to overlay this membrane, separating the epithelial cells from the stromal surface of the wounded area. The formation of the pseudomembrane, which delays the reepithelization in experimentally wounded corneas of vitamin A-deficient rats, might have resulted from the absence of PA activity in the defect region. It has been suggested that both excessive and inadequate production of PAs by the regenerating epithelium may result in impaired epithelial wound healing.[17]

III. CHANGES IN VITAMIN A LEVELS, IN TEAR PROTEIN LEVELS, AND TEAR ENZYME ACTIVITIES IN VITAMIN A DEFICIENCY

Vitamin A has been detected in the tears of man[36] and rabbit.[36,37] Experimental studies in rabbit and rat have shown the involvement of the lacrimal gland in the transport of retinol from the blood to the tears.[36,38,39] The presence of retinol in stimulated tears, in comparative studies of tear and plasma retinol levels, and a supplementation study on tear fluid levels of marginally nourished children indicate that in man the lacrimal gland is the main source of retinol at the surface of the eye.[37,40] Retinol is stored in the lacrimal gland, gets into the tears as a result of active secretion, and not due to passive transport (diffusion or transudation).[38,39,41] Retinol circulates in the blood bound to RBP, which forms a complex with thyroxine-binding prealbumin.[42] The transport mechanism by which vitamin A in the tears reaches the epithelial cells of the cornea and the conjunctiva is not fully understood. RBP could not be demonstrated in the tears of monkeys.[3] Retinol in the tears appears to be bound to protein, although the exact form of binding is currently a matter of some controversy. Some authors have suggested that retinol is bound to a specific carrier protein that cannot be distinguished chromatographically from tear-specific prealbumin,[7] while others have suggested that the binding is predominantly nonspecific.[43]

Changes in the protein composition in the tears of children with protein malnutrition and vitamin A deficiency have been studied by various authors. Immunoglobulins and enzymes are the most important among tear proteins that affect specific surface defense mechanisms in the eye. Decreased levels of secretory IgA accompanied with normal total protein levels were demonstrated in the nasopharyngeal secretions, in the saliva, and in the tears of malnourished children.[44-46] It has been suggested that lower secretory IgA levels are due to the impaired synthesis of the secretory component.[47] Watson and co-workers studied the levels of total protein, amylase, aminopeptidase, IgA, IgG, and lysozyme in the tears of children suffering from protein-energy malnutrition grades 1 to 3. Except for total protein and IgG all components decreased significantly in grades 2 and 3. The levels of lysozyme correlated with the degree of malnutrition, while the correlation for other components was not close.[48-50] Following nutrition with the missing elements (protein, minerals, vitamin A, etc.) the tear levels gradually returned into the normal range.[51] The normalization of the levels of lysozyme, lactoferrin, amylase, and secretory IgA were observed also in a controlled nutrition intervention study on Thai children.[52]

IV. THE EFFECTS OF VITAMIN A THERAPY

The treatment of vitamin A deficiency and of its eye manifestations is primarily based on the replenishment of the missing vitamin. In the presence of conjunctival or corneal xerosis, or more advanced corneal disease (ulceration or keratomalatia), vitamin A therapy should be instituted without delay. Vitamin A supplementation can be oral or parenteral. The initial dose is 200,000 IU of vitamin A oil orally or in severe cases the same dose in water-soluble form intramuscularly. This should be repeated twice within 1 week. For children under the age of 1 year, half of these doses are sufficient. In severe cases with corneal ulcers and keratomalatia, vitamin A therapy should be accompanied by locally and systemically administered antibiotics and the use of mydriatic eyedrops. There are no clinical data available, but proteinase inhibitor eyedrops may also be useful in limiting corneal ulceration and the melting of the cornea in keratomalatia. In severe cases excessive scarring is unavoidable and corneal perforation often occurs.[6]

It is well established that the ocular squamous metaplasia occurring in mild cases of xerophthalmia due to vitamin A deficiency can be reversed with topical application of retinoic acid both in experimental animals such as rabbits[23,53,54] or rats[55] and clinically in children.[56] Topical vitamin A therapy was suggested for the treatment of other disorders besides xerophthalmia such as herpetic keratitis[1] and various forms of dry eyes.[27,57]

It has been suggested that the clinical symptoms of dry eyes closely resemble those present in mild forms of xerophthalmia. It has been suggested that some of the epithelial lesions seen in dry eyes may arise from local deficiency of vitamin A, and that this deficiency may be due to insufficient tear levels of retinol resulting from impaired lacrimal gland function,[36] and/or from the fact that the tear film, the normal route of retinol to the epithelial cells, does not function properly.[58] It is very difficult, however, to prove either of these possibilities. The rationale for the treatment of dry eyes with retinoids is therefore rather speculative.

Jones and Coop[59] were the first to use vitamin A (retinol palmitate) in combination with salicylic acid topically for the treatment of epithelial desquamation and filament formation in dry eyes. They found this combination effective in reducing desquamation and filament formation, although tenderness and erythema of the lid skin prevented prolonged use. Tseng and co-workers treated dry eye patients with *all-trans* retinoic acid ointment in two different concentrations (0.01 and 0.1%). The study included 22 patients with dry eyes (6 KCS, 9 Stevens-Johnson's syndrome, 3 conjunctival pemphigoid, and 4 cases with radiation induced keratitis). The patients were followed up for

2 months in an open, uncontrolled study. Assessments were made using clinical scores, patient subjective symptoms, and impression cytology of the conjunctiva. All patients with KCS became asymptomatic as for subjective symptoms and their tear production was reported to become higher. Stevens-Johnson patients showed a reduction in clinical scores plus improved cytological findings, while patients with pemphigoid and radiation keratitis showed a reduction in clinical severity and keratinization.[27,57] The studies of Tseng and co-workers were greatly criticized by others because of the lack of adequate control (unmedicated vehicle), the small number of cases, the short follow-up, and the irritation caused by the formulation. We cannot be sure that the petrolatum ointment alone would not have produced an improvement in the symptoms as it has been reported for other nonmedicated ointments and ophthalmic gels.[60] Recently it has been proposed that water-soluble derivatives of vitamin A can easily be utilized by the epithelial cells and and may have beneficial effects on dry eyes in lower nontoxic and nonirritative concentrations.[13]

REFERENCES

1. **Sommer, A.,** Vitamin A treatment for herpes keratitis, *Arch. Ophthalmol.,* 98, 1656, 1980.
2. Report of the International Vitamin A Consultative Group (IVACG), Nutrition Foundation, Washington, DC, 1976, III-5.
3. **Rask, L., Geyer, C., Bill, A., and Peterson, P. A.,** Vitamin A supply of the cornea, *Exp. Eye Res.,* 31, 201, 1980.
4. **Wiggert, B., Bergsma, D. R., Helmsen, R. J., Aligood, J., Lewis, M., and Chader, G. J.,** Retinol receptors in corneal epithelium, stroma and endothelium, *Biochim. Biophys. Acta,* 491, 104, 1977.
5. **Wiggert, B., Bergsma, D. R., Helmsen, R. J., and Chader, G. J.,** Vitamin A receptors: retinoic acid binding in ocular tissues, *Biochem. J.,* 169, 87, 1978.
5a. **Hong, B. S.,** Demonstration of retinol-binding activity of tear specific prealbumin, *Invest. Ophthalmol.,* 27(suppl.), 129, 1986.
6. **Thoft, R. A.,** Dietary deficiency, in *The Cornea. Scientific Foundations and Clinical Practice,* Smolin, G. and Thoft, R. A., Eds., Little, Brown and Company, Boston, 1983, 401.
7. **Carter-Dawson, L. et al.,** Early corneal changes in vitamin A-deficient rats, *Exp. Eye Res.,* 30, 261, 1980.
8. **Jayaraj, A. P., Leela, R., and Rama Rao, P. B.,** Studies on cornea and conjunctiva mucus metaplasia in vitamin A deficient rats, *Exp. Eye Res.,* 12, 1, 1971.
9. **Pfister, R. R. and Renner, M. E.,** The corneal and conjunctival surface in vitamin A deficiency: a scanning electron microscope study, *Invest. Ophthalmol. Vis. Sci.,* 17, 874, 1978.
10. **Pirie, A.,** Xerophthalmia, *Invest. Ophthalmol.,* 15, 417, 1976.
11. **Hassell, J. R., Newsome, D. A., and DeLuca, L.,** Increased biosynthesis of specific glycoconjugates in rat corneal epithelium folowing treatment with vitamin A, *Invest. Ophthalmol. Vis. Sci.,* 19, 642, 1980.
12. **Kim, Y.-C. L., Wang, E., and Wolf, G.,** Vitamin A and the nucleic acids and proteins of rat corneal epithelium, *Nutr. Rep. Int.,* 11, 93, 1975.

13. **Holly, F. J.,** Ocular surface disease update: diagnosis and treatment, manuscript.

14. **Hatchell, D. L., Ubels, J. L., Stekiel, T., and Hatchell, M. C.,** Corneal epithelial wound healing in normal and diabetic rabbits treated with tretinoin, *Arch. Ophthalmol.,* 103, 98, 1985.

15. **Mahrle, G., Wilkinson, D., and Farber, E.,** Effects of retinoids on keratinocytes *in vitro,* in *Retinoids. Advances in Basic Research and Therapy,* Orfanos, C., Ed., Springer-Verlag, Berlin, 1981, 145.

15a. **Aydelotte, M. V.,** The effects of vitamin A and citral on epithelial differentiation *in vitro.* II. The chick esophageal and corneal epithelia and epidermis, *Acta Embryol. Morphol. Exp.,* 11, 621, 1963.

15b. **Beitch, I.,** The induction of keratinization in the corneal epithelium, *Invest. Ophthalmol.,* 9, 827, 1970.

15c. **Fell, H. B.,** The effects of excess vitamin A on cultures of embryonic chicken skin implanted at different stages of differentiation, *Proc. R. Soc. London,* 146, 246, 1957.

15d. **van Horn, D. L., Schutten, W. H., Hyndiuk, R. A., and Kurz, P.,** Xerophthalmia in vitamin A-deficient rabbits. Clinical and ultrastructural alterations in the cornea, *Invest. Ophthalmol. Vis. Sci.,* 19, 1067, 1980.

16. **Liau, G., Ong, D. E., and Chytil, F.,** Interaction of the retinol/cellular retinol-binding protein complex with isolated nuclei and nuclear components, *J. Cell Biol.,* 91, 63, 1981.

17. **Hayashi, K., Frangieh, G., Kenyon, K. R., Berman, M., and Wolf, G.,** Plasminogen activator activity in vitamin A-deficient rat corneas, *Invest. Ophthalmol. Vis. Sci.,* 29, 1810, 1988.

18. **Rosso, G. C., De Luca, L., Warren, C. D., and Wolf, G.,** Enzymatic synthesis of mannosyl retinyl phosphate from retinyl phosphate and guanosine diphosphate mannose, *J. Lip. Res.,* 16, 235, 1975.

19. **Smolin, G. and Okumoto, M.,** Vitamin A acid and corneal epithelial wound healing, *Ann. Ophthalmol.,* 13, 563, 1981.

20. **Sommer, A. and Emran, N.,** Topical retinoic acid in the treatment of corneal xerophthalmia, *Am. J. Ophthalmol.,* 86, 615, 1978.

21. **Sommer, A., Emran, N., and Tjakrasudjatma, S.,** Clinical characteristics of vitamin A-responsive and non-responsive Bitot's spots, *Am. J. Ophthalmol.,* 90, 160, 1980.

22. **Ubels, J. L., Edelhauser, H. F., and Austin, K. H.,** Healing of experimental corneal wounds treated with topically applied retinoids, *Am. J. Ophthalmol.,* 95, 353, 1983.

23. **Ubels, J. L., Edelhauser, H. F., Foley, K. M., Liao, J. C., and Gressel, P.,** The efficacy of retinoic acid ointment for treatment of xerophthalmia and corneal epithelial wounds, *Curr. Eye Res.,* 4, 1049, 1985.

24. **Christophers, E. and Braun-Falco, O.,** Stimulation der epidermalen DNS-Synthese durch vitamin A-saure, *Arch. Klin. Exp. Dermatol.,* 232, 427, 1968.

25. **Klein, P.,** Vitamin A and wound healing, *Acta Derm. Venereol.,* 74(suppl.), 171, 1975.

26. **Beach, R. S. and Kenney, M. C.,** Vitamin A augments collagen production by corneal endothelial cells, *Biochem. Biophys. Res. Commun.,* 114, 395, 1983.

27. **Tseng, S. C. G., Maumenee, E., Stark, W. I., Maumenee, I. H., Jensen, A. D., Green, W. R., and Kenyon, K. R.,** Topical retinoid treatment for various dry-eye disorders, *Ophthalmology,* 92, 717, 1985.

28. **Tanaka, M.,** Localization of ^{3}H-retinol in normal and sliding corneal epithelium and endothelium, *Jpn. J. Ophthalmol.,* 24, 60, 1980.

29. World Health Organization, Vitamin A Deficiency and Xerophthalmia. WHO Technical Report No. 590, World Health Organization, Geneva, 1976.

30. **Seng, W. L., Kenyon, K. R., and Wolf, G.,** Studies on the source and release of collagenase in thermally burned corneas of vitamin A-deficient and control rats, *Invest. Ophthalmol. Vis. Sci.,* 22, 62, 1982.

31. **Sommer, A., Green, W. R., and Kenyon, K. R.,** Clinicohistopathologic correlations in xerophthalmic ulceration and necrosis, *Arch. Ophthalmol.,* 100, 953, 1982.

32. **Sommer, A. and Sugana, T.,** Corneal xerophthalmia and keratomalacia, *Arch. Ophthalmol.,* 100, 404, 1982.

33. **Twining, S. S., Hatchell, D. L., Hyndiuk, R. A., and Nassif, K. F.,** Acid proteases and histologic correlations in experimental ulceration in vitamin A deficient rabbit corneas, *Invest. Ophthalmol. Vis. Sci.,* 26, 31, 1985.

33a. **Sommer, A., Hussaini, G., Tartoutjo, I., Susanto, D., and Soegiharto, T.,** Incidence, prevalence and scale of blinding malnutrition, *Lancet,* I, 1407, 1981.

34. **Sendele, D. D., Kenyon, K. R., Wolf, G., and Hanninen, L. A.,** Epithelial abrasion precipitates stromal ulceration in vitamin A-deficient rat cornea, *Invest. Ophthalmol. Vis. Sci.,* 23, 64, 1982.

35. **Seng, W. L., Glogowski, J. A., Wolf, G., Berman, M. B., Kenyon, K. R., and Kiorpes, T. C.,** The effect of thermal burns on the release of collagenase from corneas of vitamin A-deficient and control rats, *Invest. Ophthalmol. Vis. Sci.,* 19, 1461, 1980.

36. **Ubels, J. L. and MacRae, S. M.,** Vitamin A is present as retinol in the tears of human and rabbits, *Curr. Eye Res.,* 3, 815, 1984.

37. **Speek, A. J., van Agtmaala, A. J., Saowakontha, S., Schreurs, W. H. P., and van Haeringen, N. J.,** Fluorometric determination of retinol in human tear fluid using high performance liquid chromatography, *Curr. Eye Res.,* 5, 841, 1986.

38. **Ubels, J. L., Foley, K. M., and Rismondo, V.,** Retinol secretion by the lacrimal gland, *Invest. Ophthalmol. Vis. Sci.,* 8, 1261, 1986.

39. **Ubels, J. L., Osgood, T. B., and Foley, K. M.,** Metabolism of vitamin A in the rabbit: storage and secretion by the lacrimal gland, *Fed. Proc.,* 46, 1012, 1987.

40. **van Agtmaal, E. J., Bloem, M. W., Speek, A. J., Saowakontha, S., Schreurs, W. H. P., and van Haeringen, N. J.,** The effect of vitamin A supplementation on tear fluid retinol levels of marginally nourished preschool children, *Curr. Eye Res.,* 7, 43, 1988.

41. **Ubels, J. L., Osgood, T. B., and Foley, K. M.,** Storage of retinyl esters by the lacrimal gland, *Invest. Ophthalmol. Vis. Sci.,* 28(suppl.), 159, 1987.

42. **Kanai, M., Raz, A., and Goodman, D. S.,** Retinol-binding protein: the transport protein for vitamin A in human plasma, *J. Clin. Invest.,* 47, 2025, 1968.

43. **Ubels, J. L. and Rismondo, V.,** Correlation of secretion of retinol acid protein by the lacrimal gland, *Invest. Ophthalmol. Vis. Sci.,* 27(suppl.), 24, 1986.

44. **McMurray, D. N., Rey, H., Casazza, L. J., and Watson, R. R.,** Effect of moderate malnutrition on concentrations of immunoglobulins and enzymes in tears and saliva of young Columbian children, *Am. J. Clin. Nutr.,* 30, 1944, 1977.

45. **Reyes, M. A., Tye, J. G., and Watson, R. R.,** Effect of protein malnutrition on secretory enzymes and immunoglobulins of Columbian children, *Fed. Proc.,* 35, 2557, 1976.

46. **Sirisinha, S., Suskind, R., Edelman, R., Asvapaka, C., and Olson, R. E.,** Secretory and serum IgA in children with protein-calorie malnutrition, *Pediatrics,* 55, 166, 1975.

47. **Watson, R. R. and McMurray, D. N.,** The effect of undernutrition on secretory and cellular immune responses, *Crit. Rev. Food Sci. Nutr.,* 113, 1979.

48. **Watson, R. R.,** Presence and development with age of aminopeptidase(s) in tears of children and other mammals, *Ophthal. Res.,* 9, 397, 1977.

49. **Watson, R. R., Reyes, M. A., and McMurray, D. N.,** Influence of malnutrition on the concentrations of IgA, lysozyme, amylase and aminopeptidase in children's tears, *Proc. Soc. Exp. Med.,* 157, 215, 1978.

50. **Watson, R. R., Tye, G., McMurray, D. N., and Reyes, M. A.,** Pancreatic and salivary amylase activity in undernourished Columbian children, *Am. J. Clin. Nutr.,* 30, 599, 1977.

51. **Watson, R. R., McMurray, D. N., Martin, P., and Reyes, M. A.,** Effect of age, malnutrition and renutrition on free secretory component and IgA in secretions, *Am. J. Clin. Nutr.,* 42, 281, 1985.

52. **van Agtmaal, E. J.,** *Vitamin A and Proteins in Tear Fluid. A Nutritional Field Survey on Preschool Children in Northeast Thailand,* Academisch Proefschrift, Amsterdam, 1989, chap. 6.

53. **Hatchell, D. L., Faculjak, M., and Kubicek, D.,** Treatment of xerophthalmia with retinol, tretinoin and etaretinate, *Arch. Ophthalmol.,* 120, 926, 1984.

54. **van Horn, D. L., De Carlo, J. D., Schutten, W. H., and Hyndink, R. A.,** Topical retinoic acid in the treatment of experimental xerophthalmia in the rabbit, *Arch. Ophthalmol.,* 99, 317, 1981.

55. **Pirie, A.,** Effects of locally applied retinoic acid on xerophthalmia in the rat, *Exp. Eye Res.,* 25, 297, 1977.

56. **Sommer, A.,** Treatment of corneal xerophthalmia with topical retinoic acid, *Am. J. Ophthalmol.,* 95, 349, 1983.

57. **Tseng, S. C. G.,** Topical retinoid treatment for dry eye disorders, *Trans. Ophthalmol. Soc. U.K.,* 104, 489, 1985.

58. **Holly, F. J.,** personal communication.

59. **Jones, B. R. and Coop, H. V.,** The management of keratoconjunctivitis sicca, *Trans. Ophthalmol. Soc. U.K.,* 85, 379, 1985.

60. **Leibowitz, H. M, Chang, R. K., and Mandell, A. I.,** Gel tears. A new medication for the treatment of dry eyes, *Ophthalmology,* 91, 1199, 1984.

ENZYMES AND ENZYME INHIBITORS AND THE THERAPY OF CORNEAL INFLAMMATORY DISEASES

I. STEROIDS IN THE THERAPY OF DIFFERENT TYPES OF KERATITIS AND CORNEAL ULCERS

Corticosteroids are widely used in the treatment of inflammatory and immunologic diseases of the eye. They are extremely valuable due to their antiinflammatory effect; however, they must be used with caution because of the possible systemic and ocular side effects.

Some of the corticosteroids are synthesized by the adrenal cortex and play a role in neurohormonal regulatory mechanisms. Their molecule is built up of 21 carbons forming a characteristic multiring structure. All natural adrenocortical steroids and their synthetic congeners have a double bond in 4,5 position and a ketone group at C3. Prednisone and prednisolone have a 1,2 double bond that increases the ratio of carbohydrate-regulating potency to sodium-retaining potency. Dexamethasone and triamcinolone are fluorinated at C9. This property enhances all biological and pharmacological activities of corticosteroids. All presently used antiinflammatory steroids are 17-α-hydroxy compounds. An oxygen function at C11 position also seems to be necessary for a significant antiinflammatory effect.

The pharmacological effects of corticosteroids include

1. Increasing the rate of protein synthesis through interaction with cytoplasmic steroid receptors and enhancing the messenger RNA synthesis in the nucleus of steroid-sensitive cells.
2. Inhibition of the increased capillary permeability and vasodilatation induced by acute inflammation.
3. Inhibition of the degranulation of granulocytes (neutrophil, basophil, and eosinophil cells) by stabilizing intracellular membranes covering cytoplasmic granules, and prevent the release of vasoactive substances and inflammatory mediators, e.g., histamine, serotonin, bradykinin, platelet activating factor, chemotactic factors, etc.
4. Inhibition of prostaglandin synthesis by blocking the enzymatic action of phospholipase which converts phospholipids to arachidonic acid.
5. Stabilizing the cellular lysosomes in neutrophil granulocytes and preventing the deliberation of proteolytic enzymes, which once released cause extensive tissue damage through the enzymatic digestion of tissues.
6. Suppressing the proliferation, migration, cellular and humoral functions of lymphocytes, macrophages, plasma cells, monocytes, and mast cells.[1]

The appearance and severity of systemic side effects shows direct correlation with the total dose of steroids and the length of time they are given to the patient. The most important of the possible systemic side effects are osteoporosis, gastroduodenal ulceration, sodium and water retention, hyperglycemia in diabetics, endocrine disorders, steroid psychosis, susceptibility bacterial, viral, and fungal infections, and impaired or delayed wound healing. To minimize systemic side effects corticosteroids should be used in effective doses but only as long as they are absolutely necessary to have the desired therapeutic effect. Long-standing dosage of low doses should be avoided. In chronic diseases, however, corticosteroids may have to be given for a long time. The use of short-acting corticosteroids with sufficient intervals allowing recovery, alternate day steroid therapy, potassium replenishment, and regular monitoring of side effects may be helpful in avoiding serious complications.

Ocular side effects include elevation of intraocular pressure, steroid glaucoma, formation of posterior subcapsular cataract, uveitis — probably due to the activation of latent infections, delayed or impaired wound healing following ophthalmic surgery, injuries, or corneal ulceration, and the promotion of microbial eye infections by depressing natural defense mechanisms.

Systemic or intensive topical corticosteroid therapy may disturb the process of wound healing following eye surgery or ocular injuries. The topical use of dexamethasone (twice a day in 0.1% concentration) was shown to reduce tensile strength at 11 days after wounding.[1a] Prednisolone 1% eyedrops had the same effect when they were given during the first 10 postoperative days, but this effect was not observed if the drug was given only from the 10th d following surgery.[2] Topical application of corticosteroids (prednisolone acetate 1%, dexamethasone 0.1%, or fluoromethalone 1%) did not inhibit reepithelization after complete denudation of the cornea.[3]

Brown and colleagues[4] reported that both hydrocortisone and dexamethasone augment collagenase activity when incubated with cell-free crude enzyme preparations. Hook and co-workers[5] demonstrated that steroids can cause the appearance of such activity when added to cultures of corneal fibroblasts. Donshik et al.,[6] however, did not find augmentation of collagenase activity by steroids (hydrocortisone phosphate or dexamethasone) when incubated with culture media harvested from ulcerated corneas. They suggested that the mechanism by which steroids enhance corneal ulceration is not direct augmentation of collagenase activity, but probably involves the inhibition of repair processes. It has been shown by the same authors that topical dexamethasone enhances ulceration in alkali-burned rabbit corneas only in the 2nd or 3rd weeks following caustic injury. When given in the 1st 6 d or in the 4th and 5th weeks following burn, the adverse effect on the cornea was not observed.[6]

Observations that rheumatoid and autoimmune diseases afflict women more often than men suggested that sexual steroid hormones might play a role in the development of immunological diseases and in immunological reactions in general. Their effects on the production and release from different cells of proteolytic enzymes, effectors of tissue destruction, resulting from some of the immunologic and autoimmune processes was also demonstrated. Based on the observation that progesterone blocked collagenase formation in post-partum rat and rabbit uterus,[7,8] Newsome and Gross tested the effect of a progesterone derivate (medroxyprogesterone) on the ulceration of alkali-burned rabbit corneas. Three therapeutic modes were tested: (1) twice daily instillation of medroxyprogesterone eyedrops (0.5% suspension in 1% sterile methylcellulose solution), (2) subconjunctival depots of 10 mg medroxyprogesterone (Depo-Provera, 100 mg/ml, Upjohn) placed repeatedly at weekly intervals, (3) intramuscular injections of 200 mg Depo-Provera (400 mg/ml) given at 7- to 8-d intervals. Medroxyprogesterone was effective in preventing severe corneal injury (deep ulceration and perforation) in all three modes of administration. The mode of action, as suggested by the authors, is at least in part related to the suppressive effect of this hormone on the production of tissue collagenase, as indicated by considerable reduction in collagenolytic activity measured in the culture fluids of living explants of the treated corneas. Other possibilities that progesterone stimulates the production of a collagenase inhibitor or blocks the formation of an activator of procollagenase or of the latter itself also must be considered.[9]

Experiments concerning the effects of medroxyprogesterone in the treatment of herpetic interstitial (disciform, metaherpetic) keratitis lead to conflicting results. The therapeutic effect was tested on interstitial keratitis, developed in rabbits, sensitized with live herpes simplex virus type 2, and challenged intrastromally with the same virus. Subconjunctival injections of 30 mg medroxyprogesterone resulted clinically in less stromal infiltration, less neovascularization, and moderate scarring. Histology revealed a marked reduction in polymorphonuclear leukocyte infiltration; tissue explants showed a suppression of latent and active collagenase activity in the treated corneas when cultured *in vitro*. However, epithelial disease was exacerbated in prophylactically treated animals. The authors concluded that medroxyprogesterone appears to be useful in the treatment of herpetic interstitial keratitis as an antiinflammatory agent as well as an inhibitor of collagenase production but, like corticosteroids, it can lead to the exacerbation of superficial (epithelial) forms of herpetic keratitis.[10]

Medroxyprogesterone has been advocated also in the treatment of corneal ulcers, though some experimental data exist showing that this hormone may hinder corneal wound healing to a less extent but in the same manner as corticosteroids do. Phillips et al.[11] treated rabbits after perforating stromal incision, trephination, or thermal burn with 1% prednisolone and 1% medroxyprogesterone acetate eyedrops (6 times per day, topical application). After thermal burns, drug application was started right after burn, ulceration developed in none of the prednisolone treated and only in 17% of the medroxyprogesterone treated group, while 85% of the control corneas showed perforating ulcers. When drug delivery was started only from the 2nd week, severe ulceration developed in 44% of both groups. Prednisolone suppressed the tensile strength of the healing corneal wound by 20% and medroxyprogesterone by 11% (the difference was not statistically significant). Prednisolone decreased collagen formation in the trephination scar by 43% and the decrease caused by medroxyprogesterone was 39%. Both in perforating wounds and thermal burns stromal neovascularization was suppressed markedly by prednisolone, but only moderately by medroxyprogesterone.[11] These observations suggest that both the effects of progesterone on corneal ulceration and on corneal wound healing are less pronounced but very similar to that of the corticosteroids.

In conclusion, corticosteroids may disturb wound healing if they are given within 10 d after surgery. They probably have a beneficial effect within the 1st week after chemical injuries, but should be avoided after that because they may promote ulcer formation. They are beneficial in reducing tissue damage when used after perforating injuries and in cases with bacterial endophthalmitis, though to wait until the following day for the first culture results is advisable. Topical corticosteroids may be used in different forms of keratitis including corneal ulcers to suppress inflammation and minimize scarring. This form of treatment should be avoided in dentritic keratitis as corticosteroids enhance the propagation of the virus in the corneal epithelium. Steroids can be used, however, in metaherpetic keratitis, and keratouveitis to reduce scarring near the visual axis especially when the corneal epithelium is intact. Substantial antiviral coverage is important when corticosteroids are used to treat metaherpetic disease. When steroids are used in the treatment of corneal ulcers, the ulcer should be made sterile first using antibiotics. Even in cases of sterile ulcers may steroids be unfavorable in the healing stage by hindering repair processes. As for progesterone and other sexual steroids, both their therapeutic effect on corneal inflammation and ulceration, as well as their side effects resulting in delayed or disturbed corneal wound healing are less pronounced but basically of the same type as those of the corticosteroids.

II. CYSTEINE AND ACETYLCYSTEINE THERAPY

Cysteine is an amino acid occurring naturally in glutathione. It was suggested to be used topically in the form of eyedrops to prevent corneal ulceration after severe alkali burns. Brown and colleagues[12] reported that perforations of the cornea occurred only in 1 of 33 eyes severely burned with alkali which were treated with 0.2 M cysteine solution applied as two drops six times daily after 7 d routine treatment with antibiotics and cycloplegics while perforations occurred in five of seven eyes not treated with cysteine. Progress of corneal ulcers in three badly burned eyes was arrested immediately after the use of cysteine eye drops was started.[12] Cysteine eyedrops are not commercially available. They are prepared and used at the Moorefields Eye Hospital containing 15.76 g (100 mmol) cysteine hydrochloride per liter in isosmotic solution. Eye-drops containing 200 or 300 mmol/l were also used, and recommended for inactivation of collagense in ulcerative corneal diseases. Cysteine is unstable in solution forming cystine on exposure to air. It has to be stored in a cool place in airtight containers, protected from light.[13]

Acetylcysteine (*N*-acetyl-L-cysteine) is a mucolytic agent which is used, in the form of a sodium salt, as an adjunct to other therapy to reduce the viscosity of pulmonary secretions in obstructive pulmonary diseases, that of the pancreatic fluid in cystic fibrosis of the pancreas, and in other conditions where mucolytic therapy is required. Acetylcysteine is most active in concentrations of 10 to 20% at pH 7 to 9. It is also used as an effective antidote in practolol and gold poisonings.

Acetylcysteine was recommended for topical use in the form of eyedrops to treat filamentary keratitis with and without Sjögren's syndrome. Jones and colleagues[14] reported that three patients with "non-sicca filamentary keratitis" were successfully treated with eyedrops of 20% acetylcysteine sodium, applied four times daily for 2 to 3 weeks. Fifteen patients with filamentary keratitis sicca were similarly treated with relief of symptoms especially photophobia, in all patients but two.[14] In a double-blind crossover study on 30 patients with keratoconjunctivitis sicca, acetylcysteine produced significantly better objective results than treatment with artificial tears.[15] A 5 or 10% solution of acetylcysteine in Hypromellose was less irritant to the eye than 20% solution.[15a] Twenty patients with keratoconjunctivitis sicca and Sjögren's syndrome which had failed other forms of treatment and displayed mucous shreds and corneal filaments were treated with acetylcysteine eyedrops for a year. A 5% solution adjusted to pH 8.4 with sodium bicarbonate was instilled four times daily. Six patients (30%) showed objective and subjective improvement after 1 year.[16] Acetylcysteine is more stable in aqueous solution than cysteine. Eyedrops containing 5 or 10% are prepared by dilution of 20% aqueous solutions with Hypromellose eyedrops, and adjusted to pH 9 with 4% sodium hydroxide 4% solution under aseptic conditions.[13]

III. COLLAGENASE-INHIBITOR THERAPY

Ca-EDTA and cysteine inhibit collagenases by binding Zn, a metal component of these enzymes. Treatment of ^{65}Zn-labeled enzyme with inhibitors (Ca-EDTA or cysteine) at concentrations that completely inhibit collagenase activity removes all the tightly bound ^{65}Zn from the enzyme. Incubation of the enzyme with EDTA or Ca-EDTA also reduces binding of ^{14}C-cysteine by 50 and 65%, respectively. The observations that cysteine can remove all the ^{65}Zn from a labeled collagenase and that both EDTA and Ca-EDTA can, in addition to removing all ^{65}Zn, diminish cysteine binding by about 60%, suggest that cysteine inhibits collagenase at least in part by binding to and subsequently removing a Zn cofactor necessary for the enzyme action.

Ca-EDTA inhibits collagenase by exchanging the calcium for Zn, another metal cofactor, as calcium binds more tightly to EDTA than zinc. The mechanism of inhibition is thought to be similar to that of thiol inhibitors. The thiols inhibit human collagenase by binding *in situ* or stripping off from the enzyme an intrinsic cofactor needed for enzyme activity.

α-2-Macroglobulin is able to prevent ulceration and collagen degradation *in vitro* when α-2-macroglobulin has direct access to collagenase. Topical administration of α-2-macroglobulin did not prove to be enough to prevent ulceration.

EDTA and Ca-EDTA are about 100 times more effective on a molar basis than L-cysteine and its derivatives, *N*-acetylcysteine and D-penicillamine. The α-2-macroglobulin on a molar basis is superior as an inhibitor to the metal-binding agents and thiols. Calcium is thought to be a necessary cofactor of the corneal collagenases, though this requirement has not been unequivocally established. *In vitro* studies indicated a requirement for zinc. Thiols are thought to inhibit corneal collagenases by binding to or removing metal cofactors (Zn and possibly Ca) or by reducing disulfide bonds. Inhibition by both EDTA and thiols is reversible. α-2-Macroglobulin inhibits corneal collagenases irreversibly by forming complexes with them.

Ca-EDTA, cysteine, and acetylcysteine, given as eyedrops, prevented ulceration or retarded the ulcerating process in experimentally alkali-burned rabbit corneas. They were found to have some efficacy in the prevention of corneal ulceration following severe alkali burns also in humans. EDTA is quite toxic and should not be used as eye medication. Ca-EDTA has a low toxicity, and cysteine and acetylcysteine have even lower toxicity. EDTA-type compounds are stable and can be stored in the form of an aqueous solution for a long time. This is not true however for cysteine derivatives, while acetylcysteine eyedrops are more stable than cysteine solutions.[17-19]

IV. APROTININ THERAPY

Salonen and colleagues, based on the detection of plasmin in the tears of a patient suffering from therapy resistant corneal ulcer, investigated the proteolytic activity of 48 patients with different types of keratitis and corneal ulcers. Tear samples from 32 of these patients were found to be positive for plasmin. The tear samples were tested *in vitro to* determine whether the proteolytic activity can be inhibited by proteinase inhibitors. L-Cysteine (0.075 *M*) or heparin (1250 IU/ml) did not prevent lysis on the caseinolytic plates, while 20 IU/ml aprotinin almost completely inhibited the proteolytic activity in all plasmin-positive samples tested. The patients received topical aprotinin (20 or 40 IU/ml, diluted by sterile saline), one or two drops at 3-h intervals during waking hours; 2 to 3 min after the application of aprotinin fibronectin eyedrops were also administered topically, using one or two drops at a time. This combined therapy rapidly promoted corneal epithelial healing even in the otherwise nonhealing corneal erosions and torpid ulcers. Based on the favorable outcome of their cases, the authors suggested that the inhibition of proteolytic activity in tears by topical aprotinin treatment may promote epithelial wound healing and enhance the therapeutic effect of topical fibronectin treatment.[20] Tervo and van Setten also reported of beneficial effects of topical aprotinin therapy in corneal erosions and nonhealing ulcers, and suggested that aprotinin therapy is only effective in cases where plasmin activity can be detected in tears.[21]

No controlled clinical trial has been reported so far proving the efficacy of aprotinin treatment, though several corneal centers are working on the problem. Clinical criteria, the exact form of treatment, the possible parallel use of other forms of treatment, and the fact that not all forms and stages of corneal ulcers are associated with high plasmin activity in tears make these trials difficult to evaluate. The presence of other proteolytic enzymes, some of which are not inhibited by aprotinin, make the problem even more complicated. To evaluate any inhibitor eyedrop, whether it can stop the ulcerating process or prevent ulceration, might be a better approach than judging their effects on the actual healing process. Inhibitory eyedrops, in our experience, may adversely affect the process of reepithelization by inhibiting some of the enzymes involved in the process of corneal wound healing. Therefore, we use aprotinin only in the active, ulcerating stage, when proteolytic activity is detected in the tears of corneal ulcer patients, and try to promote reepithelization with fibronectin, vitamins, growth factors, etc.[22]

Very important observations concerning the efficacy of aprotinin eyedrops were made by Reim and co-workers, who in animal experiments and in a limited number of clinical cases found that aprotinin did not promote wound healing; however, it seemed to be effective in stopping the ulcerating process and preventing further ulceration.[23,24]

V. THE FUTURE PERSPECTIVES OF TEAR ENZYME DIAGNOSTICS AND INHIBITOR THERAPY

We suggest that the inhibitor therapy of corneal ulcers and the prevention of ulcer formation in various ocular surface diseases should be based on the evaluation of the levels and activities of proteolytic enzymes, activators, and inhibitors detected in the tears of these patients. Beneficial effects from the use of proteinase inhibitors can be expected only in cases in which excessive proteinase activity is present in the tears of the patients. High proteolytic activity has at least two different causes: (1) the increased level or activity of their activators and/or (2) the decreased level or activity of specific inhibitors. Proteolytic enzymes in tears form a three-level cascade in which plasminogen activators activate plasmin, which besides its broad-spectrum proteolytic activity also can activate procollagenase into collagenase. The specific inhibition of the unnecessarily active enzymes, the replenishment of the missing elements, and the use of sufficient combination of inhibitors are possibilities for more specific and probably more successful therapy.

Diagnostic methods for assessing proteolytic enzymes in tears may be used in two ways:

1. Various sophisticated laboratory methods can be used to detect, identify, and quantify different proteolytic enzymes in the tear samples. Most of these procedures require special laboratories and equipment, and several days are needed before the results are available.
2. Simple and rapid methods can be used, on the other hand, to demonstrate proteolytic activity in the tear samples, and the effect of different inhibitors can be tested in the same system as to their abilities to block the enzymes present in the samples.

The first approach, utilizing different electrophoretic methods, immunoblotting, enzymography, spectrophotometry, spectrofluorometry, etc., is indispensable for providing precise data on the proteolytic enzymes in tears, but has less of a chance to achieve widespread use in general clinical practice. The other approach, based on simple tests, such as the casein plate method, can provide useful data in a short time to be used by practicing ophthalmologists in their therapeutic decisions.

In our clinical practice we use the casein plate method of Saksela[25] with slight modifications[26] for screening patients who suffer from manifest corneal ulcers or from diseases which have a tendency for ulceration (persistent epithelial defects, recurrent erosions, severe forms of herpetic keratitis, dry eyes, chemical burns, long-standing stromal edema of different origin, bullous keratopathies, corneal dystrophies, contact lens-induced ocular pathologies, etc.). With the help of this method the presence of active proteolytic enzymes can be detected and their combined effect (total proteolytic activity of tears) can be estimated in a very short time. It is also possible to learn which of the available protease inhibitors or inhibitor combinations inhibits the proteolytic activity(ies) of the enzyme(s) present in the tears of a certain patient by testing the activity of tear samples on the same plates after mixing them with inhibitory eyedrops. By this simple method, the appropriate inhibitor combination to be used on a certain patient can be found in a reasonably short time. Tears from the same patient are kept for detailed analysis, and the levels of different proteolytic enzymes, activators, inhibitors, and enzyme-inhibitor complexes are determined with more accurate methods.

We believe that the combination of these two approaches can provide a basis for the development of new diagnostic and differential diagnostic methods as well as novel treatment modalities for the various types and stages of corneal ulcers, keratitis, and other forms of ocular surface disease. When sufficient knowledge will be available at the basic research level about the regulatory mechanism of tear proteases, emphasis will probably be placed on the second, simpler approach to provide specific data by simple diagnostic techniques for practical clinical use.

REFERENCES

1. **Mondino, B. J., Aizuss, D. H., and Farley, M. K.,** Steroids, in *Clinical Ophthalmic Pharmacology,* Lamberts, D. W. and Potter D. E., Eds., Little, Brown and Company, Boston, 1987, 157.

1a. **Gasset, A. R. et al.,** Quantitative corticosteroid effect on corneal wound healing, *Arch. Ophthalmol.,* 81, 589, 1969.

2. **Aquavella, J. V., Gasset, A. R., and Dohlman, C. H.,** Cortisteroids in corneal wound healing, *Am. J. Ophthalmol.,* 58, 621, 1964.

3. **Srinivasan, B. D. and Kulkarni, P. S.,** Effect of steroidal and nonsteroidal anti-inflammatory agents on corneal re-epithelization, *Invest. Ophthalmol. Vis. Sci.,* 20, 688, 1981.

4. **Brown, S. I., Weller, C. A., and Vidrich, A. M.,** Effect of corticosteroids on corneal collagenase of rabbits, *Am. J. Ophthalmol.,* 70, 744, 1970.

5. **Hook, R. M., Hook, C. W., and Brown, S. I.,** Fibroblastic collagenase partial purification and characterization, *Invest. Ophthalmol.,* 12, 771, 1973.

6. **Donshik, P. C., Berman, M. B., Dohlman, C. H., Gage, J., and Rose, J.,** Effect of topical corticosteroids on ulceration in alkali-burned corneas, *Arch. Ophthalmol.,* 96, 2117, 1978.

7. **Halme, J. and Woessner, J. F., Jr.,** Effect of progesterone on collagen breakdown and tissue collagenolytic activity in the involuting rat uterus, *J. Endocrinol.,* 66, 357, 1975.

8. **Jeffrey, J. J., Coffey, R. J., and Eisen, A. Z.,** Studies on uterine collagenase in tissue culture. II. Effect of steroid hormones on enzyme production, *Biochim. Biophys. Acta,* 252, 143, 1971.

9. **Newsome, D. A. and Gross, J.,** Prevention by medroxyprogesterone of perforation in the alkali-burned rabbit cornea: inhibition of collagenolytic activity, *Invest. Ophthalmol. Vis. Sci.,* 16, 21, 1977.

10. **Lass, J. H., Berman, M. B., Campbell, R. C., Pavan-Langstone, D., and Gage, J.,** Treatment of experimental herpetic interstitial keratitis with medroxyprogesterone, *Arch. Ophhtalmol.,* 98, 520, 1980.

11. **Phillips, K. et al.,** Effects of prednisolone and medroxyprogesterone on corneal wound healing, ulceration, and neovascularization, *Arch. Ophthalmol.,* 101, 640, 1983.

12. **Brown, S. I., Tragakis, M. P., and Pearce, D. B.,** Treatment of the alkali burned cornea, *Am. J. Ophthalmol.,* 74, 316, 1972.

13. **Reynolds, J. E. F.,** *Martindale. The Extra Pharmacopoeia,* 28th ed., The Pharmaceutical Press, London, 1982, 50.

14. **Jones, R. B. and Coop, H. V.,** *Trans. Ophthalmol. Soc. U.K.,* 85, 379, 1965.

15. **Absolon, M. J. and Brown, C. A.,** Acetylcysteine in kerato-conjunctivitis sicca, *Br. J. Ophthalmol.,* 52, 310, 1968.

15a. Management of the dry eye, *Drug Ther. Bull.,* 12, 81, 1974.

16. **Williamson, J., Doig, W. M., Forrester, J. V., Tham, M. H., Wilson, T., Whaley, K., and Dick, W. C.,** Management of the dry eye in Sjögren's syndrome, *Br. J. Ophthalmol.,* 58, 798, 1974.

17. **Berman, M.,** Collagenase inhibitors: rationale for their use in treating corneal ulceration, *Int. Ophthalmol. Clin.,* 15, 49, 1975.

18. **Berman, M.,** Regulation of collagenase. Therapeutic considerations, *Trans. Ophthamol. Soc. U.K.,* 98, 397, 1978.

19. **Berman, M. B.,** Collagenase and corneal ulceration, in *Collagenase in Normal and Pathological Connective Tissues,* Woolley, D. E. and Evanson, J. M., Eds., John Wiley & Sons, New York, 1980, 141.

20. **Salonen, E. M, Tervo, T., Torma, E., Tarkannen, A., and Vaheri, A.,** Plasmin in tear fluid of patients with corneal ulcers: basis of new therapy, *Acta Ophthalmol.,* 66, 3, 1987.

21. **Tervo, T. and van Setten, G.-B.,** Aprotinin for inhibition of plasmin on the ocular surface: principles and clinical observations, in *Healing Process in the Cornea,* (Advances in Applied Biotechnology Series, Vol. 1), Beuerman, R. W., Crosson, C. E., and Kaufman, H. E., Eds., Gulf, Houston, 1989, 151.

22. **Berta, A., Tözsér, J., and Holly, F. J.,** Multilevel inhibition of proteases in tears. New concept for the treatment of corneal ulcers, *Proc. 6th Int. Symp. Lacrimal System,* Kugler, Amsterdam, accepted.

23. **Reim, M., Becker, J., Borchers, H., Kohnen, S., Knott, H., Kranz, B., Leber, M., Lund, G., Marchand, M., Salla, S., and Steiger, C.,** Fibronectin, Plasmin, and Proteinase Inhibitor in Experimental and Human Severe Eye Burns, 9th Int. Congress of Eye Research, Abstracts, Helsinki, Finland, July 29–August 4, 1990, 111.

24. **Reim M.,** personal communication.
25. **Saksela, O.,** Radial caseinolysis in agarose: a simple method for the detection of plasminogen activator in the presence of inhibitory substances and serum, *Anal. Biochem.,* 111, 276, 1981.
26. **Berta, A., Tözsér, J., and Holly, F. J.,** Determination of plasminogen activator activities in normal and pathological human tears. The significance of tear plasminogen activators in the inflammatory and traumatic lesions of the cornea and the conjunctiva, *Acta Ophthalmol.,* 68, 508, 1990.

INDEX

B

Bacillus subtilis, 69
Bacterial and fungal keratitis, 171–179

C

L

Lacrimal glands
  anatomy, 1–8
  plasminogen activator-plasmin system
      contribution to tears, 134
Lacrimal pump theory, 6
Lacrimal river, 13
Lactate dehydrogenase (LDH), 31, 41,
      149–153
  contact lenses and, 256, 272
  normal ranges, 42
  plasminogen activator versus, 128
  temperature effects, 39
Lactoferrin, 64, 68–70
  and complement, 238
  contact lens wear and, 271–272
  protein determination, 22, 27
  vitamin A deficiency and, 301–302
Lagophthalmus, 17
Laminin, collagen interactions, 214
Latent collagenase, 93–94
Lattice corneal dystrophy, 94
Lecithin, 32
Lens deposits, 266
Leukemias, 16
Leukocytes, see also Lymphocytes;
      Polymorphonuclear (PMN)
      leukocytes
  and collagenase activity, 92
  in inflammation, 235
Ligases, 32
Lignotic conjunctivitis, 142–143
Limited proteolysis, 110
Lipids, 16–18
  contact lens wear and, 262–264,
      269–270
  inherited metabolic disorders, 280
  precorneal tear film, 11–13
  and precorneal tear film breakup,
      15
  secretion of, 1–3
Listeria monocytogenes, 69
Lyases, 32
Lyell's disease, 16, 235
Lymphocytes
  in herpetic keratitis, 226
  in keratitis, 172–173
  ocular surface, 235
  substance P and, 187
β-Lysin, 68
Lysosomal enzymes, 155–157

procollagenase activation, 79
proteases, 161
storage diseases, 278, 280–282
temperature effects, 39
Lysozyme
  biological function and bacteriolysis,
      47–49
  characteristics of, 51
  collagenase inhibition, 79
  contact lens wear and, 270–272
  diagnostic value, 63–64
  discovery, 45
  flow rates and, 54–56
  measurement methods, 51–54
  mechanism of action, 46–47
  nonlysozyme antibacterial factor and,
      67–69
  normal values, 59–61
  in pathological conditions, 61–63
  plasminogen activator versus, 128
  protein determination, 22, 24–25, 27
  structure, 45–46
  therapy with, 69–7
  vitamin A deficiency and, 301–302
Lys-plasminogen, 109

M

α-2-Macroglobulin, 93, 139, 161
  in herpetic keratitis, 224–225
  plasmin inhibition, 112
  therapy with, 311
Macrophages, 172–173, 226
Malate dehydrogenase, 41, 149–153
  contact lens wear and, 272
  temperature effects, 39
Manz, glands of, 3
Marangoni effect, 15
Marginal ulcer, 131
Maroteaux-Lamy syndrome, 279
Mast cells, 236–237
Matrix proteins, see Collagen; Extracellular
      matrix; Fibronectin
Measurement methods, 35–42
  cell contaminants, 40
  collagenase, 101–104
  extracellular enzymes, 41–42
  incubation period, 38
  intracellular enzymes, 41
  lysozyme, 52–56, 59–61
  measurement units, 40
  organ specificity, 41